Playing God
in Yellowstone

Books by Alston Chase

GROUP MEMORY
PLAYING GOD IN YELLOWSTONE

YELLOWSTONE
NATIONAL PARK

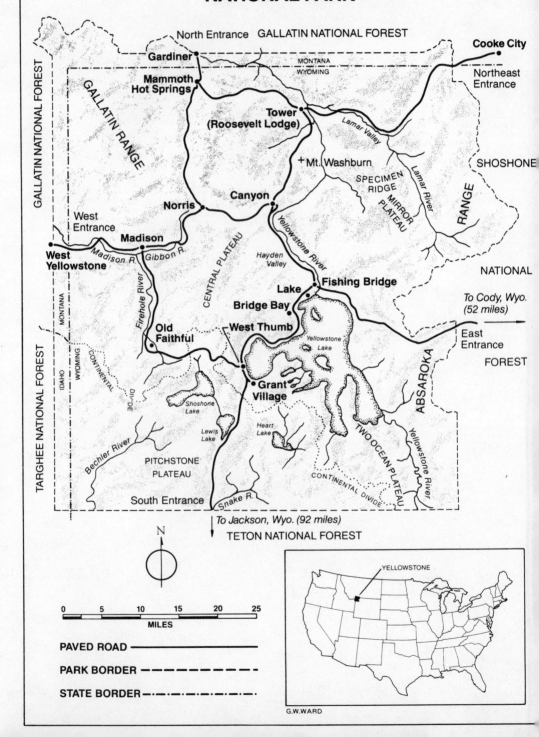

North Entrance GALLATIN NATIONAL FOREST

Cooke City

Gardiner

MONTANA
WYOMING

Northeast
Entrance

Mammoth
Hot Springs

GALLATIN NATIONAL FOREST

GALLATIN RANGE

Tower
(Roosevelt Lodge)

Lamar Valley

SHOSHONE

+ Mt. Washburn

SPECIMEN
RIDGE

MIRROR
PLATEAU

Lamar River

RANGE

Canyon

Norris

West
Entrance

Madison

Madison R.   Gibbon R.

West
Yellowstone

MONTANA
WYOMING

IDAHO   WYOMING

Firehole River

CENTRAL PLATEAU

Yellowstone River

Hayden
Valley

NATIONAL

Lake

Fishing Bridge

Bridge Bay

To Cody, Wyo.
(52 miles)

Old
Faithful

West Thumb

Yellowstone
Lake

East
Entrance

FOREST

CONTINENTAL

Divide

Grant
Village

TARGHEE NATIONAL FOREST

Shoshone
Lake

Lewis
Lake

Heart
Lake

TWO OCEAN PLATEAU

ABSAROKA

Yellowstone River

Bechler River

PITCHSTONE
PLATEAU

CONTINENTAL DIVIDE

South Entrance   Snake R.

To Jackson, Wyo. (92 miles)
TETON NATIONAL FOREST

N

| 0 | 5 | 10 | 15 | 20 | 25 |
MILES

PAVED ROAD ——————

PARK BORDER ————

STATE BORDER —··—··—··

YELLOWSTONE

G.W.WARD

Playing God in Yellowstone

The Destruction of America's First National Park

by ALSTON CHASE

With an epilogue by the author

A Harvest Book
Harcourt Brace & Company
SAN DIEGO   NEW YORK   LONDON

To the memory of my son
Lawrance

Portions of this book appeared in the *Atlantic*
and in *Outside* magazine in different form.

Library of Congress Cataloging-in-Publication Data
Chase, Alston.
Playing God in Yellowstone.
"A Harvest book."
Bibliography: p.
1. Nature conservation — Yellowstone National Park.
2. Wildlife management — Yellowstone National Park.
3. Yellowstone National Park — Management. I. Title.
QH76.5.Y45C47   1987   333.95'16'0978752   87-8446
ISBN 0-15-672036-1

Printed in the United States of America
H J K I G

All conservatism is based upon the idea that if you leave things alone you leave them as they are. But you do not. If you leave a thing alone you leave it to a torrent of change.

— G. K. Chesterton

Preface

THE two bobcats sat in the bright sun in the middle of the pasture, waiting for ground squirrels to appear. When a rodent emerged from its hole, one cat pounced, killing its prey instantly and depositing the body at the base of a tree. When the cache held six squirrels, the cats picked them up in their mouths and disappeared into the woods.

As we watched this ritual each spring, we felt like guests who had arrived at the party too early, surprising the host and hostess preparing dinner. And indeed, although this was "our" ranch, we felt as though it belonged to the bobcats and other animals — white-tail and mule deer, antelope, elk, bighorn sheep, coyote, lynx, wolverine, black bear, beaver, skunk, porcupine, badger, and others — which seemed so surprised to see us living in their midst. It was, we felt, only thanks to their patience that our presence was tolerated.

For we — my wife Diana and I — were living, or trying to live, our own version of the ecological ideal. In 1972 we had bought this land, a place so remote that no white people had ever before lived there the year around. In the middle of the canyon stretch of Montana's Smith River country, we were fifty miles from the nearest town, twenty-five from a telephone and maintained road, and ten from the nearest neighbor. We built a log cabin and lived without electricity — cooking by propane, reading by Aladdin kerosene lanterns, heating with an Ashley woodstove, and cleaning with a Hokey carpet sweeper.

Leaving tenured academic positions, we moved to this ranch,

facing payments of $7,000 a year, hoping to make ends meet by running a summer natural-history program for young people. It was a crazy thing to do, yet it was the best thing we ever did.

We loved the ranch, finding peace there we had never known elsewhere.

It was while living there that I signed the contract to write this book. It seemed like the perfect assignment. I had known Yellowstone since spending the Fourth of July there with my parents, brother, and sister in 1947. While running the natural-history program I had regularly taken classes into the Yellowstone backcountry, trekking most of the trails of the park. I held a park concessionaire's license and was a licensed Montana outfitter and guide. Having received professional training in the philosophy of science, I could decipher the often arcane language of ecologists. As a member of the Board of Directors of the Yellowstone Library and Museum Association — the official publisher of books on the natural history of the park and the sponsor of a summer educational program known as the Yellowstone Institute — I understood intimately how the park was run. And because Yellowstone was a symbol for our national ideal of natural preservation, I would be writing about a place that was the model and inspiration for conservationists around the globe.

But there was a hitch. The publisher's advance was too small to support us until the book's completion. The dozen foundations to which I sent proposals declined to support the project. The only way I could write the book was to sell the ranch. To write about wilderness, we would have to give it up. In 1981 we sold the ranch and moved to Paradise Valley.

Yet not long after beginning work I realized the book's subject would be quite different from the one I had anticipated. "I feel very good about the grizzly," Yellowstone bear specialist Gary Brown told me in 1982. "I feel very pessimistic about the grizzly," Richard Knight, leader of the Interagency Grizzly Bear Study Team told me a week later. How, I wondered, could the manager be optimistic and the scientist be pessimistic?

In asking this question, I had set off on a search that turned out to be three years long to solve unexpected mysteries. Writing the book, rather than being an inspirational experience, turned out to be exciting, to be sure, but also disillusioning, exhausting, and sometimes dangerous. I had become a detective carried along in a high-speed adventure, searching for clues to the science, history, and politics of Yellowstone. And I found more hidden traps and intrigue than I could have anticipated.

Uncovering facts was sometimes a daunting task. Occasionally I found that critical records had disappeared. Freedom of Information Act requests took months to fulfill and often, when the material arrived, revealed inexplicable gaps. And though many in government and academe were eager to recount their feelings "off the record," most were reluctant to be quoted by name. The usual fate of whistle-blowers kept many officials silent, and the threat of losing their prerogative to study in Yellowstone frequently inhibited university scholars.

As obstacles multiplied and the challenges became evident, on countless occasions I would not have continued were it not for the strong support and encouragement of many people. My editor, Upton Brady, and agent, Wendy Weil, kept their faith in the book as deadline after deadline for delivery of the manuscript passed without a written word from me. At every turn some knowledgeable and generous person was there to explain the complexities of Yellowstone science and politics, to provide me with critical documents and bibliographic assistance, or later to read portions of the manuscript, checking it for accuracy.

For helping me piece together the story of the range, I am especially indebted to Charles Kay and Leslie Pengelly. Kay supplied me with an enormous amount of bibliographic material and spent hours beyond number teaching me biology, and Pengelly was a continuing source of information, education, and support. I want to express my gratitude to former Park Service historian Aubrey Haines, former park biologists Robert Howe and Walter Kittams, former Senator Gale McGee, and former Chief Naturalist David de L. Condon, for their help in reconstructing Yellowstone's past; to Professors Robert J. Jonas and James M. Peek for the education they gave me in the intricacies of range management; to Professor Gary Wright for teaching me so much about the park's prehistory; and to Jon Swenson, Arnold Foss, LeRoy Ellig, Vincent Yannone, and Glenn L. Erickson of Montana's Department of Fish, Wildlife and Parks, for their aid in supplying me with critical historical and biological information.

For help in pursuing the story of the wolves, I want to thank David Mech, biologist with the Fish and Wildlife Service; Maurice

Hornocker, former Team Leader of the Cooperative Wildlife Research Unit at the University of Idaho; Harry Reynolds III and Richard Sideler of the Alaska Department of Fish and Game; John Weaver of the Forest Service; Professors Bart O'Gara and Robert Ream of the Wolf Recovery Team; Merlin Shoesmith, John Stelfox, and Nicholas Novakowski of the Canadian Wildlife Service; Ronald Nowak, with the Fish and Wildlife Service's Office of Endangered Species; Clare Markley, with the Department of Industry and Trade in Canada; and Eric Klinghammer, President of the North American Wolf Foundation.

In seeking the truth about bears I benefited greatly from the help of John Craighead, Director of the Wildlands–Woodlands Institute, and his brother, Frank, senior research associate (emeritus), State University of New York, Albany, who patiently supplied me with data whenever I asked for it and tutored me — for hours on end — via telephone, on the biology of grizzlies. I am also indebted to Richard Knight, Leader of the Interagency Grizzly Bear Study Team; Tony Povilitis, Director of Save the Yellowstone Grizzlies; Lewis Regenstein, Vice-President of Fund for Animals; Lee Eberhardt, with the Batelle Memorial Laboratories; and Martha Shell, former freelance writer.

In piecing together the story of Grant Village, I received invaluable help from Joe Cutter, former West Yellowstone newspaperman; Dale Snape, former budget examiner in charge of Park Service oversight with the Office of Management and Budget; Don Hummell, President of the Association of National Park Concessioners; Richard B. Bowser, former engineer with the Park Service; and William Wahlen and Conrad L. Wirth, former directors of the Park Service.

In helping me tell the story of Park Service rangers and their science, I want to thank James Kushlan, former National Park Service biologist stationed in the Everglades; George Sprugel, formerly with the National Science Foundation and the Park Service Office of Scientific Studies; Robert Linn, former Chief Scientist of the Park Service; Bruce Shaeffer, Assistant Director of the Budget for the Park Service; Carol Bickley, John Reed, and Albert G. Greene, Jr., of the Office of the Assistant Director for Natural Resources of the Park Service; Edwin Bearss, Chief Historian of the Park Service, and Robert Utley, former Chief Historian; Lowell Sumner, Ben Thompson, Russ Grater, and Victor Cahalane, formerly with the Park Service's Division of Wild Life; Mrs. Pamela Wright Lloyd and Mrs. R. A. Brichetto, daughters of George M. Wright; Raymond E. Moran, with the Office of Personnel Management, Gerald Glaser, with the National Science Foundation; John A. Warnock, Codirector of the Committee to Save Yellowstone's Heritage; and Tim Clark, Department of Biology, Idaho State University.

I could not have told the story of the discovery of the caldera without the aid of Robert Christiansen and Warren Hamilton of the U.S. Geological Survey; Francis Boyd, with the Carnegie Geophysical Laboratory in Washington; John Good, former Chief Naturalist of Yellowstone; Robert D. Smith, Professor of Geology at the University of Utah; Robert Andrews of the National Academy of Sciences; and Owen Toon, with the Ames Research Laboratory, California.

In learning some of the theoretical problems of community ecology, I benefited greatly from conversations with Professors Frank Golley, of the University of Georgia's Institute of Ecology, and James Brown, of the University of Arizona. I want to thank Professor Edward O. Wilson of Harvard University, for putting me in touch with the appropriate experts on community ecology. I was introduced to the fascinating world of deep ecology by George Sessions, Professor of Philosophy at Sierra College. In tracking the effects of recreation use on our public wildlands, I am grateful for the aid of Charles Phenicie, of the Fish and Wildlife Service, and Susan Marsh, Forest Service Researcher with Gallatin National Forest.

I am also grateful to those in Yellowstone who were forthcoming in complying with my many requests for interviews and information: Research Administrator John Varley, geologists Wayne Hamilton and Rick Hutchinson, Resource Manager Ken Czarnowski, Chief Naturalist George Robinson, Assistant Chief Naturalist Norman Bishop, Sanitarian Peter Cook, Chief of Concessions Lee Davis, Chief of Maintenance Tim Hudson, landscape architect Dan Wenk, district rangers David Spirtes, Terry Danforth, and Anthony Sisto, fisheries biologists Ron Jones and Robert E. Gresswell, and Superintendent Robert Barbee.

I owe special thanks to Yellowstone archivist Tim Manns for the many hours he spent helping me find material, to librarians Beverly Whitman and Valery Black for their patience in letting me keep books for so long, and for the many hours they spent helping me find and photocopy documents.

Finally there are also those whom I must thank but who have asked to remain anonymous. Whenever I began to feel the pressure that inevitably comes to a solitary critic of a large government bureaucracy, someone from the Park Service would appear to give me encouragement and direction. Several risked their careers to provide information. I could not have written the book without their help. Perhaps knowing that is thanks enough.

And to Diana, my partner, who worked and worried over this book with me, what can I say? Perhaps our efforts will make a difference. Perhaps they will even inspire the rescue of Yellowstone. We can always hope. Why else did we leave our sanctuary on the Smith?

Contents

1. The State of Nature in Yellowstone   3

THE RANGE

2. Jonas and the Beaver   11
3. The Perils of Playing God   14
4. Killing Animals to Save Them   31
5. An Environmental Ideal and the Biology of   38
   Desperation
6. A Solution to the Elk Problem   49
7. Nature Takes Its Course   71
8. The Secret of Yancey's Hole   85
9. Growing Apples in Eden   92

THE WILDLIFE

10. The Wolf Mystery   119
11. Rendezvous at Death Gulch   142
12. The Grizzly and the Juggernaut   170

THE RANGERS

13. Grant Village and the Politics of Tourism   197
14. Gumshoes and Posy Pickers   232

15. The Deep Hole Gap 262

THE ENVIRONMENTALISTS

16. The New Pantheists 295
17. The Subverted Science 311
18. The Hubris Commandos 326
19. The California Cosmologists 344
20. The Land Ethic 363

CONCLUSION

21. Yellowstone Elegy 371
    Epilogue 377
    Notes and Sources 395
    Index 449

Playing God
in Yellowstone

1

The State of Nature in Yellowstone

JUST outside the town of Gardiner, Montana, is a large stone monument known as the Roosevelt arch. On it is written: "For the benefit and enjoyment of the people — Yellowstone National Park, created by an act of Congress, March 1, 1872."

This arch was once the main gateway to Yellowstone. When most visitors came by rail it was the first thing they saw as they climbed from trains at the station across the road and walked to waiting stagecoaches that would take them to Mammoth Hot Springs. The arch stood at the edge of a lush meadow that was irrigated by a small stream. Between the station and the arch was a tree-lined park and pond where tourists sat to watch the tame elk and antelope.[1]

During the spring of 1903, President Theodore Roosevelt came for the ceremony dedicating the arch. It was his third visit. While there, he rode through much of the park, and was impressed with what he saw.

"During my two hours following my entry into the Park," Roosevelt wrote, "we saw several hundred — probably a thousand all told — of these antelopes." Cougars — or mountain lions, as they are known in the West, he reported, "are plentiful." He encountered bands of mountain sheep, which he found "absurdly tame." He came upon "a score of blacktails" (which we now call mule deer), and at other times came upon them "in small bunches of a dozen or so." Coyotes he found to be "plentiful." One morning, after a careful count with binoculars, "we reckoned three thousand head of elk . . . all in sight at the same time."[2]

Bears, although then in hibernation, he remembered from earlier visits, "come to all the hotels in numbers." Even the wilder species of bear, the grizzlies, he remembered "boldly hanging out around crowded hotels." Although deep snow was still on the ground, he found Yellowstone a cornucopia of smaller wildlife, including pine squirrels, rabbits, martens, geese, crows, Steller's jays, magpies, robins, finches, juncos, bluebirds, mountain chickadees, pygmy nuthatches, mountain bluebirds, blue and ruffed grouse, and one pygmy owl.[3]

Roosevelt left Yellowstone deeply moved by the rich bounty of nature that Congress, with astounding foresight, had set aside.

"This reserve," he wrote, "is a natural breeding-ground and nursery for those stately and beautiful haunters of the wilds which have now vanished from so many of the great forests, the vast lonely plains, and the high mountain ranges, where they once abounded. . . . Surely," he concluded, "our people do not understand even yet the rich heritage that is theirs. . . . Our people should see to it that they are preserved for their children and their children's children forever, with their majestic beauty all unmarred."[4]

Ever since first the white man saw it, Yellowstone was thought to be special. "I almost wished," wrote young fur trapper Osborne Russell while traveling through this region in 1835, "I would spend the remainder of my days in a place like this where happiness and contentment seemed to reign in wild romantic splendor." And today the park remains a magical and majestic place.[5]

For Roosevelt's wish, it seems, was fulfilled: this land has been preserved unmarred. The great falls of the Yellowstone River, the open meadows and ice-blue lakes, the tame and ubiquitous wildlife, the countless steaming geysers and hot pools, all seem much as they must have when Russell saw them.

Indeed, Yellowstone is like a window to our past. The tameness and visibility of wild animals keep it a place of innocence — a Garden of Eden where we might see what America was like before the Fall. And this impression is sustained by all who write about it.

"This area," wrote historian Aubrey L. Haines in 1977, "is an unusual wildlife habitat, a great outdoor zoo preserving a more representative sample of the primeval fauna of the American West than is now found anywhere else. Here, living under conditions very nearly those existing when white men first entered the area, are elk, buffalo, mule deer, moose, antelope, bighorn sheep, black and grizzly bear, cougar, coyotes, wolves, beaver and a number of smaller animals, as well as many species of resident and migratory birds."[6]

Hamilton's Guide calls Yellowstone "the greatest wildlife sanctuary in the United States." In *Wildlife in Yellowstone and Grand Teton National Parks* we find that "nowhere else can you see such a complete assemblage of native Northern American wildlife in such a spacious natural environment." The park's own research office tells us that "with the possible exception of the Rocky Mountain wolf, representative populations of all native wild animals are yearlong residents." Another official publication adds: "We may have the occasional wolf."[7]

That Yellowstone remains in original condition is an idea in which we are entitled to believe. Larger than the states of Delaware and Rhode Island put together, it was made a park when the land belonged to the Shoshone, Bannock, Blackfoot, and Crow, and before the white man could despoil it. For more than a hundred years it has been the law of the land that the United States government is to preserve the park in its original condition.[8]

"Fortunately," John Muir wrote on visiting Yellowstone in 1885, "almost as soon as it was discovered it was dedicated and set apart for the benefit of the people." Under the care of the Department of the Interior, he commented, it was protected from the "blind, ruthless destruction that is going on in adjoining regions."[9]

The Yellowstone Park Act, signed by President Grant on March 1, 1872, in setting aside the park "as a pleasuring ground for the benefit and enjoyment of the people," protected against "wanton destruction of the fish and game found within said park."[10]

The Lacey Act, "to protect the birds and animals in Yellowstone National Park," signed by President Cleveland on May 7, 1894, specifically prohibited "all hunting or the killing, wounding or capture, at any time, of any wild animal or bird, except dangerous animals when it is necessary to prevent them from destroying human life or inflicting injury."[11]

The National Park Service Act, creating the National Park Service in 1916, made it the duty of the new agency "to conserve the scenery and the natural and historic objects and the wildlife therein, and to provide for the enjoyment of the same in such a manner and by such means as will leave them unimpaired for future generations." And for nearly seven decades this agency, whose sole mission is preservation, has remained the guardian of the park.[12]

Today Yellowstone remains a symbol of our aspirations as a people to preserve the natural and cultural roots of our frontier experience. If our heritage is to be saved anywhere, then surely it can be — and has been — saved in Yellowstone. As America's first national park it was the birthplace of wilderness preservation, an ideal that one government commission said was "regarded internationally as one of the finest examples of our national spirit." For the Park Service it remains, their spokesmen are fond of saying, "the flagship of our fleet." For the world community of conservationists, the prominent biologist A. Starker Leopold told me, it is "a park which serves as a model for preservation throughout the world." Yellowstone is, in short, a place for which all Americans should be thankful and of which they can be proud."[13]

Yet this prized land, reflecting our highest ideals of natural preservation, protected for a century, is not what it appears to be. Rather than a place where time stands still, it is caught in a river of change. Rather than preserved, it is being destroyed.

Those visiting Yellowstone today will not see what Roosevelt saw. They may encounter elk, bison, and an occasional coyote, but they will see no thousands of antelope, no plentiful sheep or mule deer. They will be fortunate to see a grizzly and will find no black bears begging along the road. They will find no wolves or white-tail deer, and, in all probability, no mountain lion, wolverine, lynx, bobcat, or fisher.

As a wildlife refuge, Yellowstone is dying. Indeed, the park's reputation as a great game sanctuary is perhaps the best-sustained myth in American conservation history and the story of its decline perhaps one of our government's best-kept secrets. Several of the original species are no longer there, and many others remain in much-reduced numbers, as the capacity of the range to sustain varied wildlife diminishes. Yet this decline in Yellowstone is accelerating even while many of these species are experiencing dramatic recovery in other parts of the Rocky Mountain West.

Despite Roosevelt's heartfelt desire, we have lost much of the rich heritage of Yellowstone, and are in danger of losing much more.

Why this loss, and why have we not learned about it earlier?

The Range

Every age is fed on illusions.
— Joseph Conrad

2

Jonas and the Beaver

IN fall 1952, Robert J. Jonas, a summertime, or "seasonal ranger" in Yellowstone, came to his supervisor, Chief Naturalist David de L. Condon, with a question: Why do visitors to the park no longer see beaver?

For decades it was the custom for guests at Roosevelt Lodge at Tower Junction to walk to Yancey's Hole in the evening along the old Cooke City Highway to watch the beaver work on their dams. But beaver were there no longer, and they were gone from other places where they once abounded.[1]

Condon could not say why the beaver had gone, and so Jonas, a graduate student at the University of Idaho at the time, decided to write his master's thesis in answer to his own question.[2]

The significance of the beavers' disappearance was not lost on Jonas. Beaver had been pivotal in the settlement of the West and the discovery of Yellowstone. The search for pelts had brought white men to the Rocky Mountains even before the region's first explorers, Meriwether Lewis and William Clark, had returned to St. Louis in 1806. The first regular visitors to Yellowstone were trappers. And although trapping depleted the animals, beaver were still plentiful when Yellowstone came under protection of the Department of the Interior in 1872. Early visitors found beaverdams, surrounded by wet and willow-lined meadows, throughout the park. Walter W. deLacy, crossing the area in 1863, complained of the difficulty in crossing the Madison drainage, "owing to the numerous beaver dams." The Earl of Dunraven, traveling through the park in 1874, remarked that "all the streams are full of beaver." John

Muir, during his 1885 visit, commented on the "numerous beaver meadows." They were still plentiful in the 1920s, when Milton Skinner, Chief Naturalist at the time, estimated "the beaver population of Yellowstone National Park at about 10,000, but believe that figure to be very conservative."[3]

Yet by the 1950s beaver were scarce. Why had they disappeared? Whatever had happened, Jonas suspected, was significant for the entire park. Perhaps no animal was more important in Yellowstone ecology than the beaver. By building his dams, he slowed spring runoff in the streams, discouraging erosion and siltation, keeping the water clean for the spawning trout. By building ponds, the beaver raised the surrounding water table, adding moisture that promoted vegetation — willow and aspen, forbs (broad-leaved plants such as aster, yarrow, and clover), berries and lush grass — that were essential foods for other animals. The ponds themselves provided habitat for waterfowl, mink, and otter. The beaver themselves were, some scientists say, essential prey for predators such as wolves.[4]

Of the vegetation, forbs and berries were significant for bears and other animals, but willow and aspen, so-called browse, were essential for the survival of elk and deer. These species — hoofed mammals known as ungulates — required a diet of both nitrogen and cellulose. The stalks of plants provided cellulose, which their digestive system turned into energy. But they used nitrogen to break the cellulose down. Nitrogen fed the microflora of the ungulates' stomachs, causing these organisms to multiply; and the microflora, in turn, did the job of turning cellulose into energy.[5]

In summer, green grass had nitrogen sufficient for a complete diet. But in winter when the grass was brown it lacked nitrogen. No matter how much grass, or other forms of cellulose, a deer or elk ate in winter, unless it could also find nitrogen it would die. It was therefore possible, and indeed common, for these animals to starve to death with a full stomach. The digestive system might have been filled with grass, but without nitrogen it was useless to them.

Aspen and willow, however, stored large amounts of nitrogen in the tips of their small branches and in their bark. And as nearly the sole winter source of this vital substance, they were the ungulates' lifeline.[6]

Because of its influence on browse and water, and as food for predators, the beaver's status indicated the condition of other animals. When beaver left an area, the animals and vegetation that benefited from their presence suffered. When they left an area for good, something had gone wrong with the system.[7]

In the normal undisturbed cycle, beaver moved into an area, a small stream with banks lined by willow or aspen, and ate and built their houses with these deciduous trees. When they had eaten all the food available, they moved on to areas where food was more abundant. After they left, the raised water table encouraged quick regeneration of the willow and aspen. And when the vegetation recovered the beaver returned. Typically in Yellowstone such a cycle took twenty to thirty years.[8]

Jonas examined every watershed in the park and walked the shoreline of every lake. He found that indeed Yellowstone had only a fraction as many beaver as scientists in the twenties had reported. At Yancey's and elsewhere the thirty-year beaver cycle had been broken. Beaver, once gone from an area, had not returned. In their absence the ponds had silted in, spring runoff in the streams had increased, the water table had dropped, and the drier ground was not producing the crop of palatable browse that it had supported when the beaver had been there. The wet beaver meadows filled with aspen and willow, so obvious to early visitors, had turned into dry, grassy plains.

What had happened to the beaver?

3

The Perils of Playing God

"As a game country in those early days," wrote historian Hiram Chittenden in 1895, "it could not compare with the lower, surrounding valleys."[1]

Indeed Chittenden, a West Pointer who had worked in Yellowstone between 1883 and 1918, had surely heard stories about the bounty of primeval fauna the first white men found in the West. He had probably heard how explorer Alexander Ross found elk abundant in the Bitterroot Valley in 1823, killing twenty-seven in one spot; how in 1832 artist George Catlin saw immense herds of wild horses along the upper Missouri; how in the same year Prince Maximillian saw countless bighorn sheep along the Missouri Breaks; how between 1854 and 1856 Sir George Gore, an Englishman, killed 6,000 buffalo along the Lower Yellowstone River. He may have read the report of an army engineer named Mullan, who in 1859 wrote that game along the Jefferson River Valley was "exceedingly abundant."[2]

But Yellowstone, Chittenden and his contemporaries thought, was altogether different. A large plateau straddling the continental divide, it seemed too high, cold, and uninviting to keep most game animals. As the source of both the Snake and Missouri rivers, it averaged 3,500 feet higher than the surrounding valleys; its winters were fierce and snow prolific. About 75 percent of the plateau was covered with climax forests of Douglas fir or subclimax stands of lodgepole pine, offering little for game to eat. Many of the waters of Yellowstone, including the giant Lewis and Shoshone lakes and the upper Lewis River, were barren of fish. And certainly the herbivores that ate grass and browse had little reason to spend a winter there, when they could find warm and lush terrain just down the valleys.[3]

Elk in particular, these witnesses were convinced, were not year-round residents of Yellowstone. They had no evidence that these animals had ever wintered in the park, and they knew that nearly all early explorers of Yellowstone had trouble finding game even in summer. For although the first visitors to the plateau saw or heard many grizzlies and mountain lions, they often traveled long distances without finding elk or deer.[4]

In 1856, scout Jim Bridger warned explorer Captain William F. Raynolds not to try to traverse the Yellowstone plateau, because he would find nothing to eat. "A bird can't fly over that," he warned, "without taking a supply of grub along."[5]

In 1870, the Langford expedition found little to eat during the last four weeks of its trip through the park. The Hayden expedition of 1871, though it employed professional hunters, reported finding only one animal, a mule deer, on the entire trip. The party of Captains Barlow and Heap, army engineers who explored the park in 1871, exhausted their food eight days before the end of their trip and had to send out for more. The 1873 party led by army engineer William A. Jones also ran out of food and had to be resupplied from Fort Ellis, a hundred miles away.[6]

The Dunraven party of 1874, composed of experienced hunters, could not even find an animal track in the Hayden Valley, where many elk and bison can be seen today. "Not a fresh track," wrote Dunraven at the Hayden Valley, "and nothing whatever eatable to be seen." Army Captain Robert Ludlow, after a rapid survey of the park in 1875, wrote that "two deer were seen, the only game animals we encountered in the park." The geologist Theodore Comstock reported seeing no animals during his exploration in 1873, except "mule deer which is occasionally met in this region." Comstock, who was one of the first to suggest Yellowstone as a game refuge, urged that *game be introduced to the park*.[7]

Comstock's suggestion was not so strange at the time. In 1870, game animals seemed to be disappearing almost overnight. Dunraven commented that even the elk "was soon to be numbered among things of the past."[8]

Once common throughout North America, by the end of the

nineteenth century elk had nearly disappeared from every state. In the 1880s and 1890s thousands were shot for their eyeteeth (the so-called ivories) alone, which were sold to members of the Fraternal Order of Elks for as much as $100 apiece.[9]

Buffalo, which numbered in the millions in 1870, were nearly extinct by 1890. During the 1870s, two to four million were shot a year; 200,000 hides were sold in St. Louis in one day.[10]

Yellowstone was not spared this carnage. When Congress created the park, it did not provide for enforcement against poachers. During the 1870s and early 1880s, elk, bighorn sheep, deer, antelope, and moose in Yellowstone were killed by the thousands, perhaps, according to the park's first Superintendent, Philetus W. Norris, as many as 7,000 in spring 1875 alone.[11]

Although no early visitor to Yellowstone reported seeing large numbers of buffalo, or bison, as they are more properly called, several small herds of around four hundred were believed to have summered in the park before the white man came. These were not the plains bison, *Bison bison bison* to biologists, but cousins of this famous animal, the mountain bison, *Bison bison athabascae Rhoads*.[12]

Unfortunately the fate of the two species was similar. During the 1870s and 1880s, before the animals were adequately protected, they fell victim to poaching. And like the other game animals, they were also killed and eaten by park employees. One of the first concessionaires, ironically called the Park Improvement Company, fed its employees on game killed in the park. In 1882, to save money on beef, it signed a contract with professional hunters for 20,000 pounds of elk, deer, mountain sheep, and bison.[13]

To put a stop to this carnage, the U.S. Cavalry was sent to rescue Yellowstone in 1886, and until 1916, when the National Park Service was created, it ran the park. In 1894 the Lacey Act was passed, providing strong enforcement of antipoaching laws in Yellowstone.[14]

The army, many thought, saved Yellowstone. Within a short time it stamped out poaching. It fed elk and antelope and sent out scouts to drive animals into the park. It put out garbage for bears, and put up a fence seven feet high and several miles long to keep animals in the park. And, by nearly all accounts, animals began to multiply.[15]

"The number of elk in the Park," reported the Acting Superintendent in 1890, "is something wonderful."[16]

"The elk," he reported in 1891, "have increased enormously." In 1892, they were described as "extremely numerous, and I'm not disposed to revise the least my estimate of 25,000 made last year." In 1894, the Superintendent's Report describes "a large number of young this spring. A party sent out to Yancey's to investigate this subject in March last saw at least 3,000 of them at one time from a single point of view." In 1895, they had increased in numbers." In 1899, the Acting Superintendent reported elk "are more numerous than any other animal in the Park." In 1903, Roosevelt remarked that the elk "were certainly more numerous than when I was last through the Park twelve years before."[17]

In 1912, the first real elk census reported 30,101. By 1914, the number counted was 35,308. Apparently the Earl of Dunraven was mistaken. Rather than being counted among things of the past, they were multiplying like amoebas![18]

Likewise, protection attracted other immigrants. Although moose had been seen in the south of the park as early as 1881 (one was photographed for *National Geographic* near Yellowstone Lake by Congressman and pioneer wildlife photographer George Shiras III in 1908), the first did not appear in the northern range until 1913. By 1895, the Acting Superintendent said, "bears had increased notably."[19]

And while native trout went into decline, their exotic brethren soon began to multiply. "In passing through the Park," wrote Acting Superintendent Captain F. A. Boutelle in 1891, "I noticed with surprise the barrenness of most of the water in the Park. Besides the beautiful Shoshone and other smaller lakes there are hundreds of miles of as fine streams as any in existence without a fish of any kind. . . . I hope . . . to see all of these waters so stocked that the pleasure-seeker in the Park can enjoy fine fishing within a few rods of any hotel or camp." And Boutelle, an avid angler himself, set about to do just that.[20]

During its tenure the cavalry brought game fish from all over the world to Yellowstone: Machinaw from Lake Superior, "von Bahr" browns from Germany, Loch Levens from Scotland, brook and landlocked salmon from the eastern shores of America, coho salmon, rainbow trout and mountain whitefish from the West, black bass and yellow perch from parts unknown. They also spread native species throughout the country, sending "green" and eyed eggs of

Yellowstone cutthroat to hatcheries, fishery exhibits, and aquariums in Idaho, Wyoming, South Dakota, Oregon, Colorado, Pennsylvania, Michigan, New York, Colorado, Wyoming, and Idaho.[21]

Many of these efforts were dismal failures: salmon and rainbow dumped into Yellowstone Lake were never seen again and black bass put into the Gibbon River and into the lakes of the lower geyser basin eventually disappeared. Others succeeded too well. Yellow perch put into Goose Lake became too abundant and had to be poisoned with derris root, and in many waters brown and rainbow trout drove out the native cutthroat and grayling. But few worried about these failures. Most visitors, enjoying the fabulous angling that these efforts produced, felt as Boutelle did: the park was being improved. Exotic species might be destroying original conditions in the park, but these conditions were never so nice as those man had made.[22]

The bison also were managed by the army, perhaps not wisely, but too well. When in 1901 only twenty-five animals could be found, saving the Yellowstone bison became a national crusade. Money or support for bison restoration came from Congress, the Smithsonian, the Boone and Crockett Club, and the Sierra Club.[23]

The army wasted no time. It brought plains bison bulls from Texas in a cattle car and bred them to other plains bison purchased locally. It plowed and seeded the Lamar and Slough Creek valleys with oats and timothy. It constructed a buffalo ranch. Corrals were built, bison bred, fed, and branded, and bulls castrated.[24]

The new species, created by crossbreeding mountain and plains bison, was of course not a native. But this exoticism did not seem to matter. Turning its swords into plowshares, the U.S. Cavalry had brought the buffalo back. By 1912 the Acting Superintendent reported that the buffalo were "thriving beyond expectations."[25]

Indeed, it soon became apparent that they throve too well. The population of bison continued to grow through the teens and twenties. In fifteen years the bison, saved from extinction, were apparently growing without limit.

Teddy Roosevelt was among the first to see elk as a problem. They were, he thought, overpopulating the park and endangering the range by overgrazing. In 1915, thanks to Roosevelt's urging, the game-preservation committee of the Boone and Crockett Club (of which he was a member), urged that their numbers be controlled. "The elk of these herds," they said, "should have scientific management. The Committee believe that in addition to those killed by sportsmen, several thousand should be killed each year. The number killed must be regulated to establish a correct balance between the food supply and the numbers of elk. . . . By no other means can the problem of these elk be properly solved."[26]

The government, however, would not take this advice. Instead, the newly created National Park Service, now running Yellowstone, did everything in its power to *increase* the number of elk. Convinced that they were in danger of extinction, it continued to feed them, and when animals strayed from the park, rangers were dispatched to chase them back.[27]

During the severe winter of 1916–1917, many elk died of starvation, reinforcing Park Service fears. Two scientists, Colonel Henry S. Graves, Chief Forester of the Forest Service, and Dr. E. W. Nelson, Chief of the Bureau of Biological Survey, were sent to study the elk. These men confirmed that the animals were in trouble. "The elk situation," they concluded, "has reached a crisis." The problem, however, was not too few elk, but too many. Crowded into an area too small, they lacked winter range.[28]

It was not natural for elk, the scientists said, to winter in the park. Before the white man came, they reported, elk left Yellowstone in the fall. They

> drifted out of the mountains ahead of the storms and snow, scattering over the bordering open valleys and plains where snowfall was light and where nourishing dry grasses were plentiful. At this season they often worked their way from 100 to more than 200 miles from their summer feeding grounds. In spring they followed the melting snow back to the high mountains above the zone of the annoying flies, where the climate was cool and refreshing, the forest offering grateful cover, and where fresh and succulent feed abounded.[29]

But, Graves and Nelson reported, as civilization surrounded Yellowstone, as farmers built fences and hunters lay in ambush, the elk could no longer reach their winter feeding grounds. They were trapped in the park all year, facing harsh winters and insufficient food. Nor, moreover, could the animals ever return to their old ways. "The original conditions," the scientists wrote, "with immense herds of elk roaming freely over a vast unoccupied area, can never be restored." Instead, they urged that the herd not be allowed to increase.[30]

The Park Service, however, ignored this advice. They continued to feed the elk and do everything possible to keep the animals in the park. And during the winter of 1919–1920 the fears of Nelson and Graves seemed to come true.

The previous summer had been exceptionally dry and little grass had grown. A wet snow began to fall in September and by the end of the month more had accumulated than in any September in thirty years. October was the coldest and snowiest ever recorded in Yellowstone, and the snow froze hard, forming an impenetrable ice shield over the grass. November was the second snowiest on record and the third coldest in thirty-two years. December had the second greatest amount of snow on the ground ever recorded. In Montana at this time the towns of Butte and Dillon were completely cut off; Billings and other cities were facing severe fuel crises, and domestic livestock were dying by the thousands.[31]

"Below zero weather and lack of feed due to the protracted drouth last summer caused the death by freezing and starvation of thousands of horses and cattle," the Livingston (Montana) *Enterprise* reported on December 3. Although the weather by February was less severe, it did not let up for long. By March, stockmen in Montana had spent $30 million importing cattle feed from other states. April was the coldest ever recorded in Yellowstone. By the end of that month, more than twenty inches of snow on the level lay at Soda Butte. On May 18, the *Enterprise* reported crews clearing the roads in the park faced "unusual snow conditions." Snow was reported to be twenty-five feet deep in the canyon south of Mammoth. By June the park was still reporting "unusual snow conditions for this time of year": three and one-half feet on the level in the national forest just north of Yellowstone.[32]

Throughout this winter, elk were dying at the feeding grounds, as many as 10 percent of those present each month. Because it was impossible to count elk in the mountains, no one knew how many died there. Previous counts had estimated elk populations in 1919 to have been 25,000; yet in the following summer only 11,000 could be found. "What became of the 14,000 elk which were missing in our northern herd on June 1st?" asked the new Superintendent, Horace Albright. "The loss of nearly 60 percent in one winter is alarming and indicates most forcibly the possible danger of complete extermination of this most noble race of animals."[33]

This winter's experience convinced Yellowstone managers that the elk needed even more protection. The fledgling bureaucracy, to attract tourists, wanted to display elk and other big game as their showpiece. Predators, they had decided, threatened these animals. The new Director of the Service, Stephen Mather (a member of the Sierra Club), sent a directive to Yellowstone's Acting Superintendent, Chester Lindsley, in February 1918, ordering him to cooperate with the U.S. Biological Survey in a program of "extermination of mountain lions and other predatory animals."[34]

Soon this became a full-blown campaign of extinction, or, as Albright later urged, an "open war." Despite their efforts, though, the number of big game continued to dwindle. By the mid-1920s, the pronghorn antelope and white-tail deer had almost disappeared.

"Antelope Soon Extinct," ran a local headline in 1914, and indeed, pronghorns, numbering more than a thousand at the turn of the century, were rare by the 1920s. During winter 1916 the entire herd left the park, and scouts, sent to drive them back, were to find only two hundred.[35]

Bighorn sheep, once the most numerous animal and ubiquitous in the park, had declined drastically. Authorities guessed that by 1924 only 300 remained. The mule deer population had dropped, according to official estimates, from 2,000 in 1915 to fewer than 1,000 by 1930.[36]

The white-tail deer, once common among the willow bottoms of the Yellowstone and Gardiner rivers, began to disappear. Although a count in 1912 had found them to be as common as black bear and "puma," they were gone altogether by 1924. "I first noticed," wrote Milton Skinner, Yellowstone Chief Naturalist in 1929,

that the white-tailed deer were declining in numbers in the autumn of 1917 when I failed to see them every day and in the numbers usually seen previous to that year. From that time on, they steadily decreased. . . . A five year old buck white-tailed deer, on May 15, 1922, was the only one seen by me during that year until fall when I saw the same buck several times about Mammoth, and also another buck of similar size was seen once on November 25, 1922. I never saw this second buck again, but the first one survived until the winter of 1923–1924, and then died from eating fox-tail grass.[37]

As these animals disappeared, park managers, convinced that wolves, mountain lions, and coyotes were the culprits, redoubled their efforts at predator control. Yet the more predators they killed, the greater the decline of game; and the greater the decline of game, the more predators they killed. The strange cycle persisted.

Elk and bison meanwhile continued to thrive. For the first time it began to occur to park managers that perhaps there were too many. Each year the vegetation seemed sparser. Could the elk and bison be damaging the range? In 1928 the Park Service asked Forest Service biologist W. M. Rush to find out. His privately financed study of the range of the northern Yellowstone elk herd was completed in 1932.[38]

Rush made a startling discovery. The winter range of the elk, he found, had deteriorated 50 percent since 1914, because of over-grazing by elk and drought; on more than half this range, sheet erosion had removed from one to two inches of topsoil; cheatgrass and rabbitbrush — and unpalatable grasses whose presence indicates overgrazing — were spreading; browse was disappearing and sagebrush, a less-favored food of the elk, showed heavy use. The problem, he stated, confirming the opinions of Nelson and Graves, was inadequate winter range. Elk, confined year round in an area where once they had only summered, were eating all the aspen and willow, preventing these species from regenerating.[39]

Rush estimated the elk population of the northern herd at 12,000 to 14,000 head and recommended that they should not be allowed to increase until the range had a chance to recover. The bison herd, then around 1,000, was large enough, Rush thought. He advocated they not be allowed to increase further. In thirty years the bison had been saved from extinction only to become a nuisance.[40]

In the year in which Rush began his study, a wealthy young man named George Wright joined the Park Service in Yosemite. A student of biologist Joseph Grinnell at the University of California, he was appalled to discover that the government had made no attempt to administer the parks along scientific lines. In 1929, Wright took a leave of absence from the Service and spent two

By 1924 the herd crashed. The best estimates of those on the scene was that 60 percent of the herd died of starvation. This event was followed by a continuous, slow decline, as the decimated herd attempted to eke out a living on an impoverished range. Meanwhile the damage they did was enormous. In 1941 a thorough inventory of this damage was taken by animal ecologist D. Irwin Rasmussen. His report became a classic that found its way into nearly all the textbooks on wildlife biology. There had been, he wrote,

> no evident increase in certain animals that have been encouraged by protection such as the blue grouse . . . a very marked decrease in the band-tailed pigeon, a protected species . . . severe overgrazing took place on the plateau as a result of an overabundance of deer. Certain forage species such as Salix and Rubus have been almost exterminated by this overgrazing. The growth of young aspen was halted over a period of these 20 years and an ideal winter browse range was overutilized to such an extent that carrying capacity was reduced on a large area to an estimated 5 to 10 percent of original conditions.[52]

By the 1930s the Kaibab had become a textbook example of the perils of playing God. It was a story that greatly influenced the intellectual development of the great American conservationist (and author of *A Sand County Almanac*) Aldo Leopold. Once a believer in killing wolves and lions to save deer, he had made a visit to the Kaibab in 1941, cementing a doubt that had been growing for decades: These animals, at the apex of the biotic pyramid, were not, as his contemporaries still thought, the "Nazis of the forest"; rather they were the indispensable cornerstone for game management. The true threat was the deer, which, multiplying in the absence of natural enemies, could quickly destroy a forest.[53]

"I have lived to see state after state extirpate its wolves," Leopold wrote in his landmark confessional essay, "Thinking Like a Mountain," in 1944. "I have watched the face of many a newly wolfless mountain, and seen the south-facing slopes wrinkle with a maze of new deer trails. I have seen every edible bush and seedling browsed, first to anaemic desuetude, and then to death. I have seen every edible tree defoliated to the height of a saddlehorn."[54]

Natural predation, he came to believe, was "the only precision instrument known to deer management." With predators present, ungulate populations would go through population cycles, but the extremes of these cycles would be dampened. As deer grew in number, so too would the predators that ate them. Before deer became so numerous that they damaged the range, the combined effects of predators and a declining diet caused their decrease.[55]

Without predation, Leopold argued, prey species suffered a population boom that he called an "irruptive sequence." It had four stages.[56]

First, limited hunting and elimination of predators caused prey to multiply. Second, a "deer line" (signs of increased use) appeared on palatable browse, as the herd continued to grow. This increased grazing in turn produced a "self-aggravating" condition: it stimulated the growth of vegetation, which in turn encouraged deer to multiply even faster. Third, as numbers continued to grow, a deer line appeared on unpalatable browse as fawns died of starvation in severe winters and the herd peaked. Fourth came the crash: many adults died as the herd was attacked by disease, palatable browse continued to disappear, and unpalatable plant species spread. Eventually, a plateau, far below the preirruptive carrying capacity of the range, might be reached. This Leopold called a "downhill equilibrium," a decline so gradual that it passed unnoticed. Yet, he wrote, "under this 'downhill equilibrium' our browse will ultimately be killed off more completely than it would under an out-and-out irruption of the Kaibab type."

Such irruptions, Leopold discovered, had occurred, not just in the Kaibab, but throughout the country. Of forty-seven states with deer, he found that thirty were suffering from overpopulation. In the absence of predators, he concluded, the only way to prevent an irruptive sequence was by periodic reductions in the herd. He did not feel that this action was tampering with nature. For some time he had been convinced that truly natural conditions no longer existed. "I must confess," he wrote to the Superintendent of Glacier National Park in 1927, "that it seems to me academic to talk about maintaining the balance of nature. The balance of nature in any strict sense has been upset long ago, and there is no such thing to maintain. The only option we have is to create a new balance objectively determined upon for each area in accordance with the intended use of that area."[57]

* * *

After Murie, and with slow epiphany, Yellowstone managers came to realize the damage they had done. And they were haunted by the specter of the Kaibab, for the parallel between the two plateaus was uncomfortably close. In Yellowstone, as in the Kaibab, they noticed, predators had been eliminated, the elk herd had irrupted, and then (during the winter of 1919–1920), it crashed. Now they were in danger of destroying the range.

Provoked by Wright, Rush, and Murie, and increasingly worried that Yellowstone might be suffering the fate of the Kaibab, the National Park Service took steps to see that the herd would grow no larger. Beginning in 1934, they embarked on a program to contain the size of the herd by trapping and transplanting elk to other game ranges and wild areas, and, when necessary, by "direct control"; excess elk that could not be given away were shot.[58]

Simultaneously the Park Service, for the first time in Yellowstone history, began studying the range intensively. They hired range specialists, they invited research by independent biologists, and they began cooperative research with Montana and Wyoming and with other federal agencies.[59]

Scientists tagged elk to follow their movements and make them easier to count. They monitored grass plots, studied the spread of exotic and noxious grasses, and examined browse propagation. They measured spring runoff and soil erosion. They built exclosures, plots fenced to keep grazing animals away. By carefully comparing the vegetation within and outside of these areas, they hoped to tell the effect grazing had on the range.[60]

And the more they learned, the worse things seemed. "A detailed study of young aspen was made between 1947 and 1952," reported range wildlife biologist Walt Kittams, Yellowstone's first resident scientist. "Of 394 small trees under study," he concluded, "only 94, or 24 percent, survived the five-year period."[61]

Old photos, he remarked as well, "indicate that nearly all the bottoms of drainages were once covered by willows. Now willow thickets are almost nonexistent."[62]

Likewise the palatable grasses of the park were disappearing. "More recent and intensive analysis shows," Kittams wrote, that the late winter range "is in a seriously deteriorated condition . . . with almost complete annual removal of forage. . . . With this change in vegetation less forage is produced, and the soil has been left more vulnerable to erosion by wind and water."[63]

Exotic grasses and less palatable vegetation continued to increase. "The choice bunchgrasses have thinned out," reported Kittams, "and bluegrass and Junegrass have filled in."[64]

Elk trampling, Montana researchers discovered, was compacting the soil, diminishing its porosity — its capacity for absorbing water. When it rained, the water ran off the top of the ground rather than soaking in. As a result the soil dried out, decreasing plant growth, aggravating the condition further. The water table dropped, the spring runoff increased turbidity in the Lamar River, and the soil was eroded further.[65]

Range decline had become a vicious circle, and just when it seemed things could not get worse, Jonas solved the mystery of the beaver. These animals were dying out, he found, because their food was gone. "A result," he concluded, "more from the overpopulation of elk than from any other single cause." Once the beaver had left an area, Jonas discovered, elk moved in, eating the small shoots of willow and aspen and preventing their regeneration. Elk trampled the soil and reduced its porosity, causing greater spring runoff, which in turn caused premature siltation of the beaver ponds. As the ponds filled in, the ground dried out, discouraging further growth of browse. After the elk took an area over, they left nothing for the beaver to come back to.[66]

With Jonas's discovery that the beaver, an ecological barometer, was endangered, the last pieces of the Yellowstone puzzle fell into place. Now officials knew that, despite thirty years of trying, they had not made the elk problem go away. Between 1935 and 1961 more than 58,000 elk had been taken from the park. Yet despite this culling, elk fecundity was getting the better of them: the herd stubbornly remained at 10,000. All the while the range and animals were in decline.[67]

This was the situation that faced Lemuel "Lon" Garrison when he became Yellowstone's Superintendent in 1956 and Robert Howe when he succeeded Kittams as park biologist in 1957. Yellowstone, they were convinced, was in deep trouble. It was not good wildlife habitat: it was too high, too cold, and too covered with thick forests to support much life before the white man came. Instead, the abundance of game that Roosevelt and others saw in the park at the turn of the century was the result of protective policies and immigration of animals during the last half of the nineteenth century. The animals were there because Yellowstone was made a sanctuary at the time when game was relentlessly hunted elsewhere. Like a persecuted tribe fleeing a holocaust, they fled to Yellowstone, not because it was the best place for them, but because it was their only salvation. Once there, farmers' fences and continued hunting outside the park kept them in.[68]

Yet the park was too small; it was never a complete ecological unit for the larger animals, and it had been unalterably changed from its original condition. Wolves and larger predators had been eliminated; the Indians, who hunted the game animals, had been sent to reservations; the traditional migration routes of elk and antelope had been cut off; exotic grasses had been planted to feed the elk and bison.[69]

The elk that lived in the northern half of the park — the so-called northern range — in particular, they were convinced, were trapped in a sanctuary too small. Yellowstone had never been their winter range, and these animals, now staying in the park all year, were threatening other species. Already they had driven out the white-tail deer by consuming the dense riparian willow thickets on which these animals depended for food and cover. They had colonized the territory of the mule deer, bighorn, and pronghorn, pushing these animals into poor habitat where, because of poor nutrition, they fell prey to disease. And now Jonas had shown that even the beaver were suffering.

The problem seemed clearly defined. Never a complete habitat for the animals it now held, Yellowstone, altered by earlier misguided attempts at preservation, was even less capable of carrying them. No longer in its original condition, overuse was making it less pristine day by day.

Once man had played God, they were convinced, he could not stop his intervention. The primeval scene was gone forever. A manmade problem had struck Yellowstone, a problem that human intervention created and human intervention must solve. Natural conditions, like virginity, once lost, could never be recovered. To break the vicious circle, park managers would have to fill the void created by the departure of Indians and predators. They must now try to restore a balance between ungulates and their range. "It is National Park Service policy," wrote Superintendent Garrison at the time, to maintain a balance between Yellowstone's animal populations and their environments or reestablish such a balance where necessary. . . . It is our responsibility to manage the Northern Yellowstone Elk Herd in such a way that the vanished white-tailed deer can return, that beaver, antelope, mule deer, bison and bighorn can hold their own, and that the elk herd itself remains healthy instead of further damaging their impoverished range.[70]

By 1961 the Park Service had made plans to remove 5,000 elk. They also planned a reduction of bison. For although periodic reductions had kept the bison herd at around 1,000 animals, this total was, biologists thought, still 350 too many.[71]

Simultaneously, Garrison, sensitive to possible public criticism, launched an ambitious campaign to educate the public and to involve scientists from the university communities. He made the rounds of the luncheon circuit, seeking academic legitimacy. He sought endorsements by prominent university scientists: Biologists such as John Craighead and W. Leslie Pengelly of the University of Montana were recruited to work on the elk problem. Further cooperative studies were initiated with the Montana Cooperative Wildlife Research Unit (headed by John Craighead) and with the Montana Department of Fish and Game. More exclosures and elk traps were built. More elk were tagged, and for the first time animals were fitted with radio collars.[72]

At last it seemed the Park Service would be doing things right. Yellowstone was to be restored. Park managers had taken Aldo Leopold's advice to heart, and were to create, for Yellowstone, "a new balance of nature, objectively determined."

4

Killing Animals to Save Them

FOUR miles from Tower Junction near the Lamar River, just out of sight of the road, was the Crystal Creek trap. Constructed of high fences with thick wood planks, it stretched for hundreds of yards. Seen from the air, it looked like a large funnel — an opening a half mile wide at one end, narrowing to a bottleneck that led to a solid wood compound. The only exit to the compound was a labyrinth of chutes that seemed to lead nowhere.

Crystal Creek was one of five traps used in the 1960s. It was here, on the morning of December 12, 1961, during a snowstorm, that the first of the great reductions began, and they continued through the unusually cold and snowy winter.[1]

It was planned and executed with the precision of a military operation. Salted alfalfa was strewn at the mouth of the trap to bait the elk. Helicopters were sent to find herds. When elk or bison got near the trap's mouth, they were harried in by helicopters, Sno-Cats, and sometimes by rangers on horses.[2]

As the terrified animals reached the corral, the gate was closed behind them. Many tried to climb the eight-foot walls and fell, gashed and bleeding. Occasionally one broke a leg and had to be shot. Rarely an animal did escape. Sometimes fights broke out between bulls or between elk and bison that entered the trap with the elk, and an animal was gored to death. The others were driven through the chutes, where veterinarians stood by squeezeboxes to take blood samples. Rangers fit some with collars. Then, even in the bitter cold, they were pushed into tanks of disinfectant and loaded into open trucks to be shipped to game ranges in Montana,

Pennsylvania, New Hampshire, Texas, and seven other states. Altogether, nearly three hundred made the trip to a new home.[3]

The bison caught in the trap were transferred to the corrals at the Buffalo ranch, where they were tested for brucellosis (or undulant fever) and shipped to a bison ranch in Wyoming, to be used to breed bison for meat.[4]

The direct reductions proceeded simultaneously. Before the shooting started, orders for elk meat were solicited from Indian reservations, orphanages, prisons, and welfare agencies. Contracts were let to teams of Blackfoot Indian butchers, and meatpacking companies. Quick-freeze lockers were obtained, veterinarians, university scientists, and meat inspectors were on hand.

In the early morning, shooting crews were dispatched. Helicopters hazed the elk toward a waiting team of rangers that had moved into position by Sno-Cat. The bewildered animals, terrified by the chopper, stood still while the rangers with high-powered rifles walked within point-blank range and began to shoot. Biologists immediately went to work taking blood samples, examining fetuses, livers, and hearts. Then came the Blackfoot butchers, who skinned the animals but left the intestines to be examined later by veterinarians and meat inspectors. The carcasses were carted off in mobile refrigerators to Livingston, Montana, for inspection, butchering, and shipping. In six weeks, the direct-reduction teams had shot more than 4,300 elk.[5]

Even for hardened observers, watching these symbols of wild America that only a few decades earlier had been on the verge of extinction treated as so much meat on the hoof was upsetting. The rangers hated the work. And although every effort was made to do the job in a humane, low-key, and sanitary way, slip-ups did occur. Sometimes the veterinarians could not keep up with the killing, and butchers had to wait so long the meat spoiled. Sometimes suffering animals were not quickly put out of their misery. And although it was Park Service policy not to permit any shooting near a road, occasionally it happened. Once such a slaughter took place near Tower Junction just off the road to Cooke City, where passing motorists watched in horror from their cars.[6]

Scenes like this provided a field day for the press, which inflamed public sentiment in defense of the elk. "Yellowstone's Great Elk Slaughter," Arnold Olsen, Congressman from Montana, wrote in *Sports Afield*. "One wounded cow," he reported, "was squealing and the butchers, who were unarmed, had to dispatch the pitiful animal with a handaxe."[7]

It did not take many such stories to fuel the fire of reaction. Hunters in particular were incensed. They wanted to hunt the elk themselves and saw shooting by rangers as an example of government waste. Despite Garrison's efforts to prepare the public, demands to end the killing grew.[8]

In response to this protest, Secretary of the Interior Stewart Udall established a committee to evaluate the Park Service game-management program. This committee, the Advisory Board on Wildlife Management, was chaired by A. Starker Leopold, Professor of zoology at the University of California and son of Aldo Leopold.[9]

Leopold saw his charge as having broad implications. "We knew," he told me, "the world was looking at us. If we were to recommend public hunting of elk, parks in Africa would feel pressed to permit the public hunting of elephant. We decided that we would develop a philosophy of management that could be applied universally."[10]

In March 1963, the "Leopold Report," as it came to be known, was sent to Udall.

The report gave ringing endorsement to science in Yellowstone and Garrison's elk policy. "As models of [scientific research] that should be greatly accelerated," the committee said, "we cite some of the recent studies of elk in Yellowstone." Reductions, moreover, they urged, should continue. "The annual removal from this herd," they said, "may be in the neighborhood of 1,000 to 1,800 head . . . not a large operation when one considers that approximately 100,000 head of big game are taken annually by hunters in Wyoming and Montana."[11]

Besides these specific recommendations for Yellowstone, the report also articulated a general philosophy of wildlife management, the implications of which, its authors admitted, were "stupendous."[12]

"As a primary goal," they said, "we would recommend that the biotic associations within each park be maintained, or where necessary re-created, as nearly as possible in the condition that prevailed when the area was first visited by the white man."[13]

Managing parks as original ecosystems, they continued, required that "observable artificiality in any form must be minimized and obscured in every possible way. Wildlife should not be displayed in fence enclosures; this is the function of a zoo, not a national park. In the same category is artificial feeding of wildlife."[14]

By these means, the committee suggested, "a reasonable illusion of primitive America could be re-created, using the utmost in skill, judgment and ecological sensitivity." Unfortunately, accomplishing this goal called for "a set of ecologic skills unknown in the country today." To develop these skills, the committee urged that "a greatly expanded research program, oriented to management needs, must be developed within the National Park Service itself. Both research and the application of management methods should be in the hands of skilled park personnel." And if the primitive scene was to be accurately re-created, they observed, "the first step in park management is historical research."[15]

Udall was reluctant to accept the committee's findings. The report, Leopold told me, "was so potentially controversial it scared the Secretary to death. But . . . the environmental community received it so enthusiastically that Udall changed his mind. Within three months, it was official Park Service policy."[16]

Indeed, on May 2, 1963, the Secretary instructed the Park Service to "take such steps as are appropriate to incorporate this philosophy and the basic findings into the administration of the National Park Service."[17]

Udall might well have seen the report as potentially controversial: although attractive, it was extremely ambitious.

Park authorities had set for themselves a modest goal. They saw a specific thing that needed fixing: restoring a balance between ungulates and the range. To this end they laid plans for elk reduction and, tentatively, reintroduction of predators. Yet even these ends, they knew, might be too difficult to achieve.

By restoring a "vignette of primitive America," the Leopold Committee, by contrast, would do nothing less than reconstruct the primitive ecosystem! Yet achieving this goal, at the very least, faced enormous obstacles. To do the job properly required learning enough about the past to know how the white man had disturbed it — knowledge that we did not yet know how to obtain. It required knowing how natural processes worked, and this science was in its infancy. The work would have entailed eliminating many species of trout and perhaps the end of fishing in the park. And, although

Leopold overlooked this implication, it would have required reintroducing the Indian.

"Restoring the primitive scene," in short, required knowledge we did not have and intervention far more massive and controversial than periodic elk reductions. The son Starker was advocating restoration of a balance of nature that father Aldo — and park planners of the time — had thought was gone forever. The Leopold Report advocated that parks be windows to the past, but a past that we did not know and perhaps did not wish to recapture.

At the same time, the enormous task of re-creating and maintaining natural conditions was to be accomplished, not by independent scientists, but by government managers; and it was to be done, somehow, while keeping man away from nature. Citing studies in Yellowstone (most of which were done by independent university scientists) as models of research, the committee nevertheless advocated that all research be put under control of Park Service managers. Giving highest priority to historical studies, the committee nonetheless stipulated that "every phase of management itself be under the full jurisdiction of biologically trained personnel of the Park Service." Praising a science that used tags, traps, and exclosures, it still urged "minimizing observable artificiality in every way."[18]

The Leopold Report had, in short, inadvertently replaced science with nostalgia, subverting the goal it had set out to support. But this implication passed unnoticed. Instead, the immediate effect of the Leopold Report was to add momentum to the efforts of park managers in restoring the northern range. In 1962 they laid plans for an ambitious program to restore native grasses by reseeding. In 1965 they decided to reintroduce the mountain lion, and in 1967 laid similar plans for the wolf. They also continued to kill elk and bison, by 1965 reducing the northern elk herd to fewer than 5,000 and the bison in the same range to 85.[19]

They also killed antelope. The elk reductions of the past thirty years, park managers thought, had helped the pronghorns recover from their near disappearance in the 1920s. Apparently, though, they had recovered too well. By 1945 an estimated 800 were in the park. And that, authorities decided, was too many. The dwindling vegetation of their home range could stand no more than 300. They began direct reductions.[20]

By 1965 the herd had been reduced to 300, but park authorities decided it should be reduced even further. In the spirit of combining killing with science, the park asked a biologist from the University of Montana, Bart O'Gara, to do both. To determine the health of the herd, O'Gara decided to study the reproductive physiology of the antelope. To do so, he would have to study antelope fetuses, and to study fetuses, he must shoot the mothers-to-be.

Between 1965 and 1966, O'Gara shot and dissected 100 pregnant female pronghorns; his study tentatively concluded that coyote predation was "decimating" the park antelope population.[21]

Yet it seemed the threat to antelope came, not from the coyote, but from man. By spring 1967 a reliable census could find only 188 animals, and their number was clearly on the decline. Authorities began to worry they had so drastically culled antelope that the antelope could no longer sustain themselves.[22]

The killing of elk continued, however, and public criticism of reductions continued to grow. For unfortunately, the Park Service had little evidence with which to refute this criticism. Changes in vegetation and range came slowly. Five years, an eternity in politics, was an instant in ecology. As George Hartzog, Director of the Park Service, put it in 1967, "not enough time has passed since our massive herd reduction of 1961–1962 to recognize concrete changes in Northern Yellowstone winter range conditions." The public would simply have to wait to see the good that would come of all this killing.[23]

But the public would not wait. It was difficult to convince them that the Park Service was killing elk to save them, that this carnage was for the animals' own good.

During the winter of 1966–1967, park authorities conducted another hunt, one of the least bloody. Altogether, 1,146 elk were live-trapped and only 394 killed by rangers. Indeed, the Park Service had reached its goal. The count that year found only 3,842 elk and 397 bison. The elk herd had been reduced to the number that scientists thought the vegetation of the park could sustain. But it was, it seemed, the last straw. The public would stand for no more killing.[24]

The resistance came from the hunters. They did not want rangers to kill elk because they wanted to do it themselves, and they wanted legislation that would permit public hunting in Yellowstone.

Senate hearings on the Park Service's elk policies were convened in Casper, Wyoming in March 1967. Sportsmen's groups united in common cause to promote public hunting. The inquiry was led by Gale McGee, Senator from Wyoming. The night before the hearing, McGee and Hartzog had dinner together and agreed on the outline of a solution. According to the new plan, McGee told me recently, "The Park Service would control the elk population by live-trapping only."[25]

This "solution," however, would not work and the Park Service knew it. The demand for live elk could not keep up with the supply. Yellowstone had already saturated the market.

In eighty years the park had replenished more than an entire continent with elk. Animals had been sent to thirty-eight states and eighteen other countries. By 1967 most of the elk in the country were related to the Yellowstone species. But now these game ranges were reaching their limits. And despite promises by Montana and Wyoming at the hearings that they would each take a thousand or more elk every year, they privately told the Park Service they did not want any more. John S. McLaughlin, who had succeeded Garrison as Superintendent in 1965, was told by Montana officials to send the elk to Wyoming, and told by Wyoming to send them to Montana. Immediately after the Senate hearings, Yellowstone officials sent urgent letters to thirty-four zoos throughout the country, begging them to take elk. By June they had received orders for four bulls.[26]

Yet each winter, apparently forever, Yellowstone would need to get rid of more than a thousand animals. Clearly, live-trapping alone would not do the job indefinitely. The Park Service was between a political rock and a biological hard place.

5

An Environmental Ideal and the Biology of Desperation

THE Park Service, like any college or university, knew what to do when its team was losing: fire the coach. McLaughlin, Superintendent for only two years but a political liability following the elk controversy, was reassigned during summer 1967. That September, Jack Anderson, formerly Superintendent of Grand Teton National Park, replaced him.

A month earlier, Anderson's biologist at Grand Teton, Glen Cole, came to Yellowstone as Supervisory Research Biologist. Cole's task was daunting. Forbidden to kill elk and knowing offers of live elk went begging, he had to find a way to solve a problem that had stymied park managers for two generations. Yet within two months of arriving in Yellowstone he had conceived of a solution. It was a policy he called "natural control." The previous herd reductions, Cole said, had so lowered the number of elk that further reductions would be unnecessary.[1]

The northern Yellowstone elk herd, Cole hypothesized, was actually two: one migratory segment, the other resident. Each at that time numbered around 2,000. The migratory segment, he suggested, "would be held within a general 3,000- to 5,000-animal range by hunting outside park boundaries." The resident herd, he supposed, was living in the "ecologically intact portion of the Yellowstone ecosystem," and would be "naturally controlled by the combined actions of native predators and periodic severe winters." He predicted that, even without efforts by the Park Service, the total number of elk would not grow beyond 6,000.[2]

Elk numbers could be contained, in short, by public hunting outside the park and by mother nature within it. No longer would rangers need to kill elk. With one fell swoop the elk problem that had been dogging park managers for fifty years was made to disappear.

This was a surprising theory to come from Cole. Just a few years earlier, while working for Montana's Department of Fish and Game — an agency that derived much of its income from hunting — he had argued for elk hunting as a way to prevent ungulates from destroying their range. "Attempts to carry more deer or elk than range will support," he wrote in 1963, "only result in overuse of forage plants and a winter range which will carry fewer and fewer animals through the year." Now Cole had decided that elk could not destroy range. The "resident" herd would be left unmolested.[3]

Cole's timing, however, could not have been better. For his reappraisal of Yellowstone's elk problem fell right into step with a simultaneous service-wide reappraisal of wildlife management. In 1964 Interior Secretary Udall approved reorganization of the national park system, dividing public lands into three zones: natural, historical, and recreational. In the same year Udall directed George Hartzog, Director of the Park Service, to implement the recommendations of the Leopold Report, emphasizing that "park management shall recognize and respect wilderness as a whole environment of living things whose use and enjoyment depend on their continuing interrelationship free from man's spoliation."[4]

In 1968 Hartzog published the policies that were to guide management of natural, historical, and recreational areas (the so-called Green, Red, and Blue Books). Natural areas such as Yellowstone, the Green Book said,

> shall be managed so as to conserve, perpetuate, and portray as a composite whole the indigenous aquatic and terrestrial fauna and flora and scenic landscape. Management will minimize, give direction to, or control those changes in the native environment and scenic landscape resulting from human influences on natural processes of ecological succession. Missing native life forms may be reestablished, where practicable. Native environmental complexes will be restored, protected, and maintained, where practicable, at levels determined through historical and ecological research of plant–

animal relationships. Non-native species may not be introduced into natural areas. Where they have become established . . . an appropriate management plan should be developed to control them. . . ."[5]

Henceforth parks should be managed, in short, not as farms, but (as stated in the Leopold Report) as "biotic wholes." No longer should rangers worry about fine-tuning relations among beaver, sheep, deer, bear, and elk. "The concept of preservation of a total environment, as compared with the protection of an individual feature or species, is a distinguishing feature of national park management," the Green Book said.[6]

This outline is the one Cole set out to follow. And as the Green Book stipulated, "control through natural predation will be encouraged" for wildlife populations, Cole was to encourage natural predators to control the elk.[7]

In their enthusiasm for the new science, however, Park Service scientists went a step further. Quite on their own they *reinterpreted* the new law. Although the Green Book had stipulated that the purpose of the parks was to preserve "indigenous aquatic and terrestrial fauna and flora," Cole (and other government biologists) had construed this statement to mean "the primary purpose of Yellowstone National Park is to preserve natural ecosystems."[8]

This goal seemed a reasonable and even exciting advance in park management. There was no way, it implied, to preserve individual living things: preservation was not embalming. We could not maintain our wild heritage by spraying every tree with a protective coat of plastic, keeping animals in cocoons, or using technology to turn out bionic bear and antelope. The best we could do was preserve the system that permitted these living things to reproduce themselves.

Yet the differences between preserving species of plants and animals and preserving biotic systems was potentially great. For natural processes were perfectly capable of exterminating animals — even entire species. Sometimes, when nature takes its course, the worst happens. What would managers do if, after allowing the ecosystem to operate, some kind of animal — say the grizzly or bighorn — got into trouble?

Saving a species in these circumstances might require that the Park Service do just what the army did to save the bison — taking steps such as providing artificial feeding, manipulating the habitat, introducing non-native genetic stock, and even running a breeding program. Such efforts would clearly be tampering with the "natural ecosystem," and if the law forbade tampering, the Park Service could not do what the cavalry had done. Yet surely everyone was glad the buffalo had not disappeared. If, therefore, a mandate to preserve natural ecosystems prohibited saving another species as the army had rescued the buffalo, what would Yellowstone managers do? Would they stand silently by, wringing their collective hands, as a wild and rare species went to oblivion? And how would the public react? Was this how they wanted their parks to be run?

Underlying the commitment to ecosystems management, therefore, lay a fundamental shift in what was believed to be the purpose of the parks. Indeed, however reasonable or unreasonable, to speak of preserving an ecosystem was not just a new, more scientific way of expressing the original purpose of the parks. Rather, these biologists had quietly redefined the mission of the Park Service. They were giving a blank check to nature. And how would nature spend it? Were they, in turning over the future of wildlife in Yellowstone to the "natural ecosystem," implicitly deciding what animals — or even species — would live or die? If so, then in taking a passive role they would not have stopped playing God; they would only have changed the rules of the game.

This revision passed unnoticed. The new policy seemed at the time nothing more than a scientific rendition of the original Park Service mission.

Yet how scientific was it? What, for that matter, was an ecosystem? Few biologists, when they were pinned down, claimed to know. Every species had its own ecosystem. Some, like that of the Devil's Hole pupfish, were no bigger than an average-sized living room; others, like that of the elk, encompassed hundreds of square miles. The range of the grizzly, for instance, was huge. The home range of the average Yellowstone male grizzly was 318 square miles, and some individuals had ranges of more than 600 square miles. And we did not know what their range was in primeval times. We did not know how far the elk or bison roamed, or whether Yellowstone grizzlies in bad food years once traveled hundreds of miles in search of sustenance.[9]

Was any park in the country big enough to be a natural ecosystem for all the species it contained? According to a report of the First World Conference on National Parks, held in Seattle in July 1962, "few of the world's parks are large enough to be in fact self-regulatory ecological units." Yet if they were not, if animals such as bears and elk wandered out of them, how could these species be preserved when, once outside the park, another, entirely different policy prevailed? How could "ecosystems management" work if the park was not an ecosystem?[10]

Nevertheless, the idea of ecosystems management, conceived by Park Service scientists in 1968, eventually became Park Service policy. When administrative policies were revised again in 1978, ecosystems management was made official. In the management of wildlife, according to the new law (still in effect), "natural processes shall be relied upon to regulate populations of native species to the greatest extent possible. . . . Regulation of native animal populations in natural zones shall be permitted to occur by natural means to the greatest extent possible." For natural areas such as Yellowstone, "the concept of perpetuation of a natural environment *or ecosystem* [italics added], as compared with the protection of individual features or species, is a distinguishing aspect."[11]

To the surprise of park managers the old policy had become a victim of a revolution in the ways in which Americans viewed nature — the very revolution they had thought would foster understanding and support for their policies.

Lon Garrison had begun the reductions assuming that the public could understand what they were trying to do. Because the 1960s was a period of renaissance in ecology, people should be better able to understand the complexities of wildlife management. "We were," he wrote later, "just at the forefront of a surge of public awareness of environmental welfare (and of course, we soon had a great breakthrough in conservation literature: *Silent Spring* [Rachel Carson, 1962], *The Quiet Crisis* [Stewart Udall, 1963], and *The Greening of America* [Charles A. Reich, 1970]. The later resurgence of Aldo Leopold's *A Sand County Almanac,* published in 1946, was also part of the blooming. . . . Ecology was becoming our new watchword."[12]

Garrison was right in seeing the 1960s as a time in which the environmental consciousness of the nation was rising. And he was correct in seeing that this new movement would directly affect the Park Service. No agency benefited from the greening of America more than the Department of the Interior. As environmentalists called for protection over more areas, it was the Department of the Interior — and especially the Park Service — that was called on to administer and protect these areas. And it was only natural that environmentalists would become its constituency.

Between 1964 and 1972, thirteen new natural areas, twenty new recreation areas, and twenty-nine historical areas were created, and, during that interval, Congress passed an avalanche of environmental legislation, including the Wilderness Act of 1964; the Land and Water Conservation Fund Act of 1965; the National Historic Preservation Act of 1966; the National Wild and Scenic Rivers System in 1968; and the National Environmental Policy Act of 1970, which in turn brought into being the Council on Environmental Quality and the Environmental Protection Agency.[13]

At the same time, the Park Service opened a revolving door between itself and the environmental movement. Leading environmentalists assumed roles in the Department of the Interior and prominent bureaucrats at Interior went on to serve on the boards of the leading environmental groups. In 1967, Starker Leopold was named Director of the Office of Natural Science Studies (a position that was known as "Chief Scientist") of the National Park Service. Leopold's student at Berkeley, Mary Meagher, became Cole's assistant in Yellowstone.[14]

The environmental awakening, however, made the public less tolerant of the policy of direct reductions, not more so. For Rachel Carson, Paul Ehrlich, Barry Commoner, and others taught that nature's problems were caused by humanity. The way we lived, grew food, drank water, dumped sewage, built houses, drove cars was destroying the earth. And the silent victims of our greed and waste were the dumb and innocent animals of the world: the peregrine falcons and bald eagles poisoned by DDT, the whales and seals harpooned and clubbed for profit, the plankton and fishes of the sea smothered by oil from tankers, the jaguars, cheetahs, and ocelots skinned for high fashion, the countless species disappearing every day as the world's forests were stripped bare by hungry souls chopping trees for farms and firewood.[15]

In *Silent Spring,* Carson even specifically mentioned Yellowstone. She reminded the public of the dreadful day in spring 1957 when Park Service foresters, attacking the spruce budworm, dropped DDT from planes. Most of the pesticide, carried by the wind, landed in the Lamar and Yellowstone rivers, killing thousands of fish and destroying much of the insect life in the streams.[16]

The environmentalists were justified in their skepticism. For park managers all seemed to have black thumbs. They had been playing God for ninety-five years and everything they did seemed to make the park worse. They killed predators to save antelope, elk, and bison; then they killed the elk, bison, and antelope; then they decided they had killed too many antelope. Would they next discover they had killed too many elk? In their attempts to manage this beautiful wild area, they seemed caught in a terrible ratchet, where each mistake made the park worse off and no mistake could be corrected. Perhaps the best thing would be to do nothing at all!

This was the message summarized in Barry Commoner's "Third Law of Ecology": "Any major man-made change in a natural system," wrote Commoner, "is likely to be detrimental to that system."[17] Whatever man touches he makes worse. We must stop playing God! And what could be a more objectionable way of playing God than killing wild animals?

The new ecosystems policy thus represented a credo that coincided with a sentiment shared by a growing number of Americans — distrust of human intervention in the natural world. The parks should therefore no longer be managed as they were at the turn of the century — as game preserves, where hunting was permitted, good animals were protected, and bad animals were exterminated. Nor should parks be run as they were during many decades of this century — as farms, where trees were planted, forest fires fought, pastures plowed and harvested. Nor should they any longer be zoos, where animals were fed for the benefit of tourists. We must play a humbler role, it told us, not presuming to make judgments about which species are good and which are bad; not presuming to decide what lives and what dies. Man must stop meddling with nature.

As man was the source of all evil, expelling him became the way to restore and preserve wilderness. The Wilderness Act of 1964, which saved millions of acres of federal lands from further development, defined wilderness as "an area where the earth and its community of life are untrammeled by man." In this way the task of protecting nature was made to seem simple. By keeping man out of the woods, we would not only be protecting those woods, but also restoring them to their original condition.[18]

This was a powerfully attractive idea. All Americans desperately wanted to believe that there was wilderness, that there were still unspoiled places where primitive America still existed. No one wanted to know that it was already too late, that true wilderness had disappeared with the frontier in the last excesses of carnage and plowing that took place in the West a century ago.

That wilderness had gone with the dodo and passenger pigeon, therefore, would not serve as a rallying cry for preservation. Nor would it be as persuasive to say that preventing further spoliation was itself a worthy goal. But "Save the Wilderness!" had a ring few could resist.

Just so, wilderness was defined into existence. It did not matter that there never was a place on earth untrammeled by man. Indeed, there never had been any such thing on any continent except Antarctica. The human race has been on the earth for more than a million years. All western lands, in particular, had been trod by the early white man. Fur trappers and gold prospectors stepped on nearly every square foot of the West at one time or another. Early settlers plowed and planted in the most surprising places. They introduced viruses, microbes, plants, and animals that centuries ago spread like wildfire to every nook and cranny of the continent.[19]

In writing man out of wilderness, this new vision of original America was omitting the Indian, surprisingly even as concern for native Americans had reached new heights. For this was the time when the United Indians of All Tribes were occupying Alcatraz Island, when Russell Means and Dennis Banks had barricaded themselves at the trading post in Wounded Knee and were putting the American Indian Movement on the map, when Indians everywhere were pushing for recognition of rights and restoration for wrongs. Movies such as *Little Big Man* and books such as Dee Brown's *Bury My Heart at Wounded Knee* and Vine Deloria's *Custer Died for Your Sins* had just reawakened America's conscience to the genocide it had committed. The Park Service itself in other spheres had a renewed sensitivity to native Americans. Throughout the country, in the interest of authenticity, it had re-

named parks and natural features, replacing Anglo-Saxon or den-
igrating nomenclature with native and less derogatory ones. Thus
Mount McKinley National Park became Denali; and in Yellowstone, Buffalo Ford was renamed Nez Perce Ford and Squaw Lake
became Indian Pond.[20]

But at the very time that historians were writing the Indians into
our history, ecologists were writing them out of it. If restoring
wilderness meant re-creating a hunter-gatherer culture long since
exterminated, the task of restoration was impossible, and if it meant
giving land back to the Indians, it was undesirable. The language
of the Wilderness Act, the Leopold Report, and Ecosystems Management of the National Park Service, in turn, institutionalized this
error. They made it possible to believe there was still wilderness
to save without paying too high a cost, and to believe that the way
to manage it was to leave it alone.

As a consequence the new definition of wilderness replaced science with romance. Ecosystems management — managing a park
as a "vignette of primitive America," as stated in the Leopold
Report — was an idea that came a hundred years too late. Yet the
myth it presupposed — that man was not part of nature — would
be sustained.

By the early 1970s a team of Park Service officials and leading
environmentalists (including Sigurd Olson, former President of the
Wilderness Society) was assembled to incorporate this ideal into
official Yellowstone management policy. In 1973 this team published a grand design, the Master Plan. "The original purpose (of
Yellowstone)," it decreed, "must be translated in terms of contemporary connotations; as such it should read: to perpetuate the
natural ecosystems within the park *in as near pristine conditions as
possible for their inspirational, educational, cultural, and scientific
values for this and future generations.*"[21]

"As a natural area," it continued, "Yellowstone should be a
place where all the resources in a wild land environment are subject
to minimal management." And to implement the Leopold Report's
proscription against "observable artificiality," it urged "the ultimate removal of all artificial structures."[22]

Cole's new elk plan would be fully in accord with these ideals.
Elk numbers no longer would be altered by rangers. As "natural
conditions exist only to the extent that man does not change things

from what they would otherwise be," he wrote, "management (in Yellowstone) is restricted to protecting against, removing, or compensating for human influences that cause departures from natural conditions."[23]

Indeed once again Park Service biologists went beyond official policy — this time on the question of human intervention. For the Green Book, while discouraging intervention, had not forbidden it. It had, in fact, conceded that live-trapping and direct reduction might sometimes be necessary, and it had specifically endorsed "prescribed" (that is, intentional) burning of forests as a tool "to achieve approved vegetation and/or wildlife management objectives." Yet Park Service wish became Yellowstone's command. Park managers began to erect an impenetrable shield between humanity and nature.[24]

In this spirit, much of the paraphernalia necessary for understanding the natural world was soon prohibited. Regulations were established forbidding tags, radio-collars, helicopters, and other mechanical devices used to study wildlife. Steps were taken to dismantle the exclosures. Plans to reintroduce the mountain lion were abandoned (because these animals, as transplants from Idaho, would be "non-native").[25]

The era of direct reductions and intrusive science had come to an end. The old management had been rejected by those who wanted to kill elk and by those who wanted to save them; by those who embraced ecosystems management as scientific and by those who, in the spirit of a new environmental vision, wanted to keep man away from nature.

In its place came the policy of natural control. This was not, however, a scientific policy, but rather a policy of desperation. After the Senate hearings Yellowstone managers simply had no alternative, and Cole's hypothesis was an attempt to make a scientific virtue out of a political necessity.

The scientific virtue was provided by the idea of ecosystems management. The political necessity was the need to appease both hunters and environmentalists.

An integral part of the new policy of natural control, therefore, was to encourage hunters to kill more elk, but for them to do so out of the park. In accordance with this new policy, the National Park Service signed an agreement with the state of Montana in the summer of 1968 to organize a special elk hunt along the park boundary "to remove by sporting hunting outside the Park . . . that portion of . . . the Northern Yellowstone [elk] herd deemed surplus. . . ."[26]

And so "natural control" did not end the killing of elk, but only its location. From the animals' perspective, it was, as we shall see, a change for the worse.

6

A Solution to the Elk Problem

A MILE east of Gardiner, Montana, on the way to Jardine, site of the old gold mine the Homestake Company hopes to re-open soon, the road crosses a rough meadow known as Deckard Flat. On the far side of the flat to the south of the road is the park boundary. In summer the meadow looks as pretty and serene as a Civil War battlefield memorial. It is a nice place for a picnic. In fact, it reminds one of the park in Virginia at Manassas Junction, which commemorates the two battles of Bull Run.

This is where the special hunts took place. The locals knew it as the "firing line." Like General Hancock's men waiting for Pickett's charge at Gettysburg, hunters — more than a thousand of them — would arrive before dawn and take their stations, nearly side by side, north of the road, and cock their guns for the unsuspecting elk about to wander into the field. The road was jammed with cars, many stuck in the snow. At dawn a whistle blew, the signal to fire, while half the hunters were still in their cars fighting traffic.[1]

Sometimes people did not wait for the whistle. "I remember an opening day at Deckard Flat," one veteran of this war told writer Russell Chatham. "Just getting light, not quite shooting time. Must have been 500 elk standing out there. I said to myself, I ain't waiting another minute and I dropped one. The minute I fired the whole valley opened up. I mean those boys cut loose. Myself, I hit a dozen. When it was over all the elk were dead or run out of sight."[2]

However it began, what followed was carnage and mayhem. Bullets flew everywhere, and occasionally into elk. Sometimes bullets from more than one hunter would hit the same animal at the same time, and fights would ensue (one animal was hit by seventeen bullets). Because a hunter could not take legal possession of an elk until he had put a tag on the carcass, some opportunists would wait, unarmed, in track shoes. When someone felled an elk, the fellow in Adidas or Pumas would take off, under the hail of bullets, trying to reach the unfortunate animal first and tag it.[3]

Occasionally, hunters would surround a herd of elk and shoot into the circled animals. In the melee, once in a while hunters accidentally shot and killed one of their own. Sometimes an animal retreated wounded back into the park, where no hunter could claim it, and where it could die without any further human help. For a few years the Montana Department of Fish and Game, in an attempt to remove the firing line from park boundaries, ruled that no animals could be claimed that lay on the south side of the road. This directive sometimes caused fights between game wardens and a hunter whose quarry, shot on the north side of the road, lay bleeding to death on the south shoulder where it could not be claimed or put out of its misery.[4]

The firing line was not new. When the Lacey Act was passed and the animals were protected in the park, hunters began to wait at Deckard Flat, daring elk to exit.

An eyewitness writing in *Outdoor Life* in 1912 described it: "The bullets were falling fast and furious around the country. I hid behind the rock foundation of a near-by cabin and remained there until the battle was over. I can't figure it out how about 5,000 shots were fired and about 600 elk killed, and no one was shot. Over 4,000 shots went wild and failed to hit anyone. At least twenty stray elk were killed. I saw fifty hunters surround 200 elk and shoot into the bunch at random and no one was killed."[5]

During the 1960s the elk were so few — thanks to the policy of direct reduction — that the firing line was unnecessary. In the name of natural control, however, it was revived in 1968. Park authorities encouraged more and bigger hunts. "Huge elk kill urged," the Billings *Gazette* reported during this period. "[Park] officials want the state of Montana to issue hunting permits for a massive kill of animals when winter forces them across the park boundary." And although at first so few elk left the park that the special hunt was not thought worth the trouble by Montana officials, it gradually assumed importance as the Yellowstone herd grew.[6]

By 1984, this "harvest" amounted to more than 10 percent of the Montana statewide hunter toll for elk. During winter 1983–1984, 1,749 hunters killed 1,631 elk. In the last four years, more than 5,700 elk have been taken in the special hunt — nearly *four times* as many as were shot by rangers during the last six years of the elk-reduction program.[7]

Although the special hunt was popular with many sportsmen, Montana officials were not entirely happy with the new park policy. Being difficult to supervise, the firing line, though it brought much revenue to the state, actually ran a deficit; and despite attempts by officials to spread the hunters to ensure a "quality hunt," it did not provide sporting conditions. Even more irksome, they thought, was the policy of natural control itself. The firing line, they feared, would not keep up with the burgeoning elk population. Sooner or later these animals would multiply faster than hunters could kill them. And when they did they would leave the park, destroying the vegetation of Montana just as they were doing in Yellowstone.[8]

The ironies and inadequacies of the firing line, however, did not disturb Park Service managers. It was a successful political response to a political problem. It made hunters happy because the burgeoning population of elk in the park supplied cannon fodder for the firing line. It made preservationists happy because it ended the killing in the park.

Then, too, it was good for tourism. For ending reductions virtually guaranteed that the number of elk would expand. And because bears (as we shall see) were slated for removal, Yellowstone needed more animals for the tourists to see. "It has . . . become apparent," wrote Superintendent Anderson in 1967, "that past control operations have resulted in greatly reduced wildlife shows for visitors over large areas. We would very much like the opportunity to be sure future management actions were directed toward correcting this problem."[9]

Yellowstone managers, therefore, had many reasons for adopting Cole's hypothesis on the elk. The only reason for rejecting it: there was no evidence that the premise was true. In particular, there was no reason to believe the park was an ecologically complete habitat for one segment of the elk herd, or that elk numbers could be contained simply by public hunting outside the park. These difficulties bothered Cole's assistant, Research Biologist William Barmore. "I think," he wrote in a memo to Cole in spring 1968,

> we should recognize . . . that the factual basis for these hypotheses is far less than adequate to warrant strong belief that they are correct. In other words, I believe we could choose other data of similar factual depth and reliability and support other hypotheses that could take us most any other way we want to go with elk management. . . . We are abruptly scrapping current objectives with no stated justification or supporting information. Past objectives *were* based on supporting information that indicated relationships between elk, their habitat, and associated wildlife were different from what existed in primeval times; that ecological changes since the early 1900s *were* unnatural. . . . I think we are being less than completely scientific or completely objective and honest by proposing a drastic program switch without saying why.[10]

Barmore's pleas to base elk management on science rather than politics went unheeded. Instead, in 1970, he left Yellowstone to study for a doctorate at Utah State University. Douglas Houston, a research biologist at Grand Teton, replaced Barmore in Yellowstone.[11]

The scientific problems did not go away with Barmore's departure. For university biologists and state game managers began to raise a difficult question. The policy of natural control was based on the assumption that predators — particularly wolves and cougars — would help to limit elk numbers. How, they asked, could elk be contained by predation when wolves and cougars had become extinct in the park forty years earlier?[12]

As predators had been systematically hunted by the Park Service in the 1920s, they had, by all available evidence, been exterminated by 1930. But park managers were reluctant to admit that the wolves and cougars were gone. Without them there was no rationale for the policy of natural control. And so a curious thing happened: the biologist's office reported that predators were making a "comeback" in Yellowstone. Beginning in 1968, the number of "sightings" of wolves and mountain lions increased dramatically. Apparently these animals, through some mysterious serendipity, had returned to Yellowstone, perhaps to save the policy of natural control.[13]

Despite the increase of sightings, however, no signs appeared that wolves or mountain lions preyed on elk. The herd, in fact, was growing like Topsy. From a low of 3,172 counted in 1968 it was found to exceed 7,000 by 1971. Clearly predators — comeback or no comeback — were not doing their job. The theory of natural control was not working. The Park Service would have to think of something else.[14]

Help came from the most unlikely of places. In 1970 a wildlife biologist named Graeme Caughley at the School of Biological Sciences of the University of Sydney, Australia, published an article entitled, "Eruption of Ungulate Populations, with Emphasis on Himalayan Thar in New Zealand." The timing of this piece, from the Park Service point of view, could not have been better. The study provided the rationale for yet another shift in policy.[15]

The thar is a goatlike ungulate from the Himalayas. In 1904 five animals were released in New Zealand, near Mount Cook. Without natural enemies or competitors (no grazing mammals were native to New Zealand) they multiplied. By 1930 they numbered in the tens of thousands, and were destroying the native vegetation. A 1936 report of the Department of Internal Affairs commented, "Their great fecundity and absence of natural enemies . . . had resulted in an entire absence of natural balance with food supplies being arrived at, with the result that the country occupied by them had become denuded of vegetation to an almost indescribable extent." The government of New Zealand began a program of direct reduction. In the first twenty-five years of control actions, 24,000 thar were killed. In Mount Cook National Park alone, since 1956, rangers have killed 20,800 thar. By the time Caughley wrote his article, the number of thar had been drastically reduced. Reductions were still in progress, aided now by commercial hunting. Professional hunters in three years of the early 1970s alone exported 25,000 thar as game meat.[16]

Yet, according to Caughley, the thar population was not limited by hunters. It was, he said, self-limiting. As the thar grew in numbers, Caughley detected that mortality of kids less than a year old increased. Eventually, the death rate was sufficient to overtake the birth rate, and the herd stabilized.[17]

Based on his observations of the thar, Caughley constructed a general hypothesis: Irruptions — accompanied by a drastic decline in vegetation — naturally followed introduction of an exotic grazing species to a new area. But as the herd grew and food became scarce, their diet would decline. Gradually, density-dependent mechanisms would come into play, causing an increase in mortality. Eventually, he supposed, the death rate would overtake the birth rate, and the population would stabilize without further damage to the vegetation.[18]

Caughley's theory was another in a long debate between wildlife biologists. The issue was: What limited wildlife populations? Was it predators (those animals higher in the food chain)? Or was it the food supply (those organisms lower in the food chain)? It was obvious, as first remarked by Malthus, that populations tended to increase unless limited by their food supply. Any population would go into decline when its members starved to death. The issue between the predator people and the food-chain people, therefore, was whether some trigger limited ungulate populations *before* they destroyed the vegetation that supported them. Caughley suggested that scarcity of vegetation alone, and not predators, would eventually produce this equilibrium between ungulate and range.

Yet the Caughley hypothesis was tantalizingly incomplete. He never isolated the crucial density-dependent mechanism that was the cause of kid mortality. Moreover, Caughley never claimed to have proved a theory; he only suggested one. "This study suggests," he concluded, "but does not prove, that the thar population's rate of increase is influenced by food supply." He did not deny that the thar had done great damage to the range, concluding that "the sequence of an eruptive fluctuation reflects progressive depletion of forage by the animals themselves." And most important, he had merely hypothesized — but not established — that the thar and its food supply would eventually reach equilibrium, causing no further damage to the range.[19]

Indeed, concern over the damage done to grasslands by the thar was strong in New Zealand when Caughley wrote his article and continues to this day. The effects of the thar have been closely studied by teams of biologists in New Zealand for decades, all reporting severe damage to the alpine and subalpine vegetation. "All that remained," wrote botanist Dr. C. J. Burrows of the

University of Canterbury in 1974 (in describing thar range), "was powdery soil and dead tussock bases."[20]

Later the Yellowstone managers would find another theory that could be used to justify their policy, appearing in a study of the African buffalo done in the Serengeti National Park in Tanzania by biologist A. R. E. Sinclair.[21]

For years the numbers of buffalo had been kept in check by rindepest, an exotic disease introduced to Africa by the cattle of early Dutch farmers. In the twentieth century, however, rindepest was eradicated and the buffalo began to increase. Sinclair discovered, though, that the buffalo population stabilized, without the help of man or predators such as lions.[22]

As the buffalo increased, Sinclair theorized, both the range and the quality of their diet decreased; and as their nutritional state declined, they became prey to parasites and disease until, eventually, increased mortality stabilized the population.[23]

Sinclair, in short, had apparently found the crucial "self-limiting, density-dependent" mechanism that made natural regulation work for the African buffalo.

The Caughley and Sinclair theories were slim reeds on which to hang a policy for Yellowstone. Each was developed for a very different species in a very different ecosystem. The thar was introduced to an island that had never known grazing mammals. Its population was and continues to be heavily controlled by hunting, and so Caughley was never able to study it as a population that limited itself without human help. And the thar destroyed much vegetation in New Zealand just as the deer did in the Kaibab.[24]

The ecosystem in which the African buffalo lived was older and more complex, variegated, and stable than any in North America. The animals of the Serengeti had evolved together since Tertiary times, and had millions of years in which to secure their ecological niches. Yellowstone, by contrast, was, on the scale of geologic time, born yesterday. Periodic ice ages — the so-called "Pleistocene variations" — gave fauna little time to establish stable relations with their environment. Since the last ice age (a mere 12,000 years ago), and before the white man came, three hundred species — indeed, most of the animals of North America — became extinct. The woolly mammoth, many species of horses and camels, the sabertoothed tiger, the dire wolf, and the giant beaver disappeared within a few thousand years. And so the ecosystem the white man found in Yellowstone had far fewer animals, living more precariously, than the animals Sinclair studied in the Serengeti. The African buffalo in particular was much larger than the elk, and unlike the elk was too large ever to have been a major prey for any predator.[25]

Ironically, in a book on the Serengeti edited by Sinclair, researchers explicitly pointed out that predators had a far larger part in limiting prey in North America than they did in Africa. In many African parks, they said, "the major herbivores are too large for predators to exert much impact. . . . The American prairies, however, once contained large migratory populations of bison and pronghorn antelope . . . [and] predators were responsible for a higher proportion of mortality among these ungulates — we know that their major predator, the wolf, can indeed control prey numbers in some modern ungulate communities."[26]

The theoretical shortcomings of Caughley's model, or the irrelevance of Sinclair's, however, did not bother the Yellowstone managers. For if unproved and inappropriate, these accounts were convenient. They were the first statements in serious scientific literature of theories that could serve as justification for the elk-management policy they had already decided to adopt. The beauty for them was that, according to these accounts, elk did not need predators.

For, they reasoned, because elk had inhabited the Yellowstone plateau for millennia, then, if these animals had ever irrupted, they would have done so only when they first came to the region thousands of years ago. The elk, being native to the park, must long ago have reached equilibrium with their range. And, if limitations in food supply alone limited the ungulates of New Zealand and Tanzania, then surely the same mechanism was sufficient to control elk numbers in Yellowstone. The awkward predator problem could be ignored.

No one knew in fact what role predators had among Yellowstone fauna. In the park, wolves and mountain lions had disappeared before any studies had been done. And to sort out the relative effects of food and predation in controlling and influencing prey populations would have required ambitious studies of what biologists called three interacting "trophic levels": vegetation, herbivores, and carnivores. Yet, James M. Peek, Professor of Wildlife Biology at the University of Idaho, told me recently that "we have yet to see a good integrated study in North America where the three trophic levels have been adequately addressed. These are interlocking systems that must be studied together."[27]

Certainly in Yellowstone such a study was impossible. Without carnivores, a crucial element for understanding was missing. Their probable role in the park, therefore, had to be inferred from studies done elsewhere, in places, as we have seen, that were quite different. Yet such extrapolation, like metaphor, tended to be inexact. Truth about wildlife did not travel well. No two game ranges were alike, and none remained the same. Just as the philosopher Heraclitus had observed, "you cannot step twice into the same river," so too we could never be sure whether one place told us the truth about another, or that our learning about the past told us what we needed to know about the present. What scientists called a "controlled experiment" could never be done in wildlife ecology. Just when we might think we had found the truth, the truth would change, flying away from us, as Plato would say, like a statue of Daedalus. Mother nature would always remain one step ahead.[28]

So wildlife biologists relied on models, the similes of science, and, as with any analogy, the model chosen had much to do with the conclusions reached. "If you have no predators in Yellowstone," explained Pcek, "it is easy to minimize their influence; but if you go north where wolves are plentiful and see a lot of carnage on the calving grounds it is easy to overlook the habitat."[29]

In fact the majority of studies, elsewhere in North America, did suggest that Aldo Leopold had been correct: ungulates could destroy a range and predators could protect it.

"Of course ungulates can destroy a range," Frank Golley, Director of the University of Georgia's Institute of Ecology and former President of the Ecological Society of America, told me. "There are countless examples of this happening." Indeed, as Aldo Leopold had documented, they were out there, not just in the Kaibab but all over the map. Yet biologist George Schaller says, "the most important influence of predation is this dampening of the tendency of populations to increase beyond the carrying capacity of their range, an effect that prevents severe oscillations."[30]

Nor was predation effective just in limiting numbers, many biologists felt. It inspired evolutionary changes that made prey species better able to survive; it culled the herds, taking the weak and defective animals; it reduced competition between prey species, permitting greater niche overlap; it provided food for scavengers, including bears; it protected grasses — much like the "rest-rotation" method now preached by modern range specialists — by discouraging animals from grazing long in one spot.[31]

And if Cole and Houston had taken almost any North American study as a model, they would have found these views confirmed. Peek spoke of "a tremendous number of studies done up north where predators are still having an influence that show without question that wolf and in some cases grizzly bear predation definitely affect a population."[32]

Indeed those studies were plentiful: "High loss of caribou calves shortly after birth and throughout the first year of life to wolves is a major limitation to population growth," concluded one recent report by the Canadian Wildlife Service. A study by one of North America's best-known wolf experts, the late Douglas Pimlott, "Wolf Predation and Ungulate Populations," might have been especially appropriate reading for the Yellowstone managers. This work, published in 1967 and now regarded as a classic, suggested that deer, having evolved in the face of heavy wolf predation, had been naturally selected to be as fecund as possible. Rather than developing self-limiting biological mechanisms, they had only developed a great capacity to multiply![33]

Nearly all the studies of predation on this continent — of wolves and moose on Isle Royale, caribou in Alaska, cougars and lynx prey-switching in Newfoundland, wolf and deer declining in Superior National Forest, wolf and moose of the Pukaskwa, and wolf predation on elk in Riding Mountain, Manitoba, among others — emphasized how carnivores affected prey populations. And it seems that such studies were more appropriate role models for Yellowstone than Himalayan thar or African buffalo.[34]

But Yellowstone authorities, grasping for a rationale for natural control and finding none close to home, reached farther afield. Relying on revelations in New Zealand and Tanzania, they constructed a new hypothesis of "resource limitation." Wolves and mountain lions, according to this view, took no part in limiting numbers of elk. Yellowstone could be a complete ecosystem without them. Their absence would not, ecologically speaking, be missed; these animals could be so many sour grapes.

Accordingly in 1971 Cole introduced a new theory he called natural regulation. Elk numbers, he suggested, would be self-limiting without damaging the vegetation. They would be controlled, not by predators, not by rangers, but by their food supply. These animals, he reasoned, had *never been* limited by wolves, lions, or coyotes, but always by some density-dependent mechanism triggered by a declining diet.

Cole summarized his new theory in 1971:

> Over a series of years, naturally regulated ungulate populations were self-regulatory units. They regulated their own mortality and compensating natality in relation to available winter feed and their population size. Predation on either wintering or newborn ungulates seemed a nonessential adjunct to the natural regulation process because it did not prevent populations from being self-regulatory by competition for food.[35]

Similarly, Cole's colleague Meagher concluded that the bison population, whose historic numbers she estimated to be about one thousand, would be self-regulating. "Inasmuch as bison had apparently inhabited the Pelican area for centuries before the establishment of the park, their ability to persist without man's help seemed likely," she wrote. "Predation and disease have no apparent role in the regulation of bison numbers," she continued. "The available information does not suggest a need for future reductions."[36]

Yellowstone, moreover, she implied, was an ecologically complete habitat for this species as well. "Habitual use of particular areas in both summer and winter explains the lack of emigration of population groups from the park. . . . Past records show mixed herd groups (bulls, cows, calves) occasionally leaving near Gardiner, Montana, during a severe winter, but this exodus has happened only when artificially high numbers were maintained in the Lamar Valley by winter feeding. . . ."[37]

"Bison management in Yellowstone National Park," she concluded, "may be termed 'no management.' . . ."[38]

*   *   *

With one fell swoop Yellowstone biologists had defined the elk and bison problems out of existence. By a simple change of perspective, without one act of restoration, they had returned the park to its original condition. But natural regulation was, as Houston wrote in 1971, a "hypothesis being tested" to see whether "the park ecosystem is still enough intact that the numbers of these animals are regulated naturally." What if the experiment failed?[39]

"I guess what bothers me," wrote Les Pengelly, Professor of Wildlife Biology at the University of Montana and one of the independent scientists who worked in Yellowstone in the 1960s, "is that population ecologists are still debating over the fact of population regulation, as well as the mechanism, by using examples of lower forms such as bacteria in a Petri dish or flour beetles in a bell jar, yet [Park scientists] are willing to experiment with such valuable and immense property as Yellowstone Park and its flora and fauna."[40]

Pengelly and other university biologists who had worked in Yellowstone during the Garrison administration, were no longer required by the Park Service, however. In the spirit of the Leopold Report, calling for all science to be under government control and to be "mission oriented," no university or independent researchers were invited to take part in testing the new natural-regulation hypothesis. Those who had participated in the days of Garrison and were now associated with discarded policies no longer found themselves recipients of Park Service research funds or participants in discussions on management. Their works, expunged from official bibliographies of wildlife studies published by the research office, were no longer assigned reading for park naturalists. From the Park Service point of view, they had simply disappeared.[41]

Instead, Houston was assigned the task of testing the new theory. He was aided by Meagher, Park Service plant ecologist Don Despain, and Cole himself.

Throughout the 1970s they worked on the project, mostly unnoticed by the scientific world. Although they did produce periodic progress reports, most were never published and almost none of the original data collected were readily available to the public. Few, other than scientists like Pengelly who had long associations with Yellowstone and made an effort to keep up, knew what was going on.[42]

For the Park Service, this silence was desirable. The task facing Houston and his colleagues was daunting. Justifying the new theory required nothing less than rewriting the entire history of the park — a questionable assignment for a team of biologists — and was bound to be controversial.

To prove that Yellowstone did not have too many elk meant claiming that the plateau had always harbored large numbers of these animals, which in turn required discounting reports of the early explorers — Bridger, Langford, Hayden, Jones, Dunraven, Ludlow, Comstock, and others — who found game sparse on the plateau. It meant finding supporting archaeological evidence and discounting reports by Hiram Chittenden, Teddy Roosevelt, the early Yellowstone acting superintendents, and others who claimed to have seen many elk migrate into the park after it had been made a wildlife sanctuary.[43]

To establish that Yellowstone was an essentially intact ecosystem for the elk meant proving it was their "historic winter range," which in turn entailed rejecting the testimony of nearly every biologist for forty years — from Nelson and Graves in 1917 to John Craighead in 1967 — who thought he had established that elk had not wintered on the plateau until trapped in the park by farmers' fences and hunters' guns.[44]

To prove that predators had not limited elk numbers required denying, not only the truth of much of the scholarly literature, but also eyewitness evidence of early park administrators and leading biologists (including Skinner, Rush, Wright, and Murie) who saw — or thought they saw — elk numbers grow as predators declined.[45]

To prove that the elk did not threaten the range or other species required discounting sixty years of accumulated evidence to the contrary.

This is just what they did.

Elk, Houston said, have been present in the Yellowstone area "for millennia." The park was the "historic winter range" for all but one segment (the segment that now leaves the park to face the firing line). None had ever migrated beyond Dome Mountain, ten miles down the Yellowstone River from Gardiner. There had been no immigration of animals after the park was made a sanctuary. "The historical accounts also do not support the interpretation that large numbers of elk were compressed into a smaller area in the park." There had been "nothing resembling a population eruption [that is, irruption] and crash."[46]

Nor were predators ever a significant part of the elk's ecosystem, Houston argued. Elk were so abundant in primeval times, he said, and wolves and Indians so scarce, that "predation did not prevent elk, as the most abundant, large migratory ungulate, from being resource limited."[47]

Yellowstone, in short, was largely an "intact ecosystem" for elk. And if they had been living on the plateau for millennia, they could do it no damage. *It was impossible for there to be too many or for them to threaten either range or other species.*[48]

There could never be too many elk. The number in the northern range, they insisted, had always fluctuated around 12,000. The early censuses that found more than 30,000 had been inflated. Nor was it possible for a crash to have taken place in 1919–1920. Self-regulatory, density-dependent, compensatory mechanisms ensured that numbers would never exceed what Houston and Cole called, following Caughley, the "ecological carrying capacity" (or *"K"*) of the range, the number that nature determined was optimum.[49]

The range had never been damaged by ungulate grazing. What earlier biologists saw as overgrazing, they said, was natural — and being natural *it could not be overgrazing.* "An elk population with a low reproductive rate," Houston wrote in 1971, "on ranges that appear temporarily 'overgrazed' . . . is an excellent relationship in a park ecosystem if it is essentially natural."[50]

In the past, they said, evidence of range damage was derived from exclosures, and exclosures themselves were unnatural. "Like old cemeteries or isolated relict areas," they wrote, exclosures could not tell us what natural conditions should be like. "Interpretation that such exclosures illustrate how things should be in a park," he added, "would more often than not be confusing artificial with natural conditions."[51]

Those sites which *seemed* overgrazed Houston and Cole called "zootic climax" sites. They were, to be sure, places where elk naturally tended to congregate and therefore graze heavily. But they had done so for millennia, and so we need not worry about them.[52]

There was no interspecific competition and no native species had been driven out by elk. The white-tail deer had not fled because of the elk; rather, Houston suggested, "the park and adjacent areas represented the extreme upper limit of marginal winter range" of this species. "A combination of land clearing, livestock grazing and human predation outside the park were primarily responsible for the decline. . . ." Nor, moreover, were any animals currently suffering from competition: "The elk population and the populations of the less abundant ungulates were not closely linked by interspecific competition. . . ."[53]

Elk did not overgraze browse. Instead, the decline of browse, which Houston and Despain admitted had occurred, was in part the fault of mother nature and in part human, for dry weather and fire-suppression policies during the last hundred years had inhibited its growth. Since the "little ice age" of the 1880s the climate in the park had been hotter and drier, they observed, gradually depleting the water table and driving out both browse and beaver.[54]

Aspen and willow, they stated, were "seral" species — they grew only during one stage of the biological progression of the range. Mature aspen plants produced chemicals that inhibited new growth. If undisturbed, they would gradually die out and be replaced by pine, which happened when range was allowed to grow old — to reach "climax." To encourage new growth, therefore, old plants had to be burned. Burning, therefore, kept the range young. Tree-ring studies by Houston suggested that before the white man the northern range had burned on the average every twenty years. Yet the Park Service "has been able to almost completely suppress fires on the grasslands of the accessible study area for at least 30 years, and has substantially reduced the frequency of natural fire for as long as 80 years. . . ."[55]

"A variety of information," Houston concluded, "suggests the hypothesis that reduction of aspen has been primarily a function of reduced fire frequency, rather than of forage by ungulates, particularly of elk."[56]

By brilliant argument in a single equation, natural regulation had captured the scientific high ground: As an intact ecosystem Yellowstone could come to no harm so long as we did not interfere. It was a simple, compelling idea with which to dismiss criticism: range damage and interspecific competition could not exist. If potential critics believed otherwise, it was up to them to prove that these conditions existed, not up to the Park Service to show that they did not. The burden of proof had shifted from the Park Service to the public.

Yet how could the Park Service be so sure? Many university or state scientists who had worked in the park during the Garrison and McLaughlin administrations, but who now found themselves on the outside looking in, remained puzzled. Jonas, Peek, O'Gara, Pengelly, Arnold Foss (biologist for the Montana Department of Fish and Game), Alan Beetle (biologist at the University of Wyoming), John Craighead and his brother Frank, and the many students of the Craigheads who went on to earn distinguished reputations — Maurice Hornocker, Unit Leader of the Cooperative Wildlife Research Unit at the University of Idaho and now the country's leading expert on mountain lions; Vince Yannone, biologist for the state of Montana; Merlin Shoesmith, elk expert working for the Canadian Wildlife Service; Harry Reynolds III, biologist for the state of Alaska; and many others — having long associations with Yellowstone, knowing it intimately and far longer than almost any Park Service official, could not understand the sudden certainty that park biologists attached to their revisionism.[57]

For the idea of natural regulation, although appealing, did not rest on overwhelming evidence. The historical claims, almost without exception, were not supported by new evidence but were made only by ignoring or reinterpreting the old. The literature was cited selectively — especially on the subject of predation — almost to a bizarre extent. On many crucial points Yellowstone biologists merely cited each other, referring to work that was unpublished and in many cases withheld from the public.[58]

How did they know that carnivores were few in prehistoric times, or that they did not deplete elk numbers, these men wondered? Park biologists offered no archaeological evidence and admitted that early historical accounts were scanty. "The early accounts of carnivore abundance are generally too limited to determine whether or not elk were more intensely or frequently predator limited," Houston wrote.[59]

Yet that wolves could have depleted elk in Yellowstone was suggested by one of Houston's own colleagues. Barmore, together with Charles W. Fowler, a biologist from Utah State University, had conducted a computer-simulation study to estimate the probable effect of predation and had found that it could have been considerable. A wolf, they pointed out, can eat ten pounds of meat a day. But even assuming that they ate half that amount, 90 wolves would consume 13,500 pounds of meat a month — equivalent to 1,200 elk calves a year! This regime, Barmore and Fowler concluded, would, in time, reduce the elk herd to half the range's carrying capacity. "It seems," they concluded, "entirely conceivable that predation could influence the size of the northern Yellowstone elk herd."[60]

Houston rejected this conclusion, however. In their computer model they had assumed, he said, that the wolves had been helped by human hunting. But there were too few Indians in Yellowstone, he suggested, to significantly limit numbers of elk. " 'It seems improbable,' " he wrote, quoting archaeologist Larry Lahren to make his point, " '. . . that the hunter-gatherer populations ever operated at a level which significantly affected the evidently large biomass (of their main prey in the upper Yellowstone Valley)." Yet this reply was puzzling. Lahren's study, from which Houston took the quotation, was not done in the park, but in the valley of the lower Yellowstone, in Montana.[61]

How did they know elk had lived on the plateau in numbers for millennia? "Remains of elk," Houston wrote, "occur in archaeological sites from the Yellowstone area dated in excess of 9,000 years B.P. [before present], continuously to historic times (Lahren 1971, 1976)." Lahren, however, had never found an elk bone in Yellowstone. His principal excavation, at a place known as the Myers-Hindman site, was not in the park but in the lower Yellowstone Valley between Livingston and Big Timber, Montana — sixty-five miles from, and 3,500 feet below, the Yellowstone plateau in land that Lahren described as a "plains grassland life zone" — hardly "the Yellowstone area."[62]

How did they know elk wintered on the Yellowstone plateau before it was made a park, that these animals never migrated beyond Dome Mountain, and that the park was their "historic winter range"? Houston's evidence was: "M. Meagher, personal communication." How then did his colleague know this information? When asked she would not say. This evidence represented, she insisted, part of her work in progress, and as such was exempt from the Freedom of Information Act. How work done by a public employee at public expense could be private information she would not say. Yet when Professor Pengelly requested the information in a letter to Robert Barbee, Yellowstone's Superintendent, his request was denied.[63]

"Since you seem to be questioning the validity of Dr. Meagher's personal communication statements used in Dr. Houston's book," Barbee wrote Pengelly in refusing his request, "it seems unlikely that you would be satisfied with a personal communication from Dr. Meagher to you."[64]

How did they know the early elk censuses were inaccurate? Park biologists, it turned out, had taken sides in an old controversy.

In 1914 Congress gave the U.S. Biological Survey a mission: to eradicate carnivores in the West. During the summer of 1915, Vernon Bailey, an agent of the Survey, came to Yellowstone to convince the cavalry officers then administering the park of the need for predator control. His most persuasive argument was that wolves were a threat to elk. "Mr. Bailey," wrote Yellowstone's Acting Superintendent Colonel Lloyd Brett on September 18, 1915,

> has found and reported that the wolves which have recently appeared in the park, have become so numerous as to seriously interfere with the young elk, deer and other game. . . . His Department is about to hire and put in the field, a number of expert hunters to devote their whole time to the extermination as far as possible of wolves, coyotes and mountain lions, and has offered to place two of them in the park and possibly others not far from the border.[65]

As part of his effort to determine their vulnerability to wolves, Bailey made an elk count in spring 1916. And to the dismay and disbelief of the army, he found far fewer animals than they had supposed were there. The northern herd, the army had estimated in 1915, numbered 37,000. Bailey could find only 13,700. This was persuasive evidence in favor of predator control.[66]

Nevertheless, the army was skeptical. "So far as his report is concerned," wrote Colonel Brett's assistant, Captain F. T. Arnold, on April 5, 1916, "it is not considered of any value, as it would have taken several times the amount of work done by him and his assistants at that time to have made anything like an accurate count of the elk." The army conducted a recount on April 27, 1916, finding more than 28,000. Bailey, they thought, had missed a large portion of the herd. "It developed later," wrote Brett, "that a very unusual number of elk were still outside the park" at the time that Bailey made his count.[67]

The architects of natural regulation, however, chose to accept Bailey's count as accurate, a count that coincidentally fitted their theory. All the earlier estimates, from 1886 to 1920, which had reported herd sizes of up to 40,000, were, they insisted, flawed.[68]

Yet their certainty in revising past counts was surprising. Were the scouts who took the census so careless? Aubrey Haines, retired Yellowstone historian who knew many of these people, insisted to me that they were not. "Many of them were intelligent, experienced, and dedicated men," he told me. "They would not have made stupid mistakes."[69]

Any game census, moreover, was more guesswork than science. Weather conditions, location, the extent to which animals were dispersed all affected the count. Double counting could not be ruled out, nor could one know how many animals were missed.

Without tags or collars, no good way was known to measure the "census efficiency" of a count: if 10,000 animals were found, were these 50 percent of the herd, or 80 percent, or even 100 percent? Even today, using the most sophisticated techniques — airplanes, helicopters, aerial photography — park biologists dare not guess the efficiency of their own counts. Yet paradoxically these biologists were not reluctant to guess the efficiency of counts made sixty years before on horseback.[70]

How did they know a crash had not occurred during the winter of 1919–1920? Following Caughley and Sinclair, Yellowstone biologists believed that only ungulates in disturbed ecosystems such as New Zealand (where the thar was introduced) and Africa (where the exotic parasite rindepest upset nature's balance) could irrupt. Elk, living in an "ecologically complete habitat," could not irrupt. And if they could not irrupt, they could not crash. Bailey's count, therefore, fit the government hypothesis. If his numbers were correct, then there had been no irruption, and if there had been no irruption, there could not have been a crash.[71]

Besides, Houston suggested, the winter of 1919–1920 was too mild to produce a big elk kill. "December to early March," he reported (failing to notice that the winter set five records for snow and cold and that April was the coldest in Yellowstone's history), "were unseasonably mild," and "spring growth of vegetation apparently began in April" — even though on April 30 Tower Junction and Soda Butte Creek (both places where elk wintered) still had between ten and twenty inches on the ground.[72]

*   *   *

How did they know that no animals had suffered from competition with elk? Their claim puzzled many biologists, yet was hard to challenge. It was easy to prove the potential for competition but nearly impossible to *prove* it existed, at least until it was too late. Professor Peek told me recently,

> Anywhere you have a lot of elk, you see few other ungulates. This is now occurring in northeastern British Columbia, where there has been a dramatic increase in elk and a decrease of moose. In the Selway [Idaho], which is a major elk range, you do not see deer, nor do you see moose. This is not hard to understand. The elk is a strong competitor; it is gregarious, its forage habits overlap all the other major species. When you have that kind of competition from an animal so much more adaptable than any of the others, the question of competition does come into it.[73]

But proving that the competition was there was another matter. Without collars or tags the actual number of any species could not be known. Counting thousands of animals, some hidden among trees, while flying in a Piper Cub at ninety miles an hour, even in perfect weather, involved as much guesswork as calculation. Besides, sometimes competition showed itself indirectly. Often the immediate cause of a decline was disease. How could we know when an epidemic was "natural" and when it was caused by poor nutrition or stress from competition?

How did they know the decline of browse was due entirely to dry weather and suppression of fire and not to grazing?

No one doubted the importance of weather or the benefits of fire to rangeland. Some scientists were surprised by the certainty with which park biologists concluded that the decline of browse was *entirely* caused by these factors.

Although some years in the late nineteenth century were slightly colder and damper than average, long-term studies showed that Yellowstone was not drying out. Indeed, according to Richard A. Dirks of the Department of Atmospheric Science of the University of Wyoming, "No long-term trends are evident in the data from Yellowstone Park."[74]

The record for fire was similar. Forests in the entire West had the same fire history as Yellowstone. Why had aspen and willow not disappeared everywhere? Park biologists provided little evidence to support their view. Houston, J. R. Habeck, Professor of

Botany at the University of Montana, commented, "makes a lot of statements . . . without providing enough, if any, real facts to back them up." There were, Habeck said, "abundant field 'data' to demonstrate the elk have indeed been the cause of aspen reductions."[75]

How did they know? The question persisted. As a revisionist history written by biologists, the new official science left these university scholars unconvinced. Park Service certainty in a fledgling discipline, while groping with a complex mystery, seemed to them out of place. Yet these brave new government scientists seemed to be without doubt.

Previous records, Houston stated with finality, "reinforced a largely incorrect historical account of the northern Yellowstone elk in a manner analogous to that reported by Caughley for the Kaibab deer of Arizona." Indeed, their unearned conviction seemed epitomized by their rush to judge the Kaibab as myth. The Kaibab, to be sure, as a textbook example of how loss of predators led to destruction of the range, was an historical embarrassment to the government hypothesis of resource limitation. But how did they know it was "incorrect"? Caughley himself was far less certain than they. Reinterpreting the histories of other game ranges, he admitted, "is something of a dangerous exercise. . . . I have not been on the ground [in the Kaibab]. . . . My interpretation may therefore be wrong."[76]

Yet perhaps Park Service scientists were not as certain as they appeared to be. In implementing the policy of natural regulation, a government planning document suggested in 1974, "the general trend of park range conditions may go downward and other species dependent on grass may suffer as a result."[77]

That the pristine condition of Yellowstone's ecosystem remained unproven mainly escaped notice. The private doubts of government biologists lay buried in obscure government reports while the general public and larger scholarly community interpreted criticism by scientists such as the Craigheads and Pengelly as the natural but biased expressions of resentment by those who had become pariahs in the park. And the very inaccessibility of data for and against the theory worked to its advantage. Natural regulation had developed an aura of the occult: only the initiate understood it. Independent scientists who did not know kept silent.

There was, besides, no stopping a simple idea whose time had come. The ecosystem, it seems, had been restored; and for many Americans the illusion of the park as a primitive ecosystem was too attractive to be denied. "Natural," according to a recent report on marketing, by the 1970s had become the most popular word in America, preferred even to "new" and "improved." Natural regulation was a triumph of packaging. It was a policy containing nothing artificial. It was more nutritious than gorp and safer than granola. Therefore the thought that the grass and animals might suffer, that Yellowstone's capacity to sustain life might decline if nature were left to its own devices, was inconceivable.[78]

Cole, Houston, and their colleagues had thus made Yellowstone safe for natural regulation. The Park Service could continue its policy of benign neglect. In 1972 it applied for inclusion of Yellowstone in the National Wilderness Preservation System, so that it would be officially recognized as a place "retaining its primeval character . . . affected primarily by the forces of nature, with the imprint of man's work substantially unnoticeable." In the same year the Service put a natural-burn policy into effect: All fires that started naturally (without human help) would be permitted to burn, so long as they did not endanger human settlements. This practice, they hoped, would in time solve the browse problem, and with that, the last remaining step in restoring Yellowstone would be complete.[79]

Things were not to be so simple, however. The elk problem could not be defined out of existence. Difficulties, like the elk themselves, quietly multiplied.

7

Nature Takes Its Course

THE week after Christmas 1983 was one of the coldest ever recorded in Montana. Temperatures in Paradise Valley reached fifty below zero. Yet if you stood at the bar of the Old Saloon near Emigrant with a spotting scope and looked up at the hillside by Six Mile Creek you could see around 3,000 elk pawing through the snow for grass.

At that time I was two miles from the Old Saloon, standing in the forty-below cold with Allen Nelson, a local rancher whose family had farmed this valley for a hundred years. His summer range at Six Mile Creek was twenty miles from Yellowstone.

The elk were on his land. We watched as a helicopter hovered above them. The pilot maneuvered until the herd stood still in the snow. Jon Swenson, a biologist for Montana's Department of Fish, Wildlife and Parks, rode shotgun. In fact, he aimed a gun loaded with a tranquilizer dart and fired. The chopper climbed and hovered above until the animal dropped.

But the animal would not drop. The act was repeated, with the same result. No elk would go down.

The trouble was a bad batch of tranquilizer. Trying to save money, the gunners had used an inexpensive kind. After a futile morning the helicopter returned to Helena; they would try again another day.

After years of pleading with officials, Nelson had finally gotten the state interested in his elk problem.

During recent winters, Nelson told me, thousands of elk left Yellowstone Park and moved up Joe Brown Trail or over Slip and

Slide Creek past Dome Mountain to Six Mile. Despite the depredations of the firing line, elk had been multiplying. More came out
each year to spend the winter on Nelson's pasture.[1]

"They are eating me alive," he said to anyone who would listen.
But he has had trouble finding a sympathetic ear.

According to Park Service biologists, there could be no park elk
on his land. Their elk, they said, never migrated beyond Dome
Mountain, several miles to the south. If Yellowstone elk were at
Six Mile or farther north, Park Service policy, which assumed elk
never migrated past Dome Mountain, was in error. Then they could
not claim the park was "historic winter range" for the elk. Thus
these animals must belong to the state.[2]

Eventually Nelson was able to enlist the interest of the Forest
Service. They owned land next to him and also did not like to see
it denuded of vegetation by foraging elk. They in turn persuaded
Montana officials that it was in their interest to identify the elk on
Nelson's land. These animals, they suspected, were of two separate
groups: one small resident herd that stayed at the Nelson ranch
and a larger migratory herd that came out of Yellowstone in the
winter. Unless they were able to distinguish the two, hunters might
eliminate all the residents by mistake. Then the state of Montana
would lose a segment of "their" elk.

Consequently, the state and the Forest Service agreed to put
radio-collars on elk at Nelson's, some on animals they thought
were members of the resident herd, and some on those they thought
were members of the Yellowstone herd, and follow the animals by
radio to see where they went in the spring. This tracing should
settle, once and for all, whose elk were whose.

Eventually, Swenson collared a dozen elk, and by late spring he
found some of them had traveled far into the park. They were
Yellowstone elk after all. But in the meantime they wintered at
Nelson's. By early April, several hundred remained and Nelson
was frantic. "They have eaten the grass to the ground," he told
me. "They are leaving my cattle nothing but dirt." He called Montana game officials. Something had to be done, he told them.[3]

Something was done. The state sent a helicopter to drive the elk
back into the park. Natural regulation, after all, ended at the
Yellowstone line.

*   *   *

Why had these animals strayed so far from the park? Swenson told me that the elk "are reestablishing traditional migration patterns." If so, then the earlier generation of biologists had been right all along. Just as Nelson and Graves had theorized in 1917, in primeval times elk had not wintered in the park. Instead, they came out — and how far did they go? If Yellowstone was not "historic winter range" for these animals, neither was it an "essentially intact ecosystem" for them.[4]

The growing evidence that Yellowstone was not an intact ecosystem was just one of many problems for natural regulation. Throughout the 1970s and into the 1980s the animals and range seemed to go into a steep decline. But the bad news turned up piecemeal and seldom attracted attention. Often it was something that happened slowly, such as soil erosion, or was on a subject about which wildlife biologists knew little, such as the population trends of the mule deer.

Local citizens who had spent their lives around the park — like Nelson — saw, and were disturbed by, these changes. But although they knew much about local fauna and had a historical perspective that park managers lacked, they were not trained scientists.

And so few noticed when the elk herd began to grow. The 1968 census had found 3,100 animals; by 1975 the count was 12,607 — a 400 percent increase in seven years.[5]

What, then, was K, the theoretical carrying capacity of the range? As the herd grew, estimates of K escalated. In fall 1967, Cole expected "to hold fall herd numbers within a 4,000 to 6,000 range." In 1968, Yellowstone biologists had written that "plans are to allow the late winter size of the Northern Yellowstone elk herd to range from about 5,000 to 7,000, as it has since 1962." By 1969, Cole was suggesting that the carrying capacity was between 5,500 and 8,500. In 1971 Houston commented that "calculations suggest that 8,500 to 9,000 elk should occur in the 1971 fall herd. The elk population is approaching levels where these hypotheses [of natural regulation] will be tested."[6]

Park managers, however, did not use these numbers to test their theory. Instead, they reinflated their estimate of K again. By 1974, Houston conceded that earlier estimates of the carrying capacity of the range were "an underestimate of K and of environmental variability." Instead, he hypothesized that the elk population would stabilize "at prereduction levels of [around] 12,000."[7]

For a while it seemed Houston was right. The winter of 1974–1975 was a long one, bringing unusually heavy spring snows. Elk began to starve in large numbers. Emaciated and mangy animals flooded the roadside, dropping dead in front of photographers, and in the town of Gardiner. By summertime, perhaps as many as 2,000 elk had died.[8]

This large winterkill was just what the Park Service had hoped might happen. They thought it would end the population boom. But starving elk were not popular with the public. There were calls to feed the elk. The Western Association of Game and Fish Commissioners, saying " 'self regulation' all too often means death by starvation and disease," signed a resolution reminding the Park Service that "no National Park is a separate and distinct ecosystem," and urged the Park Service to recognize the "needs and desires . . . of other resource users."[9]

"Other resource users," of course, was a code word for hunters. The specter of public hunting in Yellowstone was rising again. As soon as elk died without the supreme unction of a hunter's bullet, all the old wounds reopened.

In 1976 Glen Cole left Yellowstone to become Supervisory Biologist at Voyageurs National Park, Minnesota. His task there, he told me, was to "reestablish the caribou," an original species that had been wiped out by brainworm carried by deer. "The wolves," he said, "had done such a good job on the deer" that a vacuum had been created in the habitat, and he hoped the caribou could fill it.[10]

Meagher succeeded Cole. At the same time, Jack Anderson retired and was replaced as Superintendent by John Townsley. In 1978, Houston left for Olympic National Park. John Townsley died of cancer in summer 1982 and was replaced by Robert Barbee, Superintendent of Redwoods National Park. In January 1984, Meagher was replaced by John Varley, a fisheries biologist with the Fish and Wildlife Service. Meagher stayed in Yellowstone as a research biologist.

In the year Cole left, the United Nations Educational, Scientific and Cultural Organization (UNESCO) designated Yellowstone as a "World Biosphere Reserve for the global value of its ecosystem and gene pool." In the year Houston left, the park was one of twelve places in the world designated by UNESCO as a "World Heritage Site for its universal natural significance."[11]

As the architects of natural regulation went on to other things and as the shrinking gene pool of world wildlife made Yellowstone's resources all the more irreplaceable, successive park managers continued to struggle with the implications of their policy. The proscription of collars and tags for animals remained in effect, an exception finally being made for the grizzly bear when this species was declared threatened in 1975. "We wanted to keep the animals as anonymous as possible," one official told me. Collars were still worn by alligators in the Everglades, caribou in Denali, and raccoons in Rock Creek Park, but the fauna of Yellowstone were left to their anonymity.[12]

The end of trapping brought other worries. Stockmen began to fear that if left unculled, bison would become so numerous they would leave the park in search of better range. Yet bison carried brucellosis (also known as Bang's disease or undulant fever), an ailment causing domestic cows to miscarry. If they left the park they could infect domestic livestock. State officials, fearing an epidemic, urged the Park Service to eradicate or at least contain brucellosis in Yellowstone. Such a program had been in effect before the era of natural regulation: rangers had regularly trapped bison, inoculating those without the disease and shooting those which tested positive.[13]

But now park managers balked. Trapping animals, they argued, was not only inconsistent with a policy of natural regulation, it was also against the law. "Wilderness designation precludes trapping," Assistant Secretary Reed wrote Congressman John Dingell in 1973. A brucellosis eradication campaign, declared Meagher in 1974, "fails to recognize that wilderness designation precludes use of trapping operations." Yet these were curious replies, for Yellowstone, as it turned out, was never granted wilderness status.[14]

While government biologists continued to search for reasons not to interfere with nature, the Fish and Wildlife Service, responsible for fisheries in Yellowstone, was only too willing to intervene. Remaining obedient to its mandate to restore original conditions, it embarked on a "native species restoration project," seeking to reintroduce native species — cutthroat and grayling — to waters where they once had lived. But to do so they had first to eliminate the so-called exotic species: rainbow, brown, brook, and lake trout.[15]

Taken literally, such "restoration" would also require eliminating *all fish* from waters — such as Lewis and Shoshone Lakes — that were originally barren. Yet this was an ironic turn, for Lewis Lake contained a genetic gem: some of the last remaining Lake Superior lake trout. This species, having been severely reduced in numbers in its original home, was abundant in Yellowstone. The Lewis Lake gene pool was therefore an irreplaceable reservoir: if lake trout were ever to be revived in the Great Lakes, they would have to come from Yellowstone. Consequently, in "recognition of the importance of this sustained genetic integrity," the Fish and Wildlife Service in 1982 took fertilized lake trout eggs from Lewis Lake to be used as brood stock for reintroduction into Lake Superior. Genetically speaking, the lake trout would be going home. But in the meantime public policy called for their eventual removal from the park.[16]

Did Yellowstone contain any other irreplaceable genetic pools? If so, would the policy of restoring original conditions threaten them? Poisoning trout in the park in any case was not a prospect to please anglers. Yellowstone had some of the finest rainbow, brown, and brook trout fishing in America. Some waters supported trophy-sized specimens. The fisheries biologists did not relish the thought of destroying these fish, but the law was the law. In 1976, therefore, beginning at a little stream called Canyon Creek, after poisoning the resident rainbows, they introduced 2,863 nine-inch grayling.[17]

Neither the poisoning nor the transplant was successful. The grayling disappeared and the rainbows reappeared. Restoring the primitive scene was turning out to be more difficult than had been thought. In recognition of these and similar difficulties, an Ad Hoc Fisheries Task Force of the Fish and Wildlife Service was established "to review and evaluate the effectiveness of the National Park Service's fisheries policies and practices."[18]

This committee urged that the policy be changed. In February 1980 it proposed, through an insertion in the *Federal Register,* that "where appropriate and after careful analysis the Director, National Park Service, should be empowered to declare certain introduced species 'naturalized' and managed as natural components of the ecosystem." This proposal, they hoped, would achieve through the magic of legislation the result that was proving impossible for biology to do. But it was rejected; natural regulation, it seemed, was here to stay.[19]

While park administrators came and went and government departments struggled with the legal niceties of managing ecosystems, Montana state officials were increasingly alarmed over the elk problem. The theory of natural regulation, James W. Flynn, Director of Montana's Department of Fish, Wildlife and Parks said in a letter to Yellowstone Superintendent Barbee in February 1983, "will not be endorsed by our department." Their experience "provided some clearly documented situations relating to the capability of elk to materially influence the environment they live in. . . . Intense vegetation and population studies have very adequately documented significant and positive changes in composition and conditions of vegetation on winter ranges associated with segments of these populations."[20]

But, Flynn continued, "our most paramount objection to the natural regulation philosophy . . . is that the basis from which it was initiated is totally lacking in sound ecological documentation. . . . Yet the current management direction for elk in the park indicates a complete willingness to disregard some of the most glaring examples of deterioration of vegetation under the guise of natural regulation."[21]

The firing line also remained unpopular with state officials. "The manner in which these hunts are conducted," Flynn commented in 1985, "is not what we would generally term as the ideal hunting opportunities we prefer to provide sportsmen."[22]

Local residents with the perspective of time began to fear that the range was dying. There seemed to be less ground cover. Pedestaling — soil washing from around a clump of grass — was more obvious, a sure sign of soil erosion and overgrazing.

The boundary-line area, the part of the park between Mammoth and Gardiner, had such sparse vegetation it looked like a desert. "I get nearly sick," Montana state biologist Arnold Foss told me, "every time I drive through the northern range now. I would not turn a jackrabbit loose there without packing it a lunch first."[23]

Robert Howe, who retired as park biologist in 1966, returned to the winter range recently and was distressed by what he found, especially the signs of elk trampling around Soda Butte Creek in their winter range. "It almost looked like they had taken a Rototiller to it," he told me, "leaving it a perfect setup for a lot of erosion."[24]

"There will be a day of reckoning," he added, "and when it comes, the Park Service will deserve what it gets."

Former researcher Walt Kittams was equally distraught. "I am so disgusted with the management of the northern Yellowstone elk herd," he told me recently, "that I don't want to visit the park any more."[25]

Exclosures showed a surprising trend: natural regulation was encouraging unnatural plants. Inside the exclosures native species, such as Idaho fescue and bluebunch wheatgrass, could still be found. But outside, these plants, preferred by the elk, were giving way to exotics such as Kentucky bluegrass and timothy, which were more robust or received less grazing pressure.[26]

As evidence from exclosures mounted, so too did Park Service antipathy to them. A five-acre exclosure near Tower was taken down early in the era of natural regulation. That was the last straw, said Kittams, who retired from the National Park Service in 1973, "when you do a thing like that you need to have a good reason. The only reason the park had was that the exclosure made management look bad."[27]

When many university scholars objected, this demolition temporarily ended. Instead, the fences were not maintained and began to collapse on their own. As they collapsed the Park Service removed more of them, saying they conflicted with wilderness values or were too expensive to maintain. During summer 1983 two exclosures on the southern range, at Red Mountain and Chicken Ridge, were removed.[28]

As vegetation deteriorated, many began to suspect that soil erosion was increasing. On Big Game Ridge, south of the park — an area heavily grazed by the southern herd — erosion was found to be up to two-tenths of an inch a year, a rate that geologists considered phenomenal. It would mean that the ridge, which had been there tens of millions of years, would be level in less than another 200,000 years.[29]

Government biologist George E. Gruell took this measurement, but he considered the "rate not greater than presettlement times."[30]

Likewise, after a rain, the Lamar River turned dark chocolate, and stayed that way longer, oldtimers felt, than in the past. John Bailey, son of the famous fly fisherman Dan Bailey and lifelong resident of the area, was so bothered by the muddy waters of the Yellowstone that he chartered a plane to fly the drainage, well up into the park, during summer 1984. "You wouldn't believe how brown and bare the range looked," he reported to me. "Elk and bison were everywhere. The range looked awful!" Was more of Yellowstone washing downstream?[31]

Continuing archaeological research on the Yellowstone plateau by Professor Gary Wright of the State University of New York at Albany had failed to turn up evidence to establish the Park Service assumption that elk were native to the region. After excavating more than two dozen sites, Wright had failed to find one elk bone. "Where," he asked, "are those thousands of prehistoric elk?"[32]

Despite the natural-fire regime, willow and aspen seemed ever scarcer. What was going on? Charles Kay, an independent researcher and former student of Pengelly's, decided to find out. During summer 1983 Kay took a census of aspen in and around the park. He studied 3,082 stands outside Yellowstone and 799 stands in it, and his study confirmed that unless something was done there would soon be no browse left. "Within Yellowstone," he reported at a symposium on western elk management at Utah State University in April 1984, "only 4 percent of the aspen stands had reproduction and this was limited to snowbank sites or steep rocky outcrops, both of which restrict or prevent elk use." Outside the park he found aspen reproduction averaged 60 percent. "Outside Yellowstone," Kay concluded, "aspen is clearly regenerating itself in the absence of fire, and this difference cannot be due to variations in climate." Rather, he concluded, the decline was caused by "repeated browsing by elk."[33]

Yet the areas around Yellowstone had been under a policy of fire prevention as long as the park itself had been. If, as Houston had said, fire prevention was the cause of aspen decline in the park, why was it not the cause of aspen decline outside of the park?

Indeed, the park natural-fire policy was not working. Unlike some parks, such as Everglades and Sequoia, Yellowstone did not have a policy of "prescribed burning." That is, managers were not permitted to ignite fires intentionally. Instead they were required to wait for lightning to strike. Yet apparently lightning was not striking often enough. After twelve years of the "natural-burn" policy, not one significant fire had burned in the northern range. Although since 1972 thirteen natural fires in Yellowstone had burned

160 acres or more, not one acre of the northern range had burned. Yet Houston's tree-ring study had discovered that before the white man the entire range had burned on the average every twenty years. Why would this land not burn?[34]

Fire needs fuel, ecologists suggested. It was buildup of dead grasses that fed a fire when lightning struck. But in Yellowstone, there is no fuel. "The lack of fire is no surprise," Professor Peek told me, "the elk ate all the fuel."[35] Charles Kay was more emphatic: "The northern range would not burn if you napalmed it."[36]

As the range declined, so too, many thought, did the animals that depended on it.

"The situation in Yellowstone is sad, sad," Scotty Chapman said to me. Chapman had been a ranger in Yellowstone from 1930 until retirement in 1962. Since then he had lived on a small farm next to the park.[37]

"I used to be able to watch the mountain sheep from my window," he went on. "But no more."

"A myth about Yellowstone," Joe Gaab told me, "is that it is a tremendous reservoir for wildlife. It isn't." Perhaps no one alive knew Yellowstone better than Gaab, who had spent a lifetime in the Yellowstone backcountry, first aiding in elk studies and later patrolling its borders for Montana Fish and Game.[38]

"Take moose, for instance," he went on.

There are hardly any left. They can't compete with the elk. And mountain lion. People think there are mountain lion in the park. More were killed by hunters between Gardiner and Livingston this summer than exist in the entire park. Cougars feed on deer, and there are almost no deer left. And in the fifties, when I was assigned the job of following the elk herd, there was a place on the Mirror Plateau where I knew and could distinguish twenty-seven grizzlies. There is not one grizzly now where there were more than twenty then.

Indeed, many scientists also worried about the animals. Antelope numbers remained around three hundred. No one knew about the beaver, for no studies had been done, but it was universally believed that their numbers were down. And no one could guess how the more secretive moose was faring, but there were reasons to worry. Because moose required browse to live it did not seem they could have much to eat.[39]

Mule-deer numbers were not good. The most recent census, conducted by Glenn Erickson of the Montana Department of Fish, Wildlife and Parks in 1979, could find only 126 in the park. "And they," Erickson told me, "had the worst fawn-to-doe ratio I have ever seen." Whereas biologists estimated that a ratio of thirty-five fawns for every hundred does was necessary for a population to survive, Erickson could find only fifteen fawns for every hundred does in Yellowstone.[40]

The bighorn was in trouble. In 1979, Kim Keating, a student of wildlife biology at Montana State University, conducted the first and only independent study of Yellowstone's bighorn population since the era of natural regulation began. Wanting to put radio-collars on the animals so that he could follow their movements, he could not do his research in the park. But luckily the Yellowstone bighorns lived on high ridges that lay along the boundary with Montana. When the sheep were out of the park they did not belong to the federal government but to the state, and the state would give him permission to place collars and marks on the animals. Keating climbed the mountains and waited for sheep to leave the park. When they did, he put collars on them.[41]

"One ranger gave me a hard time," he told me. "He threatened to take the collars off when the sheep crossed the line. But I told him he had no right to do that, and finally he backed down."

Keating, therefore, was able to complete his study, and he found something surprising. The bighorn population had made a dramatic comeback after the reductions of the elk in the 1960s. Originally common throughout Yellowstone, bighorns had remained very few between 1930 and 1960, never, apparently, exceeding two hundred. But following the massive elk reductions of the 1960s, Keating found, their population had zoomed up to five hundred. This come-back had not been noticed because a four-year lag had separated the elk reduction and the surge in numbers of bighorn. When the population boom showed itself, the policy of elk reductions had come to an end.[42]

The elk, Keating's study implied, had hurt the sheep. Nor, as it turned out, had they completed their damage. During the winter of 1981–1982 the herd became infected with chlamydia, or pinkeye, a disease that caused temporary blindness. It was mating season and rams, butting heads in fights over females, spread it quickly. Although chlamydia was not serious and was easily treatable, the Park Service, following the policies of natural regulation, would not intervene to medicate the sheep.[43]

Nature was allowed to take its course. Although a few rams were saved by local ranchers who, defying park regulations, trapped an animal and put drops in its eyes, the majority, blinded by the disease, unable to keep their balance or eat, disoriented, walking ceaselessly in circles, fell from cliffs or starved to death. Some suffering animals were shot by rangers to spare them further misery. For park officials, though not permitted to save the animals, were permitted to kill them.[44]

By spring only 180 bighorns were left in the park. Sixty percent of its rare bighorn population had disappeared in one winter. What really killed them? Chlamydia, Jon Swenson explained to me, was a native organism, to which sheep were probably exposed frequently. Why then, did they suddenly contract it in so virulent a form? "Perhaps," said Swenson, "it was due to nutritional stress. Whenever an animal population succumbs like that you have to suspect that they were being stressed in one way or another."[45]

Certainly the pinkeye epidemic would not have surprised the old park scientists. "Biological studies of bighorn," Robert Howe had written in 1961, "have indicated that poor range conditions are a basic difficulty in maintenance of sheep bands, and though the sheep may die of various diseases, these fatalities are brought on by weakened physical condition of the animals due to poor range conditions."[46]

The architects of natural regulation, however, did not blame the elk; rather they blamed man. Following the epidemic, Meagher and Assistant Resource Manager Gary Brown closed portions of the northern range to hikers for "the protection of bighorn sheep and elk during a period when both species are most seriously stressed by human activity." This action, they reasoned, would "provide the opportunity for the sheep population to recover from the high mortality of the winter of 1981–2 . . ." and ". . . for this segment of the elk population to utilize this important area of the northern winter range and to move from the park allowing hunter harvest in the state of Montana."[47]

By 1981, a census of the northern range taken by Meagher found 16,019 elk. Assuming 80 percent census efficiency, the herd had topped 20,000, growing 500 percent in thirteen years.[48]

During her flight, covering the entire northern part of the park, Meagher saw *one* moose.[49]

Meanwhile, contrary to Meagher's prediction, the bison herd had grown as well, from fewer than 400 in 1967 to more than *2,000* in 1981. And as they proliferated they began to leave the park.[50]

Groups of several dozen or so came into the town of Gardiner, wandering the streets or grazing in the schoolyard. To Montana officials this visit was not a laughing matter. The fear of brucellosis grew as these emigrating bison threatened to infect the state's population of domestic livestock.[51]

But the state was reluctant to let hunters kill them. Because they carried undulant fever, a hunter who dressed a buffalo without wearing gloves could get very sick. Instead townspeople, rangers, and ranchers rounded them up and drove them back. But bison could not be herded like cattle. Some refused to return. They wanted to stay where food was more plentiful.[52]

As more animals left the park the situation became difficult to control. "Buffalo," said Ron Marcoux, Assistant Director of Montana's Department of Fish, Wildlife and Parks, "can move a long way in a short time."[53]

Indeed, some soon did. Three wandered past Gardiner, where they were shot by state officials. All three tested positive for brucellosis.[54]

Local stockmen were almost literally up in arms. It seemed to state officials that they would have to maintain a posse at the park line indefinitely, ready at a moment's notice to blow away any bison that left the park.[55]

According to the theory of natural regulation, this growth and exodus was not supposed to happen. If Yellowstone were an intact ecosystem for bison, why were they leaving? And where were the density-dependent self-limiting factors now that we needed them?

They had never been found. Houston had searched vigorously but in vain for them. "The density-dependence shown by the population's natality rate was less than might be expected," he concluded, adding, "density-dependence of male survival, although suggested, could not be demonstrated." He thought he had detected a trend of increased calf mortality as the herd increased, but reaching that conclusion, he admitted, was "actually a sticky blend of data and guesswork."[56]

By 1984, park biologists had given up looking for the mysterious density-dependent self-limiting mechanism. The best they could say was that it was "inanition." That is, as elk multiplied, more would starve to death until eventually the number starving equaled the number born.[57]

Park biologists were saying, in short, that they no longer knew how their theory was supposed to work. Instead, they hoped for winters severe enough to kill elk. "If we would only have a decent winter as we are supposed to," one park biologist said to me recently, "you will see the number of elk stabilize."

Indeed, natural regulation had become nothing more than the policy of waiting for bad weather.

The Secret of Yancey's Hole

A ONE-MILE walk from the Roosevelt Lodge at Tower Junction
is a place called Yancey's Hole. "Uncle John" Yancey ran
the Pleasant Valley Hotel and Saloon there a hundred years ago.
He built it for the teamsters from the mines of Cooke City, who
had no choice but to pass by on their way to the toll bridge across
the Yellowstone run by another early Yellowstone entrepreneur,
Jack Baronett.[1]

The Pleasant Valley Hotel was probably the most uncomfortable
establishment in the West, a two-story log building that accom-
modated twenty guests in five rooms. Yet many people — Owen
Wister and Earnest Thompson Seton among others — stayed there.
The guests would sit on Yancey's front porch in the evening, swat-
ting mosquitos and watching the beaver carry willow branches
across the ponds that filled the field below the buildings.[2]

Seton studied the beaver there in 1897. "The best opportunity
I ever had to study Beaver work," he wrote, "was in the Yellowstone Park. . . . On Lost Creek, not far from Yancey's, where I
stayed some months, was a family of Beavers with their usual
contrivances to make a great pond of a very little stream." There
were thirteen ponds in the meadow, most of them created by one
giant dam more than 300 feet long and containing up to 200 tons
of stone, mud, and willow.[3]

One of America's great experts on the beaver, Edward Warren,
wrote a book about the beaver at Yancey's Hole in 1922. By Warren's time the beaver had left the meadow and retreated a mile up

Lost Creek, but were still abundant. Jonas began his study there in 1953. By then the beaver had gone from the entire area.[4]

The hotel, ponds, willow, and beaver were all gone now. Yancey's Hole was no longer the brushy thicket it had been in Seton's day, nor the lush hay meadow it was when Warren saw it. It was an open field split by a serpentine stream. Yet for scientists this place still had special value. It was one of the few places in wild America that have been studied with minute care over a long period. The work of Seton, Warren, Jonas, and others gave a unique view of ecological change that was too slow to be perceived in a person's lifetime.

Last summer I visited the Hole with Les Pengelly and Charles Kay. We went to look into Yellowstone's past.

Kay led the way. The current park concessionaire, TW Services, was preparing a cookout below us. About thirty tables were under a marquee, their gingham covers blowing in the wind. Two workmen were building a hitching post for the miniature conestoga wagon that would carry visitors there from Roosevelt Lodge.

We walked into the woods behind the marquee. Although the trees were thick, we could see for some distance, for all the limbs below six feet had been eaten away by elk — or "high-lined," as it was called. A half mile up the hill we found the remains of the old beaverdams that Warren studied, filled in now, and dry. Where, I wondered, had the water gone?

Deeper into the woods Kay stooped over.

"Look," he said, touching a bare twig about six inches high, nearly buried in the grass. "That's chokecherry. See how far down it's been eaten? A stem that short can't produce berries for animals like bears to eat. Compare that with the lush stands of berry we saw out of the park in Yankee Jim Canyon. Plus, it won't reproduce this year." In fact, we could find no edible vegetation at all, neither berries nor forbs, in the understory.

We left the woods, heading across the meadow. Kay stopped again. "Look at that pedestaling," he shook his head. "See how the clumps of grass stand above the surrounding ground? That means more than an inch of topsoil is gone. That's not supposed to happen!"

At the other end of the Hole we reached an exclosure. It was tiny, perhaps only fifty feet to a side. When it was put up in 1936 a corral was on the spot, and the park scientists noticed that where the corral fence had kept the elk out, aspen was growing. To test whether the aspen was growing because it had escaped grazing pressure, the fence was torn down and the exclosure put up.[5]

Kay showed me a picture taken of the exclosure in 1953. Half the aspen lay inside the exclosure and half outside.[6]

Today we saw something that was almost comical. Inside, aspen of every age filled the exclosure so densely we could not see daylight through it. Outside was nothing but grass, cut as close and smooth as a golf fairway. The impression was of a potted plant sitting on a lawn.

Kay bent over again. "Take a good look at this grass," he said. "This is Kentucky bluegrass, brought in by the Park Service fifty years ago to seed the lawns at Mammoth."

When elk were grazing, Kay explained, bluegrass had a competitive advantage over the natives. The exotic propagated itself by the root system, but natives spread seed from above. Because the elk preferred native grasses and ate them before they could go to seed, they disappeared and the bluegrass prospered.

"Is this range damage?" I asked Pengelly.

"What is truth?" he replied. "You and I look at this scene and see range damage. The Park Service sees nature at work."

Park managers, indeed, did not seem troubled by this evidence. Conditions that others took as signs of decline were, they insisted, natural, and being natural they could not be decline.

Instead, giving explanations that seemed suspiciously ad hoc, they dismissed all the cries of alarm. The apparent range damage, they insisted, was merely "zootic climax" vegetation. There was no pedestaling. The decline of browse was "natural seral succession." Evidence from exclosures was skewed, for fences trapped moisture, creating a "microclime" that made for artificially lush growth of grasses. There was no sign of soil erosion. The mud in the Lamar and Yellowstone rivers was put there by "geologic instability." No fires had started in the northern range because sagebrush had not grown tall enough to sustain a fire.[7]

There was still no sign of interspecific competition, they insisted. The pronghorn population, Houston said, was suffering not from elk but from poaching and coyote predation. Mule-deer numbers, he claimed, were the same as in "primeval times."[8]

As evidence, they turned to numbers. For elk and bison, whose proliferation proved an embarrassment, park authorities reported the actual (and therefore minimal) number counted by Meagher during aerial censuses. But when they published figures on species that some feared were threatened by these ungulates, they reported an estimate, based on some (optimistic) multiplier.[9]

"I estimated that the winter population (of mule deer) was 2,000 during the late 1970s," Houston reported. Yet he derived this number from *the same census* (conducted by Glenn Erickson, mentioned in Chapter 7), which found only 126 mule deer in Yellowstone. Rather than reporting the actual number of deer found in the park (126), Houston counted all 1,088 deer that Erickson found in an entire Montana hunting district (stretching past Dome Mountain to Point of Rocks, twenty road miles from the park). Then, to reach the figure 2,000, he assumed that, because "counting conditions were poor," and many animals had been missed, he *doubled this number*. In this way he managed to multiply the actual count almost twenty times and reach a number that seemed consistent with the "historic population."[10]

With such arithmetic the government reassured the public and buoyed the confidence of the seasonal rangers who faced questions from summer visitors. Conversely, they called hunters' attention to the bounty of elk the park provided. If the herd were reduced, former Yellowstone naturalist Paul Schullery said in an official defense of natural regulation, "this would dry up the great hunting now available in the Gardiner Hunt."[11]

Though such responses may have satisfied some, they left others uneasy. What was the truth in Yellowstone? Were elk even a native species? Anthropologist Wright was coming to the conclusion that they were not. He had still not found any of their bones in middens. Yet he could not convince the Park Service.

"I have been battling wildlife biologists from Grand Teton and Yellowstone Parks for some years," he wrote to Professor Pengelly. "One told me after a seminar that I gave at the Jackson Hole Biological Research Station on the faunal resources of the region, 'even if you demonstrate that no elk were here, we will still argue for them because our management policies require a herd of at least 10,000 by the end of the Pinedale Ice.' "[12]

As elk and bison fecundity slowly ticked skyward, unease grew, in and around Yellowstone. Park officials and local citizens alike wondered if they were sitting on a time bomb. No one believed the "elk problem" had gone away. Natural regulation seemed to have reached a dead end. Without intensive research, Yellowstone remained *terra incognita* even to its own biologists. The extent of soil erosion, pedestaling, the real causes of browse decline, and the population status of all the large mammals remained obscure. Introduced as a hypothesis to be tested, it had not been tested.[13]

In 1982, Douglas Houston's book, *The Northern Yellowstone Elk* — the first complete statement of natural regulation — was published. Paradoxically, the work had been presented to the public fourteen years after the policy had been applied. Natural regulation was all dressed up in time to have nowhere to go. It was making its debut just as it was dying of old age.[14]

Yet the power of the idea had not died. Unaware of conditions in Yellowstone, many greeted the "new" idea with enthusiasm. A. R. E. Sinclair, author of the study on the African buffalo in the Serengeti, in his foreword to Houston's book, praised it as "one of the first indications that an objective and experimental approach is entering the field of wildlife management."[15]

Gary Blonson, writing in the November 1983 issue of *Science 83*, described natural regulation as "a brave and determined policy, an attempt to apply scientific knowledge in an arena where politics and presumption often have dominated."[16]

But the question would not go away: How did they know? Mysteries crucial to the future of the park steadfastly refused resolution: Were the mule deer in trouble? What kept the pronghorn numbers so low? Why had the bighorn population crashed? Would the whitetail deer, which was now regularly seen just north of the park, ever return? Where were the beaver? What was destroying the aspen and willow? Why would the range not burn? How serious was soil erosion?

If the hypotheses of natural regulation were truly scientific, they had to be verifiable; there had to be some tests that would determine their truth or falsity. But such testability was not easy, for it seemed that theory had outpaced experiment in Yellowstone. Hypotheses such as natural regulation were based on highly abstract and mathematical analyses, and it was not clear what experiments would verify them. Authorities would not or could not make clear what they would accept as evidence for a test. The decline of browse, increase in elk, crash of the bighorns — all the conditions that worried their critics — had no relevance, government scientists suggested, to the truth or falsity of their theory.[17]

Then what evidence would be relevant? No one would say. Debate became uncertain and abstracted as critics, not knowing the rules by which they were playing, could not be sure what studies might resolve the issue. They seemed to be boxing at shadows, and all they hit was air.

"Is what I saw at Yancey's Hole range damage?" I asked Dr. Habeck recently.[18]

"It is change," he replied. "Whether it is bad change or good change depends on whether it is natural or not."

"Is it natural, then?" I insisted.

"That is unlikely in an open ecosystem like Yellowstone's," he told me. "Two hundred years ago buffalo or elk might have come in there and eaten it down to the ground. But then they might move on and not return for ten years. The grass had time to recover. Now the elk simply camp in there every winter, year after year."

In just this way a question that seemed to be about biology — what was the status of the park ecosystem? — became a question of history: What was Yellowstone like before the white man came?

Here was the reason biologists had so much difficulty judging natural regulation: historical assumptions were clothed in the language of biology. Was the range deteriorating? "No, that was just *zootic climax vegetation.*" (These places always looked as they do now.) Did the numbers of elk exceed the carrying capacity of the range? "No, they would reach equilibrium at the *ecological carrying capacity.*" (The population would stabilize when it reached *historic* levels.) Was browse declining? "No, what we saw was *seral climax vegetation.*" (The browse was going through a cycle it had been through many times *before.*) Should we worry about the bighorn sheep? "No, pinkeye was a *natural disease.*" (The sheep have been through this cycle many times *in the past.*)

Despite signs of decline, the park biologists' optimism was sustained by a simple belief: So long as Yellowstone was an intact ecosystem, these bad things could not happen.

Here was the crux of the debate. For if this belief had any meaning at all, it was historical, not scientific. It was a claim that the apparent decline — irruption of elk and bison, decline of vegetation and other animal species — was part of a cycle of change that had gone on for millennia.

The earlier generation of park observers, from Chittenden and Skinner to Garrison and Craighead, had been convinced that the park was no longer in its original condition. The architects of natural regulation, however, now insisted that, for the ungulates, it remained whole. The survival of Yellowstone depended on who was right. What was happening was either slow decay or natural change. Which it was depended on whether conditions in the park were "original" or not.

The key to the future of Yellowstone, therefore, lay in its past.

9

Growing Apples in Eden

DURING his first day in Yellowstone, Lieutenant Gustavus C. Doane saw a fire. A member of the Langford Expedition of 1870, he had just crossed the Gardiner River (near the park's present north entrance) when he noticed "the woods were . . . on fire in every direction." The Indians, he suggested, had set the fires "to drive away the game."[1]

Lieutenant Doane was half right. Indians probably had set the fire, but almost certainly they had not done so to drive the game away. It was more likely that they had set fires in a circle to surround and concentrate game, making it easier to kill. "Fire hunting," as Thomas Jefferson called it, was a common practice, not only among the Indians, but among aboriginal peoples throughout the world.[2]

Doane had seen one of the last examples of an activity that had gone on for millennia. Yellowstone had been Indian country for 11,000 years. At the end of the last ice age, the first settlers camped at the foot of the glaciers that shortly before had covered the entire plateau to a depth of 3,000 feet. They soon settled in along the shores of Yellowstone Lake, at the edges of the thermal areas, and along the riparian valley bottoms, hunting woolly mammoths and bighorn sheep. And no sooner did they arrive than they began to burn.[3]

Indeed, Pleistocene people, archaeologists tell us, used fire drives ten to twenty thousand years ago. Today natives in South America, Africa, Australia, Tasmania, New Guinea, China, and Turkestan, among other places, find their game in this way. Nor was hunting the only use to which fire was put. Fire, the earliest human tool, had been used by people throughout the world for every conceivable purpose. It must have been nearly a million years ago that our Promethean ancestors, seeing lightning strike a tree, rushed to ignite a torch. And once they had stolen the flame, they used it with abandon. They kept fires in their camps for warmth, cooking, communication, and for warding off predators in the night. They burned around their campsites to keep the brush away and the bugs down. They burned woods to open trails and make roads. They burned to prevent the buildup of combustibles that would later cause a climactic and dangerous forest fire. And they set fires accidentally.[4]

They also used fire for farming. "It is often assumed," historian Stephen J. Pyne writes, "that the American Indian was incapable of greatly modifying his environment and that he would not have been much interested in doing so if he did have the capabilities. In fact, he possessed both the tool and the will to use it. That tool was fire."[5]

Fire improves nature in many ways. It changes the chemical composition of the soil, promoting more varied vegetation and supporting more diverse wildlife. It releases soluble mineral salts in plant tissue, and increases the nitrogen, calcium, potassium, and phosporus in the soil that act as fertilizers. It decreases soil acidity, neutralizing the effects of acid rain, not only in the earth but in the streams and lakes as well. It blackens the ground, permitting more of the sun's heat to warm it and promoting faster growth of vegetation. Burning forests stimulates germination in some trees, such as lodgepole pine; it produces habitat for nesting birds, such as the mountain bluebird. But the most important result of burning is that it arrests or reverses the normally inexorable seral succession.[6]

Left undisturbed, areas such as Yellowstone would progress from grasslands, to shrub, to deciduous stands of aspen, then to lodgepole pine, and finally to spruce and fir, a succession that takes around three hundred years. As succession progresses, the variety and volume of plant life diminishes, and thus the capacity of the land to support wildlife declines as well. Fire is the only way to interrupt seral succession, and human beings apparently learned this lesson long ago.[7]

Burning stimulates the growth of browse, forbs, grasses, and berries, promoting greater diversity of wildlife. (The number of animal species in an ecosystem, studies have shown, peaks about twenty-five years after a fire.) Repeated burning turns a forested area into a grassland plain, simulating growth of grasses that support bison and elk. "In North America," ethnohistorian Lewis wrote, "the most important resources of Indian hunter-gatherers are the early succession species commonly found in recently burned areas: bison, moose, deer, elk, rabbits, grouse, grass seeds, legumes, berries, bulbs. However, natural fires are too irregular in occurrence and distribution to be completely relied upon." To ensure that their food resources would flourish, the Indians, rather than waiting for nature to take its course, ignited the range themselves.[8]

By contrast, without fire, succession progresses and the number of species of plants — and animals that live on these plants — declines. Sagebrush spreads in the grasslands, berries and forbs disappear, forests advance into open areas, and much of the aspen and willow age and die. As the forest spreads, trees, which lock up greater amounts of water, lower the water table, drying the soil; as the forests age, insects and diseases spread. Entropy overtakes nature and eventually, as the forest becomes a monoculture, the animal kingdom becomes one as well. The open savannah that once supported bison, elk, deer, antelope, beaver, bears, birds, and wolves, becomes the closed boreal forest inhabited by squirrels, ravens, and pine martens, but little else.[9]

Accounts by early settlers in America showed that they were well aware of the Indians' penchant for burning, and that they knew what the Indians were doing. The Puritan chronicler Thomas Morton wrote in 1637 that "the Savages are accustomed to set fire of the Country in all places where they come, and to burne it twize a year, vixe at the Spring, and the fall of the leafe. The reason that mooves them to doe so, is because it would other wise be a coppice wood, and the people would not be able in any wise to passe through the Country out of a beaten path." Thomas Jefferson commented on the Indians' practice of fire hunting. "They make their circle," he wrote to John Adams, "by firing the leaves fallen on the ground, which gradually forcing animals to the center, they there slaughter them with arrows, darts and other missiles." Lewis and Clark reported that the Indians set fires to improve pasturage for their horses. Botanist Edwin James wrote (in 1819) that the

Indians set fires "in order to attract herbivourous animals, by the growth of tender and nutritious herbage which springs up soon after the burning." Army engineer Mullan, exploring Montana in 1860, was surprised to find the Coeur d'Alene Indians remarkably sophisticated in their use of fire. "In returning," he wrote, "the Indian [Mullan's guide] set fire to the woods himself, and informed us that he did it with the view to destroy a certain kind of long moss which is a parasite to the pine trees in this region, and which the deer feed on in the winter season. By burning this moss the deer are obligated to descend into the valley for food and thus they [the Indians] have a chance to kill them."[10]

"To simply note that all Indians used fire to modify their environments," wrote Lewis, "is no more an ecological generalization than to note that all farmers used plows." With fire, they tilled the earth. It has been documented that at least one hundred tribes of North America used fire for at least fifteen purposes, but nearly all these uses dramatically affected the landscape and ecosystem. The "forest primeval" that Longfellow described was not the scene the first settlers on this continent saw. Rather, forests were created by the European settler. Whereas to the Indian trees were an obstacle to be removed, to the white man they were property to be preserved and used for building ships and houses. "The virgin forest was not encountered in the sixteenth and seventeenth century," writes historian Pyne, "it was invented in the late eighteenth and early nineteenth centuries. For this condition Indian fire practices were largely responsible."[11]

Virgin America, rather than a place where a squirrel could swing from tree to tree from the Atlantic to the Mississippi without touching the ground, as some said, was a place, Pyne reported, "where it was nearly possible to drive a stagecoach from the eastern seaboard to Saint Louis without benefit of a cleared road." Indeed, it was not the primeval forest that impressed the first Europeans about America, but the absence of trees. Near Narragansett Bay, explorer Verrazano found "open plains twenty-five or thirty leagues in extent, entirely free from trees or other hinderances." In Salem, Massachusetts in 1630 it was reported that "a man may stand on a little hilly place and see diverse thousands of acres of ground as good as need be, and not a tree in the same."[12]

Using fires, the Indians virtually alone created the great prairies of the Midwest, and were still extending their range when the white

man came. The so-called prairie peninsula — an arm of open grassland extending through Ohio to Pennsylvania that the first settlers
found — was created by Indian fires to extend the range of the
buffalo, and it was successful.[13]

Evidence abounded that the Indians of the northern Rockies,
including the Yellowstone plateau, engaged in broadcast burning,
and that they changed the area dramatically. The Cree, Slave,
Flathead, Salish Kootenai, Pend Oreille, and Coeur d'Alene all
practiced it. So too did the Shoshone and Blackfoot, both tribes
living in parts of Yellowstone. Blackfoot Indians, it is said, were
so named because their feet were blackened by running through
burned areas. Lewis and Clark frequently ran into fires set by
Indians. The journals of dozens of other explorers — including
George Catlin, John Wesley Powell, and Father Pierre deSmet —
also remarked on the numerous fires they saw set by the natives.
Osbourne Russell, traveling across the plateau in 1835, described
how the Blackfoot tried to flush his party of trappers by setting
fire to the brush. In 1842 countless fires started by the Indians
raged out of control throughout land that is now Montana. At that
time, reported historian Pyne, "the landscape around major Indian
trails resembled the wasteland around rights-of-ways of railroads."[14]

The prospector deLacy, reaching the Firehole in 1863, recorded
that "the grass was partly burnt," and suggested this "indicates
that Indians might be near." The Hayden Survey expedition to
Yellowstone in 1872 found large expanses of the plateau had been
recently burned. According to a report of the Acting Superintendent of Yellowstone in 1886, a group of Lemhi Indians (a Shoshone tribe), set two fires in the western part of the park.[15]

Recent research has established that the Indian practice of burning around Yellowstone was not only widespread, but had been
practiced for millennia. "Broadcast burning by the native peoples
of the inland Pacific Northwest," biologist Dean A. Shinn said
recently, "was widespread and persisted over an extended period
primevally. It may have dominated, perhaps largely pre-empted,
natural burning in shaping aboriginal environments." Fire-scar data
taken from trees in western Montana, according to biologist Stephen Barrett of the University of Montana, indicated "substantially
shorter Mean Fire Intervals in Indian habitation zones than in the
remote zones before European settlement in 1860," and that the

frequency of fires was about twice the rate at which the forests caught naturally from lightning. A study of charcoal deposits in a bog on Montana's Lost Trail Pass found frequent broadcast burning by Indians over the last 2,000 years. On the Yellowstone plateau near Jackson Hole, anthropologist Wright discovered more than three dozen early campsites occupied for 4,000 years and more, and that the area around these sites had been repeatedly burned by natives. "Fire was an important tool," wrote Wright, "in the regulation of the human ecosystem, and ethnographic evidence clearly demonstrates that Native Americans understood this."[16]

This activity profoundly affected the environment around and on the Yellowstone plateau. Human burning kept large areas in open grassland, forests from reaching climax, sagebrush from spreading, and many edible plants prolific. The many open parks, the riparian thickets of willow, the aspen colonies, the lush patches of berries that early explorers wrote about, were helped along by the kinds of fires that Doane encountered.

The land that the early settlers found in the West, in short, was not made that way by God alone, but partly by man. The Indians, too, knew how to play God, and one of the tools with which they made their Eden was fire. They were, however, capable of diminishing nature as well as improving upon it, and they did so by hunting.

"The true hunter loves and reveres his quarry," Isak Dinesen is said to have remarked. "Alas," she added, "these feelings are unrequited." Indians, indeed, worshiped the game they hunted, but they also took their toll. For them the chase was not an activity done in season to give the game a sporting chance, but a deadly serious undertaking, done whenever possible and under conditions that gave the hunter every advantage. They knew their quarry and practiced sophisticated techniques for killing.[17]

Antelope were especially easy to outwit. Knowing that they would not jump a fence, the Indians learned to construct false fences of sagebrush, often trapping dozens at once. Bison were caught in traps as well, or driven off cliffs. Moose and elk, too weak to move quickly in winter and in deep snow, were hunted on snowshoes, and they were easily run down. Large numbers of elk, suggested archaeologist George Frison, could be killed at once by the simple technique of shooting the bunch leader, usually a female, first. The remainder of the herd, rather than running, would mill about in circles. Sheepeater Indians lured bighorn sheep (which they dispatched with poisoned spears) by banging two logs together, imitating the sound of rams butting heads. Because mule deer had the fatal habit, when startled, of stopping after they had run a few hundred yards, Indians had no trouble following. White-tail deer took the same route day after day, and hunters knowing this route had only to wait to find their prey.[18]

Throughout the park are archaeological sites that testify to extensive hunting. Running through the park from Duck Creek in the west to Cooke City in the east is the Great Bannock Trail, an Indian hunter's superhighway. Although the trail has been unused since 1878, the ruts made by travois are still visible in many places. Obsidian Cliff, named for the glasslike rock formed by rapidly cooling lava, was the site of a quarry for thousands of years. Obsidian is ideal material for stone weapons, and this quarry was one of the most important sources for it in North America. Rock taken from this cliff has been found throughout the country. An obsidian core from this cliff, more than 3,000 years old, was found in a Hopewell Indian burial ground in Illinois, more than 1,500 miles away![19]

Clubs, spears, bows, arrows, darts, jumps, and traps — every conceivable tool for killing — have been uncovered in various parts of the park. Remains of an early hunting camp, including a rock fence compound where bison were captured, is visible at Slough Creek campground. Stone and cedar sheep traps remain northeast of Mount Everts near Turkey Pen Creek; more traps and hunters' blinds lie on the ridges of the Absaroka, along the park's eastern boundary. Remains of early settlements line the shores of Yellowstone Lake and lie in river bottoms and around thermal areas. The West Thumb thermal area had been occupied continuously since the last ice age; so too was a similar site near the Grant Village marina. At Fishing Bridge are many places that had been occupied since 6,500 B.C. Tepee rings dot the Blacktail Deer Creek drainage. Large hunting camps were found near Mammoth, Lava Creek, Swan Creek Flat, Sheepeater Cliffs, the Gardiner River, and so on. In fact, although the park has never been systematically surveyed by archaeologists, at last count 247 sites of previous human occupation have already been identified, and more are discovered each year.[20]

And lying just east of the park boundary, along the North Fork of the Shoshone River, is perhaps the most dramatic evidence of Indian hunting in the area. It is a place called Mummy Cave, which was used by hunters for 9,000 years. In the cave archaeologists have found a remarkable record of early hunters' success: remains of sheep, deer, antelope, rabbit, beaver, marmot, duck, goose, grouse, porcupine, chipmunk, ground squirrel, dog, several species of carnivores, and other small mammals, birds, and rodents.[21]

As the age of this evidence attests, Indians were remarkable hunters long before they acquired horses or guns. In fact, many early hunting techniques and cultural traditions encouraged overkill. Ceremonial dresses for women of the Crow tribe (whose territory included the eastern part of the park) often sported 700 elk teeth. Only two teeth, the "ivories," are taken from one animal, and so each dress represents 350 elk killed. The Sheepeaters used dogs to drive the sheep uphill to waiting traps, sometimes capturing a dozen or more at one time.[22]

Jumps and traps, used particularly for bison and antelope, were especially wasteful. Used for thousands of years they killed many animals at a time. Frison discovered a jump near the Green River in Wyoming, where early Indians killed 212 antelope in just a few days. "Antelope trapping was so successful in the Great Basin," he wrote, "that several years were required to regenerate the population."[23]

Frison found another killing ground near Jackson Hole containing remains of an "extensive number of animals [bison and mule deer] that were killed and butchered." The Vore buffalo jump site in Wyoming, also excavated by Frison, was used five times between 1550 and 1690, and holds remains of 20,000 buffalo. That means 4,000 or more buffalo were killed each time the jump was used. Other buffalo jumps in the West display the remains of as many as 300,000 buffalo. These sites were so numerous, in fact, and held such large deposits of bone, that for many years they were mined as a source of phosphorus for fertilizer![24]

These killing techniques were often so effective that large amounts of meat were left to rot and herds of animals were decimated. Buffalo and antelope traps killed so many that it took the animals decades to recover. An average jump site was used only once or twice in a person's lifetime, but archaeologists surmise that it was the limitation of animal numbers, not Indian ingenuity or sense of self-restraint, that determined how often these jumps could be used. "One successful kill of a number of adult animals," wrote Wright, describing the effects on the ecosystem of a jump near Jackson Hole, "would have reduced the breeding potential of the local [bison] herd to a level where it was no longer a significant part of the valley ecosystem. Only immigration from Yellowstone and/or the Green River Basin could have returned the population to a level at which killing could begin anew."[25]

Yet this overkill was not an isolated phenomenon. Human hunters all over the globe were the most effective predators the world had ever known, and had been so for thousands of years. Archaeological evidence suggests that the hunter-gatherer societies of North America as well as elsewhere were not only exceptional killers, but actually may have pushed to extinction countless animal species.

Until ten thousand years ago an incredible bestiary of mammals roamed North America. These were the so-called megafauna, an exotic menagerie that included the woolly mammoth, saber-toothed tiger, giant sloth, giant beaver, camel, horse, two-toed horse, and dire wolf. These were the dominant fauna on this continent for tens of millions of years. Then suddenly and mysteriously they disappeared.[26]

Apparently not the victims of climatic changes, these species had already survived three ice ages, times of great extremes of temperature; they were also highly mobile, able to follow the expanding tundra north as the ice melted. At the time they became extinct there had been no changes in vegetation to indicate that variations in either climate or habitat caused their disappearance. Also, strangely, only mammals disappeared, a class of life biologists consider one of the most robust; yet these vanished from every kind of habitat: camels disappeared from the desert and woolly mammoths from the northern boreal forests. Nor was their disappearance caused by nonsurvival of the unfittest. They were not pushed out by other animals that were better able to compete, for they left behind empty habitat. The horse, once a native species, disappeared 10,000 years ago. But when the Spaniards "introduced" a similar species to this hemisphere in 1519, it flourished, spreading rapidly over the entire continent, a sure sign that a niche was just waiting to be filled that had remained empty for ten thousand years.[27]

For these and other reasons, many scientists believe the megafauna were exterminated by early hunters. Coming to this continent where game was unused to man, they found the giant herbivores easy prey. And as the herbivores were eliminated, the carnivores that depended on them disappeared as well.

Whether or not early people alone exterminated these species, most scientists believe they at least executed the coup de grâce. We know that early societies were eating these giant animals at the time they disappeared. Human fecal samples have been found containing animal parts; bone chips and charred remains of various species of megafauna have been found around ancient campfires, and radiocarbon dating suggests that some of the animals being eaten had become extinct at around the same time.[28]

The disappearance of the megafauna left behind a fragile and empty ecosystem, with many vacant niches and less diversity of wildlife. "The Pleistocene faunas may represent the last natural climax," wrote paleontologist Paul S. Martin:

> The pristine range of the American West "where the buffalo roam, where the deer and the antelope play," must contain many empty niches, space once shared by elephants, camels, horses, sloths, extinct bison, and four-horned antelope. . . . Was late-Pleistocene extinction so effective in upsetting the ecosystem that our National Parks, wilderness areas, and wild lands are an illusion? On a continent where herbivore herds evolved and thrived for tens of millions of years, can there be a natural community without them?[29]

After the megafauna disappeared, there was less food to eat. And for the next ten thousand years, up to the moment when the white man entered the scene, the native technology of hunting improved, putting additional pressure on the remaining species. Game throughout the country went into decline, suggesting to a growing number of anthropologists and ecologists that these early people were such successful hunters that they were annihilating their food supply.[30]

As wildlife disappeared, the quality of the Indians' diet declined as well. The archaeological evidence suggests that although Pleistocene peoples often ate woolly mammoths and other giant species, their descendants, first making do with mule deer and bighorn sheep, eventually were often forced to subsist on rodents. Still later tribes frequently had to sustain themselves almost entirely on roots and berries.[31]

The period since the last ice age was, in short, anthropologist Mark N. Cohen writes, "one of gradually growing population intensities accompanied by increasingly broad-spectrum use of localized resources." Having a difficult time making ends meet, early Indians coped by diversifying their diet. Apparently their population, not limited by "density-dependent mechanisms," was outgrowing the "carrying capacity" of its resources. They were forced to run harder to stay in the same place.[32]

By the beginning of the nineteenth century, Indians were enormously influencing animal populations. Their practice of hunting by fire and with traps and jumps had kept overall animal numbers low for millennia, and more recent events further improved their odds against game. Once they acquired the horse (around 1700 in the Yellowstone region), they gained a tremendous advantage over the buffalo, and began to burn and hunt over wider areas. Their acquisition of guns (around 1800) once again improved their prowess. The result of these improvements in the Yellowstone region was overkill.[33]

Hunting practices may also have affected the relative populations of different species, giving an advantage to those which the Indians found less desirable, or more difficult to hunt. Overkill at the trap sites kept bison numbers low, but elk and moose, being especially easy quarry, were apparently hunted nearly to extinction. The first moose to appear in Yellowstone's northern range did not arrive, as we saw in Chapter 3, until 1913, apparently, many biologists believed, having been decimated previously by the Indians. Mummy Cave, despite a complete chronological record of hunting results for 9,000 years, held not one elk or bison bone. Nor, after more than a decade of research, could Professor Wright find evidence of elk. In fact, he declared, no one had ever found prehistoric evidence of elk or moose on the Yellowstone plateau.[34]

"We have excavated several deep stratified sites," he wrote to Professor Pengelly recently. "There are plenty of elk (and moose) around these two sites today, but no elk (or moose) from the excavations. However, we have recovered animal bone in Swan Lake Flat above Mammoth. It is mule deer and associated with a date of 2965 B.C. Swan Lake Flat [and Willow Park just to the south] are [today] areas of substantial elk and moose populations."[35]

Historical evidence, moreover, Wright said, supported the prehistoric record. "We have good descriptions of [Chief] Washakie complaining to government officials about the fact that the entire tribe was starving. If there were 15,000 to 35,000 elk moving down toward the winter ranges why didn't the Shoshone ride a few days northward and secure more than enough meat for the winter?"[36]

Their numbers affected by Indian hunting, elk and moose without question did not live on the Yellowstone–Grand Teton plateau in the numbers they do today. In descending order of abundance, the animals that once lived there were bighorn sheep, mule deer, antelope, and bison: almost exactly the reverse of their present numbers.[37]

Thanks to their hunting prowess, the Indians of the Yellowstone region — the Shoshone and their cousins, the Bannock and Lemhi — had eaten themselves out of house and home. When Lewis and Clark first met the Shoshone in 1805, they were starving. Their chief told the explorers that they had "nothing but berries to eat." A year later, they offered the explorers a horse for slaughter, for they had nothing else to offer. Another explorer, visiting the Lemhi (a Shoshone tribe, northwest of the park) in 1811, described them as the "poorest and most miserable nation I ever beheld; having scarcely anything to subsist on except berries and a few fish." By 1840, the Shoshone and the Bannock, with the help of a few trappers, had wiped out the buffalo in Idaho (the Great Bannock Trail through Yellowstone, in fact, was opened by the Bannock in 1840 in a desperate search for better hunting grounds in Wyoming).[38]

Wildlife populations were still abundant in other parts of the West, as all the early white settlers reported, but these animals were not always accessible to the Indians. The people were nomadic and lived by following their prey, but often enemy tribes blocked their way. The Shoshone were hemmed in by the Ute on the south, the Nez Perce on the west, the Crow and Sioux on the east, and the Blackfoot on the north, and when their food gave out in Idaho, the Yellowstone area and the Bannock Trail was their only route to better hunting. Indeed, as Lewis and Clark discovered, the best places to hunt lay in the areas between warring tribes. Game congregates where it is safest, and, these explorers discovered, it quickly found a sanctuary in no-man's-land.[39]

The explorers' observations, moreover, are confirmed by recent research. Intertribal buffer zones were important in early ecology.

"Warfare between members of the two tribes," writes anthropologist Harold Hickerson in studying the effects of the buffer zone between Chippewa and Sioux on the red deer, "had the effect of preventing competing hunters from occupying the best game regions intensively enough to deplete the supply." Moreover, he added, "in the one instance in which a lengthy truce was maintained between certain Chippewa and Sioux, the buffer, in effect a protective zone for the deer, was destroyed and famine ensued." Just so the shifting boundaries between nomadic peoples became vital niches in the ecosystem.[40]

These were the complex and subtle conditions that prevailed before Lewis and Clark reached the upper Missouri in 1804. But they may not have been the conditions that the first explorers found in Yellowstone sixty-five years later. In the meantime smallpox, the white man's disease, swept the West. Indians, living on a continent remarkably germ free, were exceptionally vulnerable to the various organisms the white men brought with them.

Tuberculosis, pneumonia, dysentery, bubonic plague, and probably syphilis were European diseases that became chronic among the Indians, but smallpox nearly destroyed them. No one knows just how many Indians died from these diseases, but the best estimates are that they killed between 50 and 90 percent of the Indians in America. The first plague epidemic swept New England in 1619 and 1620, taking more than 90 percent of the Indians with it. The first smallpox epidemic probably occurred in the same area around 1617, and was followed every twenty-five years with successors that eliminated several Indian tribes along the East Coast. On Nantucket, for instance, four epidemics between 1659 and 1792 killed all but twenty out of 3,000 in one tribe.[41]

No one knows when smallpox first reached the Rocky Mountain West, for it preceded the white man — carried, probably, by Indians fleeing epidemics — but it had done great damage before Lewis and Clark arrived. A plague in 1781 decimated the Gros Ventre Indians, one of the tribes that hunted on the Yellowstone plateau. "In 1781," write historians E. Wagner and Allen E. Stearns, "a war party of Kenistenos, Assinboin and Ojibways . . . came to the village of the Gros Ventre, which they attacked. Resistance made to their attack was very feeble, so that they soon rushed forward to secure their scalps. They found the lodges of the village filled with dead, and the stench so terrible that they quickly re-treated."[42]

In 1804, Lewis and Clark found that the Omaha Indians had just been ravaged by smallpox, only a remnant of the tribe remaining. During the next forty years periodic epidemics swept the Yellow-stone region, each successive one taking an appalling toll in the Indian population. But it was the great epidemic of 1836–1840 that did most damage. It devastated the Arikaree, Mandan, and Gros Ventre (again); it swept through the Dakota, Pend Oreille, Kal-ispel, Spokane, Colvile, Salish, and Kiowa.[43]

"Entire villages," write the Stearns, "were completely exter-minated." And it again attacked all the tribes that lived in and around Yellowstone. Between 1836 and 1840 the Blackfoot lost half their number; the Crow population was reduced from 41,000 to 16,000; the Gros Ventre were virtually exterminated. No one knows how many Shoshone were lost, but interviews with Sheep-eaters, the tribe found living in Yellowstone when the first explorers arrived, suggest that they too were decimated by smallpox. "By and by, Sheep Eater not many," said an old Sheepeater woman interviewed in 1913 (reputed to have been 115 years old at the time), describing the smallpox epidemic, "then Sheep Eater no more, no more papoose, no more squaw, all gone."[44]

These tragic epidemics changed not only history but the ecosys-tem as well. The Indian population never recovered, and their depleted numbers were never again able to defend or feed them-selves adequately. Their decline encouraged both early settlers and later historians to underestimate their population and their im-portance to the region, and allowed ecologists to underestimate their influence on wildlife. The reduced hunting activity that fol-lowed the pox reduced pressure on game populations, which had twenty-five years to recover before the first large waves of white settlers reached the upper Missouri after the Civil War.

"It is a new phase in the natural world," Nathaniel Pitt Langford wrote on visiting Yellowstone in 1870, "a fresh exhibition of the handiwork of the Great Architect." When the early exploring par-ties ascended the Yellowstone plateau they thought they had found a place of pristine beauty, fashioned only by God. It was indeed a place still unspoiled, but it was also a place changed by man.

The Indian burning and hunting had immeasurably altered the ecosystem. Yellowstone, their country for eleven thousand years, had been changed by what they did and what happened to them. By introducing the horse, the gun, and disease, white men too had intruded by proxy, enhancing and then diminishing the Indians' prowess in hunting and their influence on the range by burning, centuries before they saw the Yellowstone. The chain of events stretching more than ten millennia — megafaunal disappearance, changes in the Indians' diet, use of traps and jumps, invention of the bow and arrow, broadcast burning, acquisition of the horse and gun, contraction of smallpox — all helped to make the original conditions that so impressed Langford and his contemporaries.[45]

The early explorers were, of course, keenly aware that they were in Indian country. Many saw Shoshone, Crow, Bannock, or Blackfoot Indians on their forays into the park, and nearly all saw Sheepeater. Osbourne Russell, on a trapping expedition into Yellowstone in 1835, remarked that "the woods seemed to be completely filled with Blackfeet, who rent the air with their horrid yells." The explorer de Lacy, traveling through the park in 1863, saw many signs of the Blackfoot and Crow. "We are in an Indian country," he wrote uneasily, "admirably fitted for their warfare." After the park was founded, early staff and visitors discovered signs of Indians everywhere. The park's second Superintendent, Phelitus Norris, recorded in 1880 that "skin covered lodges, or circular upright brush-heaps called wickiups, decaying evidences of which are abundant . . . in nearly all of the sheltered glens and valleys of the Park." For many years concessionaires did a thriving business selling arrowheads and other Indian artifacts.[46]

Even today, near Lava Creek Campground, hidden for more than one hundred and fifty years in an aspen grove, sit four wickiups. Incredibly well preserved, they remain silent, seldom-seen signs of Yellowstone's Indian past. Made of long thin poles of aspen and pine leaning together and rotting at their base, they have stood for a century as reminders to the white man that the Indians did once make their living here.[47]

The newly arrived white man did not know, however, that the signs of man were not simply the constructions and artifacts once ubiquitous in the "sheltered glens" of the park; the landscape and the wildlife too were evidence. These effects, however, were less obvious and little noticed. Instead, sadly for both the Indian and

Yellowstone, no sooner was the park created than the Indian, and his contribution in shaping the conditions the white man loved so well, was forgotten.

Not long ago a local businessman was given a concession to open a gift shop at Roosevelt Lodge. His plan was to place his shop where it would be clearly visible to visitors as they entered the Lodge grounds. The man was an Indian, however, and his shop was to be a tepee. According to Indian custom it could be put up only in certain ways at certain locations. In fact, only one place satisfied all the cultural and commercial criteria, yet park authorities would not permit the tepee to be put there. It would, they said, detract from the appearance of Roosevelt Lodge, which had been nominated by the United States government for listing on the National Register of Historic Places.[48]

The Indian was incensed: "How can a white man's building be more historical than an Indian's tepee?"

His point was lost on authorities. For, according to conventional wisdom and official doctrine, the Indian had little past in the park. In nearly all service publications for a century and more, Indians, when mentioned at all, were called early "visitors" to Yellowstone. A booklet sold today at the visitors' centers, the *National Parkways Comprehensive Guide to Yellowstone National Park,* tells the reader that "Yellowstone proved to be an infrequent habitation for most Indian tribes," explaining that "the natural phenomena of the region, particularly the geysers, deterred their [the Indians'] visits," and later reassures us that "the majority of the Park is in its natural state, exactly as it was before man arrived." Another guide tells us that "man's presence affects only two percent of Yellowstone's two million plus acres of wild lands. The rest is as it was."[49]

Scientists whose success in managing the park depended on understanding the original ecosystem, and therefore the Indians' influence in that system, also neglected them. Instead, they portrayed them as ecological eunuchs, perhaps living in the park, but never significantly affecting the ecosystem. "We do not know in detail the role of primitive man in Yellowstone," park biologist Meagher wrote in 1975, "but as a predator in this particular area that role must of necessity have been minor."[50]

"Yellowstone Park was virtually uninhabited in 1872," wrote former Yellowstone naturalist Paul Schullery in 1984, adding that "a small band of Shoshone sheepeaters occasionally occupied the park. . . ."[51]

"Man-caused fires," wrote biologist Dale Taylor, "have been superimposed upon the natural history of Yellowstone only since the late nineteenth century."[52]

"Archaeological evidence and historical accounts suggest," biologist Houston wrote in 1982, "that densities of man were comparatively low in the upper Yellowstone River Valley and the area that became the park" — too low, he suggested, to affect the ecosystem.[53]

As these statements imply, the managers of Yellowstone, and independent scientists as well, despite abundant archaeological, anthropological, and biological evidence that the Indian was a big factor in Yellowstone's past, overlooked him. How could these people, responsible for keeping the park as an original ecosystem, overlook such a vital member of that system?

The Indian's official past began to suffer shortly after Yellowstone was made a park. During summer 1877, one year after the battle of the Little Big Horn and shortly after the natives of Yellowstone — the Shoshone, Sheepeater, and Bannock — had been sent to reservations in Idaho, Chief Joseph and the Nez Perce, trying to escape the cavalry, which had chased them off ancestral lands in Oregon, traveled through the park. A pathetic band of 1,000 men, women, and children and 2,000 horses, they captured a group of tourists and required one of them to serve as their guide. It was a case of the blind leading the blind. Following their captive, who did not know the way, they were promptly lost. They spent more than two weeks in the park, cooking in the hot springs and pasturing their horses while their braves looked for a safe exit, raided park headquarters at Mammoth, and burned a ranch near Gardiner. They narrowly missed capturing the Civil War hero, William Tecumseh Sherman, who was touring the West at the time.[54]

A year later the Bannock made their last visit to Yellowstone. Starving on the Fort Hall Reservation, to which they had been consigned in 1868, they left in search of food. Following their old trail, they set out across the plateau for Wyoming. They did not make it. Pursued through the park by the cavalry, they were defeated in battle near Clark's Fork and were returned to Fort Hall.[55]

These events greatly dismayed the recently appointed Superintendent, Phelitus Norris. Congress had given him no money to run the park, expecting it to pay for itself with tourism. But stories of Indians capturing visitors would be bad for business. He had lobbied hard for expulsion of the natives from the park, thus, he said, "averting in future all danger of conflict between these tribes and laborers or tourists." Now he needed to undo the damage that the Nez Perce and Bannock had done. He therefore conceived a myth that would help the park's image. Indians, he wrote in his reports, avoided Yellowstone because they feared the geysers. He wrote in 1880,

> Although the Crow Indians upon the north, the Shoshone upon the east and south and the Bannocks upon the west might have, during the brief summers, traversed the few difficult passes to the park, there is little evidence to show that they did so. It is probable that they were deterred less by those natural obstacles than by a superstitious awe concerning the rumbling and hissing sulphur flames of the spouting geysers and other hot springs, which they imagined to be the wails and groans of departed Indian warriors who were suffering punishment for their earthly sins.[56]

These were surprising statements from one who, only two years previously, had, in a panic, ordered park staff to refurbish their one Gatling gun, preparing for a Bannock attack that never came; he must have known that the Nez Perce were so unafraid of the hot springs that they cooked their food in them. Yet, contrived as it was, the myth never died. Ever since, the story of the Indians and the geysers has been part of Yellowstone lore. Four generations of park naturalists were taught to tell visitors that "Indians never lived in Yellowstone because they were afraid of the geysers." In 1952, historian Ake Hultkrantz stated that, for the Indian, Yellowstone was "almost frightening — evil spirits lived in the geysers." And as we have seen, the story persists today.[57]

The myth of Yellowstone's Indian past persisted partly because it reinforced the conventional wisdom of white America. Indians, we have always believed, lived in harmony with their environment, but did not change it. "We have a tradition," wrote Peter Steinhart in *National Audubon* magazine recently, "of regarding Indians as ecological sages." It never occurred to us that they, being human, could be capable of evil as well as good; that they might affect the earth, changing it sometimes for better, sometimes for worse; that the difference between primitive and modern people might be one of degree, not kind. Much less did it occur that, thanks to population pressure, humans may have been living beyond their means long before they learned to write.[58]

Surely, we thought, they could not burn forests or cause the extinction of animals. They were noble savages, the innocents who, ever since Columbus landed in the Antilles, were viewed by whites as children of nature. Arthur Barlowe wrote in 1584 that Indians lived "after the manner of the Golden Age." Indeed, writer Peter Mathiessen has even suggested, the name *Indian* was given to them, not because the Italian explorer supposed he had landed in India, but because he saw these people as *una gente en Dios,* a people of God. And Indians see themselves, not as exploiters of the environment, but as people who revere nature; they are animists who take the names of animals as their own, and see something sacred in the smallest forms of life.[59]

This view of Indians as innocent children of nature was conceived as soon as the first white men set foot on this continent, and it has persisted because it has been convenient for many: the Indians' enemies did not want to believe they might have been sufficiently civilized to improve upon their surroundings; the Indians' friends did not want to believe they could, like the rest of us, exploit nature.

"In the beginning," wrote English philosopher John Locke in his *Second Treatise on Civil Government* in 1690, "all the world was America." Writing these words in a chapter entitled "On Property," Locke, whose ideas dominated the thinking of America's early settlers, meant that America, when the first Europeans arrived, was in a state of nature wherein land belonged to no one. To call something our own, he said (repeating Saint Thomas's labor theory of value), we must mix our labor with it. By working the land, we make it ours. But the Indians, he suggested by implication, never worked the land. Therefore it was not truly theirs, and the Europeans could make it their own with a free conscience simply by tilling the soil.[60]

For this reason it was convenient for the first settlers in America to believe that the Indians did not improve upon nature, did not farm, fertilize, or set fires. According to European law and political theory, people owned those things with which they mixed their labor; the settlers' claims for taking land from the Indians was justified in part, therefore, by their assumption that the Indians never worked the land.[61]

The early settlers were aware, of course, as we have seen, that the Indians did use fire, often to their advantage. But this memory was quickly forgotten. In a market economy where lumber was inventory and housing an industry, trees had value and were to be preserved. By the eighteenth century the white man found it strange that the Indian might want to burn valuable property; by the nineteenth century, it seemed inconceivable.[62]

The Indian as burner of the land was forgotten. In this century an entire bureaucracy — the Forest Service — was built around a national agenda calling on people to fight forest fires. Since 1945 a benevolent national symbol — Smokey the Bear — has been used to enshrine prevention of fire as one of the world's greatest goods. And in this atmosphere the idea of intentional burning seemed not only un-Indian, but sheer lunacy. In 1940, when the Forest Service received reports that people in Appalachia were intentionally burning the woods, they sent *a psychologist* to investigate; they could not believe such behavior was rational. Neither could the psychologist. He reported that the only reason they acted as they did was blind acceptance of authority. The reason they gave him for starting fires, he reported, was that "our pappies burned the woods"![63]

Similarly, the Department of Agriculture Yearbook for 1949, entitled *Trees*, refused to accept the possibility of Indian incendiarism. "Tradition says that [Indians burned] to drive game, but we have no positive proof that they did this. . . . The Indians had no matches and they used small campfires that they tended carefully; so it is improbable that they set many fires."[64]

If early settlers and later forest managers did not want to believe the Indians could do any good for the land, environmentalists and other friends of the Indian have not wanted to believe they could do any harm to it. Just as the Indians in the tales of James Fenimore Cooper could walk through the forest without breaking a twig, the Indians of the environmentalists could live off the land and not endanger moose or buffalo. And, as long as burning was considered bizarre behavior, they could not be guilty of these activities, either. "With the emergence of ecology," wrote Lewis, "or at least popularized ecology, the hunter-gatherer has been seen by some writers to represent a kind of *ecological hero* [his italics], a revival of the 'noble savage.' This view is linked inextricably to the idea of a passive relationship to nature."[65]

This view, Pyne said, was reflected in our changing views of Indian incendiarism. Although this role is still ignored in Yellowstone, it is now generally acknowledged elsewhere. But we are now willing to credit the Indian with burning only because we finally recognize fire as good: "As the child of nature, the American Indian could not deliberately have damaged his environment, and by mid-century fire was generally considered as an environmental evil. In more contemporary times, when prescribed fire has again been accepted as an appropriate tool in the management of natural systems, it has been discovered that, indeed, the Indians burned."[66]

Actually, this reverence for the Indian began long before ecology was born. For a hundred years, conservationists have made the Indian their hero. John Muir thought the Indians were pantheists like himself who saw themselves as part of God. "To the Indian mind," he wrote, "all nature was instinct with deity." George B. Grinnell saw the Indian as a great conservationist. "He protected the beasts on which he depended," Grinnell wrote, "and practiced methods of economy in hunting that American sportsmen may well take to heart." Stewart Udall, calling Indians "pioneer ecologists," saw them as models for all environmentalists. "Today the conservation movement finds itself turning back to ancient Indian land ideas," Udall wrote in 1963, "to the Indian understanding that we are not outside of nature but part of it." Justice William O. Douglas in 1974 saw the Indian as one who lives with nature without changing it: "Although the Indians took their living from the wilderness, they left that wilderness virtually intact." Sigurd Olson saw the Indian as one who refused to interfere with nature as a matter of theological principle. In 1976 he wrote, "Indians believe the land is God's and no one has the right to manipulate it."[67]

These were lofty and well-intentioned sentiments, but they did not tell the whole story. In truth environmental activists' attitude toward the Indian was ambivalent. As ecologists, they desperately needed a model, an example somewhere in history of people who lived in harmony with nature. Yet, admitting that the Indian was an essential part of primitive ecology would mean accepting that, without the Indian, wilderness was an illusion. As recreationists who hiked, rode, and fished in national forests and parks, they were users of the wilderness; and as preservationists legitimately

caring about the future of our natural areas, they were their pro-
tectors. Both positions occasionally put them in conflict with those
contemporary Indians who attempted to use the land and water as
their forefathers once did. Traditional native economies depended
on harvesting the fruits of nature; yet when Indians attempted to
exercise their traditional rights to net salmon, harpoon whales, or
club seals, they found themselves at odds with environmentalists.
Hunter-gatherers, it seemed, were attractive to white people only
so long as they were no longer hunting or gathering.[68]

The history of Yellowstone embodied this ambivalence, a symbol
of the underlying conflict between Indians and well-intentioned
preservationists. Congress, in creating the park in 1872 "for the
benefit and enjoyment of the people," destroyed the livelihood of
another people. From the beginning this sad fact was a skeleton
in the closet of our park system, a past too embarrassing to con-
template, for it demonstrated that the heritage of our national park
system and environmental movement rested not only on the lofty
ideal of preservation, but also on exploitation. Instead, almost
immediately, thanks to the imagination of Phelitus Norris, the past
was denied and the Indian mythologized. The U.S. Cavalry, following Norris as park administrators, sustained this myth.

Park Service attitudes were products of these military and preservationist traditions and reflected both in their ambivalence toward the Indian. The first twenty-two rangers hired by the new
service in 1916 were cavalry soldiers who resigned their army com-
missions on the day they donned the new green uniforms. As direct
descendants of the same agency that hunted down the Nez Perce
and Bannock, they, too, kept Norris's myth alive. As members of
an agency dedicated to keeping the park unimpaired for future
generations, they were reluctant to admit that, without the Indian,
it was already impaired.

Instead they ignored the Indian and thereby further impaired
the park. Rather than employ broadcast burning, the Park Service
spent millions to prevent and fight forest fires. Rather than encourage predation, they stamped it out. Rather than cull the elk
and bison herds, they gave these animals sanctuary.

The policy of natural regulation was born in the heat of enthusiasm for the new environmental awareness, a policy conceived by
Starker Leopold and dedicated to a new ecology, but an ecology
that had no place for humanity. Yellowstone biologists conceived

a theory that sustained this view: Primitive people may have been predators, they argued, but predation had no significance in the ecosystem. As with the wolf, park scientists defined the Indian's ecological niche out of existence.

In just this way, minimizing the Indian, which earlier had been justified by military pride and prejudice, became a scientific principle: If primitive people had no biological significance in the ecosystem, then pristine conditions could be defined without them. If the park were in original condition, if man had never counted in the ecosystem, then the way to preserve the park would be to keep man and nature apart. In so defining their mission, park scientists were, of course, ignoring volumes of evidence about the Indian's past, but this evidence was historical and archaeological, and the scientists — all biologists — were not trained to understand it.

Indeed, all those responsible for the future of Yellowstone worked within a time horizon too short to see nature's direction. Because it took the forest three hundred years to reach climax, change was slow to show itself. Scientists, whose entire careers spanned less than a tenth of that time, and park managers, whose average tenure in the park was less than a decade, lacked the perspective to see the past. For these reasons the Indian and his works were forgotten.

Throughout Yellowstone we can see signs of this forgetfulness. By comparing present conditions with photographs taken by W. H. Jackson and others a century ago, we can see how the range has changed. In the hundred years since the last Sheepeater, Shoshone, Crow, Blackfoot, Bannock, Piegan, and Lemhi left, Yellowstone, managed with an animal husbandry the Indians would have deplored, has been in steady decline. The forests, which now cover more than 80 percent of the park, are reaching climax and dying of old age. Stands of lodgepole pine, Douglas fir, Engelmann spruce, and subalpine fir have encroached perceptibly on the open grasslands. Tens of thousands of acres of the old, densely packed coniferous trees, attacked by spruce budworm and pine-bark beetle, have died and turned brown.

By contrast, the open areas are shrinking. Deciduous trees — cottonwood, mountain alder, red birch, aspen, and willow — have declined, the latter two drastically. Sagebrush has spread. Grasses, no longer "remarkably luxuriant, indicating great fertility of soil," as Hayden found them in 1871, are in many places short and sparse. Forbs have decreased, and exotic plants have multiplied. As forests have grown, the soil has dried and the plants that require moisture have given way to other species. Much of the blue camas — a plant once common to wet meadows around Yellowstone — whose root was an important food source for the Indians have disappeared.[69]

We can never know just what Yellowstone was like before the white man came. Many of these "original conditions" will always be a mystery to us. Boundaries of the early ecosystem were surely huge and fluctuating, a product not only of climate and wildlife, but of Indian activities as well. But we can say the area that is now a park is far different from what Doane and Langford found in 1870. The elk and bison that now dominate the fauna of the park were in all probability not native and certainly not abundant. Animals that were once common are now rare, and the Indian is gone.

Once we know that present conditions do not represent the park, we must conclude that conditions we now see at Yancey's Hole and elsewhere are not natural, not a window to our past, but symptoms of slow decline. Yellowstone, having long ago ceased to be an original ecosystem, is becoming, thanks to the policy of natural regulation, less pristine every day.

These are the sad ironies of Yellowstone. Created for the benefit and enjoyment of the people, it destroyed a people. Dedicated to preservation, it evicted those who had preserved it. Touted as pristine, the policy required that we forget those whose absence diminished it. Denied its Indian past, it deprived us of the knowledge needed to keep it pristine. As it turns out, ignoring the Indian was not only bad history, but bad ecology as well.

In the next part we shall discover what this ecology did for the wolf and the bear.

The Wildlife

"It's too late to correct it," said the Red Queen: "when you've once said a thing, that fixes it, and you must take the consequences."

— Lewis Carroll
Through the Looking Glass

10

The Wolf Mystery

IN December 1967, Marshall Gates, a summer seasonal ranger in Yellowstone, had returned to his cabin in Cooke City, near the northeast entrance, for a vacation. Shortly after Christmas, Gates, his wife, and another couple started on a drive through the park to Livingston. While they were passing through the Lamar Valley near Soda Butte Creek, something jumped onto the road in front of them. It was a wolf. Gates grabbed a Super Eight movie camera that he luckily had along and kept it aimed at the animal as it sped away from him.[1]

If Gates had not taken his picture, surely no one would have believed him. For he photographed the wolf that many believed was the first seen in Yellowstone in more than thirty years. Yet it was not the last. In the following months wolves were seen frequently, and in the four years after Gates's trip to the park, nearly four hundred sightings of wolves were reported, and although many were dismissed as erroneous, park authorities and independent experts agreed that many were solid evidence that wolves were indeed in Yellowstone.[2]

Where did the wolves come from?

The Park Service's official answer is that they were *Canis lupus irremotis,* a remnant of Yellowstone's native species.[3]

To the scientific community, however, this answer seemed incredible. *Irremotis,* they thought, was extinct, exterminated with the help of the Park Service itself. And nearly everyone familiar with Yellowstone — the scientists studying wildlife, the rangers who lived in the backcountry, and the maintenance people who maintained the trails — were convinced that no wolves of any kind had been in the park for decades. If wolves had been there, surely, they thought, they would have seen some sign of them, and they had not. Instead, they began to suspect that the Park Service had secretly planted the wolves in the park, and then falsified the historical record to make it appear that they had been there all along.[4]

Thus did the rumor of the wolf conspiracy begin. It spread to the gateway communities surrounding Yellowstone. Some said that the wolves came from Canada. Others said Minnesota. Still others believed they were purchased from individuals who maintained captive populations. There were stories of trucks, loaded with crates carrying wolves, entering the park in the middle of the night in winter and disgorging their woolly contents in the Lamar Valley. According to other versions, they came by train.

The rumor also spread through the worldwide community of wildlife biologists. In his 1973 report on *The Status of the Wolf in the United States,* renowned wolf expert Dr. David Mech wrote, "There are persistent rumors that the Yellowstone wolves are imported from Canada and released." Indeed, many biologists — including Mech — investigated the rumor, but without success. And throughout they met with Park Service denials.[5]

Yet the suspicion lingers to this day, an unsolved mystery with momentous implications that piques the curiosity of nearly every wolf biologist in North America.

This is the story of that mystery, about how wolves returned to Yellowstone, and why they did not stay.

In 1907 a federal agency dedicated to preserving wildlife underwent an abrupt and curious role reversal. The U.S. Biological Survey, originally called the "Division of Economic Ornithology and Mammalogy," was created by Congress in 1885 to promote research for the protection of wildlife. Run by C. Hart Merriam, a distinguished scientist from the American Ornithological Union, it had been dedicated, for most of its first twenty-two years, to protecting waterfowl. But this year it set its sights on something else.[6]

A study published by the Survey in 1907 urged an all-out national campaign to exterminate wolves, especially by "killing young in breeding dens."[7]

This transformation was prompted by an apparent plague of

wolves in the West. As sheep and cattle replaced the buffalo on the Great Plains, wolves multiplied, apparently thriving on the domestic stock. Cattlemen called on the federal government to help, and the Biological Survey answered that call. In 1907 alone, Biological Survey officers killed more than 1,800 wolves and 23,000 coyotes in thirty-nine national forests.[8]

Up to this time the U.S. Cavalry, then running Yellowstone, had often resisted demands by farmers and ranchers that it kill predators. After all, killing any animal in the park was against the law. The park enabling act of 1872 had forbidden the "wanton destruction of fish and game," and the Lacey Act of 1894 had specifically prohibited "killing, wounding or capture, at any time, of any wild animal or bird, except dangerous animals when it is necessary to prevent them from destroying human life or inflicting injury." Although an occasional coyote or mountain lion had been killed by authorities or by employees working in the park, the cavalry for the most part followed the letter of the law.[9]

Captain Moses Harris, in his first annual report as Acting Superintendent in 1887, wrote

> I have heard considerable anxiety expressed by those who profess interest in the Park lest the rule which protects equally all animals in the Park should work to the detriment of the game proper by causing an undue increase in the carnivora. But while it is true that there are some noxious animals that are not worthy of protection, chief among which is the skunk or polecat, yet I am convinced that at the present time more injury would result to the game from the use of firearms or traps in the Park than from any ravages which may be feared from carnivorous animals.[10]

But thanks to pressure from the Biological Survey and farm lobbies, during the first decade of the twentieth century the cavalry began to permit destruction of predators. Cougars were occasionally hunted, and were killed in large numbers. By 1911 the park had hired part-time staff "assisting in the extermination of coyotes along the north line of the park." Wolves, however, remained unmolested. The cavalry was convinced that they were not a menace to the park. Indeed, not until 1914, according to official records, was the first wolf killed in Yellowstone.[11]

Around the West, however, the demand grew that predators be eliminated. In 1914 Congress appointed the Biological Survey as chief predator-control agency, and charged it with the task of "destroying wolves, prairie dogs, and other animals injurious to agriculture and animal husbandry." By 1916, the West had been arranged by the Biological Survey into control districts, preparing for systematic extermination.[12]

Coincidentally, 1916 was also the year in which the National Park Service was formed, and administration of Yellowstone passed from military to civilian control. The new agency was charged by Congress "to conserve the scenery and the natural and historic objects and wild life therein, and to provide for the enjoyment of the same in such manner and by such means as will leave them unimpaired for the enjoyment of future generations."[13]

Few noticed, however, that the new agency was also granted a license to kill. The Park Service enabling act also authorized the Secretary of the Interior to "provide in his discretion for the destruction of such animals and of such plant life as may be detrimental to the use of said parks, monuments or reservations."[14]

With the Park Service created and the Biological Survey in the ascendant, Yellowstone wildlife was in deep trouble. For now two bureaucracies saw that they could build budgets by demonstrating the threat these animals posed. For the Biological Survey this was an opportunity to grow from a small department of scientists to a huge organization with damage-control districts throughout the country. For the Park Service it was an opportunity to grow as well, adding new national parks by convincing the public that it cared for game animals while reassuring neighboring farmers that it would not be a threat to their economies.

In 1915, Merriam's son-in-law, Vernon Bailey, was (as we saw in Chapter 6), sent to Yellowstone to develop a program of predator control for the park. Predictably, therefore, Bailey found that wolves and other predators had put Yellowstone in trouble. Although almost no one else had ever paid attention to wolves in the park, he "found wolves common, feeding on young elk." At the end of the year he reported, "their numbers have become alarming. . . . It is strongly recommended that the Biological Survey continue their campaign in this region without abatement until these pests are greatly reduced in numbers."[15]

At the same time he urged the army to step up predator control. He wrote Yellowstone authorities in October 1915 that "the case seems so urgent that I feel strongly that no time should be lost in

getting wolves and coyotes out of that region." In the following
spring he conducted his controversial elk count, which, contrary
to earlier ones, suggested that elk were far less numerous and more
vulnerable to predation than the cavalry had thought.[16]

When the Park Service was organized a year later, more men
were put to work on predator control. A professional lion hunter
was hired and a system of payment was set up whereby rangers
would receive part of their wages in pelts. The Park Service entered
into a cooperative agreement with the Biological Survey, permitting their men to work within Yellowstone. Park Service hunters
were ordered by the acting director of the new Park Service in
Washington to work with the Biological Survey "in the extermination of wolves, coyotes, and mountain lions in Yellowstone National Park."[17]

The public, however, was unaware of these plans. For the same
month in which he issued the confidential directive to exterminate
predators, the acting director also announced (publicly, in his first
annual report) that "the killing of wild animals, except predatory
animals when absolutely necessary, is strictly forbidden in Yellowstone National Park by law."[18]

In taking its first step, the new agency had set off on the wrong
foot. Its first rangers were hired for predator control, and this task
would become one of the highest priorities of the entire Service.
As extermination augmented rangers' wages and increased the Service budget, it provided economic incentive for its own perpetuation. Yet, because wholesale extermination of species was clearly
against the intent of the law, the program had to be kept from the
public. And so a pattern was set. Killing was to increase, but it
would be done in the the the name of conservation.

During its first eight years the Park Service campaign against
predators gained momentum. Although the cavalry had managed
to kill only fourteen wolves in its thirty-two years in the park, the
new agency, according to the records in the Yellowstone archives,
had killed at least 122 by 1926. During the same time they eliminated nearly 1,300 coyotes. Mountain lions also were killed in large
numbers. Nearly every employee — from ranger to plumber, elec-
trician, and mechanic — was encouraged to shoot or trap animals
and allowed to keep or sell the hides.[19]

Each year, superintendents of the national parks met to coordinate the policy of predator control, and one of their jobs was to define "predator."

"What is a predatory animal?" they asked at their conference in 1922.[20]

"We must judge an animal's destructive nature by the results," they concluded. Because nearly every animal was destructive in some way, each year they defined the word more loosely. Foxes, lynx, and bobcats soon were added. Later, fisher and marten were put on the list. When rangers discovered that the otter at Fish Lake were eating trout that had been artificially planted there, the otter were eliminated. When they discovered that pelicans carried a parasite that infested the trout of Yellowstone Lake, a plan was conceived to dispatch rangers to their breeding grounds on the Molly Islands to stomp on eggs.[21]

Yet while this war was being waged, Park Service spokesmen continued to preach the rhetoric of conservation. "The national parks," said Secretary of the Interior Franklin K. Lane in 1918, "must be maintained in absolutely unimpaired form for the use of future generations."[22]

"It is contrary to the policy of the Service," wrote Park Service Director Steven T. Mather in 1926, "to exterminate any species native to a park area."[23]

These remarks reflected a confused sense of institutional purpose, rather than personal hypocrisies. In a sense, they felt that conservation and extermination were both goals of the new agency. For they were charged not only with preserving the parks, but also with providing for the "benefit and enjoyment of the people." And killing wolves and lions, they thought, was a way to please.

Yellowstone Superintendent Toll said in 1932,

We have always assumed that the elk and the deer and the antelope were the type of animals the park was for. We have had the support of the game associations only on the basis that the parks would act as reservoirs for the game and the increase would overflow and form legitimate hunting. If we change that policy and say there is to be no killing, coyotes will increase to balance the increase of the deer and elk, there will be no hunting and we would have no support whatsoever from the sportsmen's associations of the adjoining states. To me a herd of antelope and deer is more valuable than a herd of coyotes.[24]

Thanks to the Service's rhetoric of preservation, however, this policy did not become public until 1928. Through the efforts of Berkeley Professor Joseph Grinnell and his former student, George Wright, among others, the whistle was blown at last. The Service received a barrage of protests from environmental groups condemning their predator-control policy.[25]

"No so-called 'vermin,' " Grinnell insisted at a superintendent's conference on predator control in 1928, "such as wildcats, coyotes, weasels, hawks, or owls should as a rule ever be killed inside of National Park boundaries."[26]

Wright urged that "the rare predators shall be considered special charges of the national parks in proportion that they are persecuted everywhere else."[27]

More than a hundred environmental and game organizations signed a joint resolution urging that "each park shall be a sanctuary for . . . all wild plants and animal life." The New York Zoological Society sent the Secretary of the Interior a letter announcing that it was "strongly opposed to the extermination of any single species of our American wild life." The Boone and Crockett Club sent Director Albright a resolution stating that it "is opposed to the practice in some of the national parks of drawing a line between game animals and those of predatory habits, to the detriment of the latter."[28]

In response, Park Service officials publicly denied that they were exterminating predators while privately resisting all attempts to end the killings.

Albright, then Director of the Park Service, wrote to the American Society of Mammalogists in 1930, "We have avoided all wholesale campaigns of destruction. Such campaigns certainly have no place in a national park."[29]

Director Mather replied to Professor Grinnell's plea for predators, "That brings in a great many problems. In Yellowstone, if Mr. Albright didn't kill off his 200 to 300 coyotes a year it might result in being the developing ground for the coyotes and wolves spreading out over the country and the cattle or sheep men getting much greater losses than they ordinarily would."[30]

Albright's successor, Arno Cammerer, told the Ecological Society of America that "it is very seldom necessary to kill any animals within the national parks." And by the time predator control became a political issue it was no longer necessary to kill wolves and mountain lions. For although the public did not yet know it, these animals were already extinct in Yellowstone. Only coyotes, then being killed at the rate of nearly one a day, could be saved by a reversal of policy.[31]

Wolves apparently were gone by 1926. Indeed, two years earlier Yellowstone officials reported that "wolves and cougars [were] so nearly exterminated as to make hiring of special hunters unprofitable." Although predator control lasted more than another decade, the Service could find no more to kill, nor could they find any physical evidence that these large predators remained in the park.[32]

George Wright and his fellow biologists in the Park Service's Wild Life Division were painfully aware what damage had been done. In 1939 Victor H. Cahalane, succeeding Wright as Director of the division, conducted a thorough in-house review of the history of predator control in Yellowstone. "We knew the cougar and wolf were gone by 1930," he told me recently.[33]

Because Park Service managers had resorted to the language of preservation to mollify the public, though, it was difficult for them to admit that these animals were gone. If they did not countenance "wholesale destruction" of animals, if it was "seldom necessary" to kill, if Park Service policy was to preserve nature intact, then where were the wolves and cougars?

These animals, they insisted, were still in Yellowstone. Mysterious, unconfirmed reports of "sightings" of wolves came in from rangers or their families at convenient times during the predator-control controversy and the investigations that followed. To reassure the visitor who cared about the fate of predators, Albright admitted at a Superintendents' conference in 1932, "They had a coyote specially trained [to parade in public on command] when I was out there in Yellowstone."[34]

Continued public pressure led, in 1934, to a temporary decision to prohibit the killing of any predator in a national park without written authorization by the director of the Park Service. Three years later, Adolphe Murie was invited to study the role of coyotes in Yellowstone. Yet both Murie and the new policy faced fierce Park Service resistance, even, sometimes, from the Director himself.[35]

"Dr. Murie," Yellowstone Superintendent Edmund Rogers suggested, "would prolong his studies as long as possible because it was a soft job and . . . when completed the report would show the coyotes do not kill game animals." Albright (who by then had resigned from the Park Service to join the American Potash Company, which had mining interests in Death Valley National Monument), urged the Director "that the studies of coyotes on the antelope range should be brought to a conclusion and I suggest that open war be declared on the coyotes." And the Director was inclined to agree. "My opinion," he wrote, "is that a very strict coyote control will have to be undertaken this winter."[36]

And when Murie's study — confirming that the wolf and mountain lion had disappeared, exonerating the coyote, and calling for the end of all predator control in Yellowstone — was completed in 1938, it was greeted with howls of protest from the Park Service.[37]

The Yellowstone administration pleaded with the Director to delay announcing or amend Murie's conclusions and to reinstate predator control. Rogers complained of inaccuracies and "shaded statements." John R. White, Yellowstone's Chief of Operations, commenting that "there was no great haste to publish Adolph Murie's coyote manuscript," urged Cammerer "that there should not be a presentation to the public which will enable unfavorable comments to be made about the National Park Service, whether in the Yellowstone National Park or elsewhere." Albright, following the issue closely, chafed at the support Murie received from environmentalists.[38]

He wrote Cammerer that "the National Association of Audubon Societies and perhaps other organizations are more interested today in saving the predatory species of birds and mammals than giving reasonable consideration to the species that are regarded as very important by the general public. I hope this is not true of the National Park Service."[39]

Opposition within the Service remained strong in part because some feared the moratorium would incite ranchers to oppose new parks. "In connection with the establishment of the John Muir King's River National Park," wrote the Supervisor of Fish Resources, "I found that the greatest opposition to the establishment of this park was based on what was believed to be our policy regarding predatory animals. . . . The same arguments were used against the establishment of the Olympic National Park. . . . It is not likely that this opposition is fully appreciated by men like Dr. Murie. . . ."[40]

Fortunately, Murie himself was appreciated by preservationists. Grinnell wrote to the director at the time, "I urge that Murie's recommendations, every one of them, be adopted by the National Park Service, for all national Parks, and kept in fullest extent." After a year's delay, the Service relented, and the work was published.[41]

With publication of Murie's study the era of predator control in Yellowstone had come to an end. Not all the killing stopped, however. Superintendent Rogers, frustrated by the new ruling that protected coyotes, continued, on occasion, to order their destruction when he felt they endangered park personnel. Beaver that chose to make their dams above the Mammoth water supply were dynamited and trapped. Pocket gophers that tunneled through the bluegrass lawns of Mammoth were asphyxiated with carbon monoxide. A campaign was carried out against ground squirrels, which, it was thought, carried the plague.[42]

Systematic extermination, moreover, was continued at park borders by the states and the Fish and Wildlife Service (formerly the U.S. Biological Survey) Division of Predator and Rodent Control. The states continued to offer bounties for predators (sometimes including magpies and other birds) until 1975. Packets of rotten meat baited with cyanide — so-called mickey finns — were dropped from planes around the park. "Coyote-getters" — shotguns attached to bait and loaded with cyanide — were placed around the park. And compound 1080 poison stations sprang up throughout the West. Although the cost to the taxpayer of this predator control was a hundred times the economic losses stockmen suffered from coyotes, the war continued. The park became an island surrounded by cyanide. Any carnivorous animal leaving the park did so at great risk. Local ranchers did not dare let their dogs go out alone. Many pets met a coyote's fate.[43]

By the 1960s, Park Service personnel as well as independent scientists knew that wolves and cougars — and in all probability lynxes, bobcats, and wolverines as well — had become extinct in the park. Garrison began to explore with John Craighead a plan for reintroducing wolves; his successor McLaughlin asked Hornocker to do the same for the cougar.[44]

The public, however, was not told that wolves and lions were gone. Every ranger-naturalist and every wildlife checklist listed them as among the native species still present. The 1964 *Manual of General Information on Yellowstone National Park,* "compiled by the Division of Interpretation for the use of Ranger-Naturalists" (who are charged with informing the public on the flora, fauna, and natural history of the park), listed the wolf, cougar, wolverine, lynx, and bobcat as "fairly common, at least in certain localities."[45]

Basing management decisions on the presumption that large predators were gone while publicly claiming they were still present was a difficult position for park authorities to sustain, however. And this dilemma deepened when the Service adopted the policy of ecosystems management. If our parks were to be restored to their original condition and managed as self-sustaining ecosystems, the absence of major predators presented a dilemma. Without them the park could hardly be called an original ecosystem. Yet to follow the advice of the Leopold Report — to reestablish original conditions — would require a public program to reintroduce predators into the park. And reintroduction would mean not only confessing that these animals were gone and that the Service had deceived the public for decades; it was also bound to arouse the wrath of sheepmen, hunters, and the surrounding states, not to mention the government's own Division of Predator and Rodent Control.

By 1967 the Park Service seemed hoist by its own petard. To run Yellowstone as an intact ecosystem required wolves. But getting them into the park would damage its public credibility and would face strong political resistance from well-financed and organized agricultural and sporting interests.[46]

The dilemma deepened after the Senate elk hearings in Casper in spring 1967. Once the Park Service had been denied permission to cull the elk herds it seemed even more necessary that large predators be there, doing their ecological job, killing elk. In 1968, the Green Book made predators central to policy, stating, "control [of wildlife] through natural predation will be encouraged." Now they were needed as an instrument of management.[47]

Shortly after Anderson and Cole arrived in Yellowstone from Grand Teton, during early fall 1967, a rumor began to spread among the rangers. A decision had been made, they told one another, to reintroduce wolves to the park. In fact the idea seemed to be on everyone's mind. Gordon Eastman, a professional photographer living in Cody who kept wolves as pets, remembers visiting Mammoth at that time. Two wolves were in his station wagon. "Would you consider releasing those animals of yours in the park?" Anderson asked him.[48]

"No, Superintendent," Eastman replied. "Animals have a way of disappearing in Yellowstone, and mine are much too valuable."

A few weeks later, according to Richard Sideler, now a biologist with the Alaska Department of Fish and Game, Cole drove to the Charles Russell Game Refuge near Great Falls, Montana, to talk with Sideler's boss, Bob Berkholter, a renowned wolf expert. "Cole was there," Sideler said, "to talk about a wolf transplant. He very much wanted to do it. But I remember Berkholter cautioning him. He kept telling Cole the political resistance would be too great. Livestock interests would never permit it."[49]

Nevertheless, on December 5, 1967, Cole announced his policy of natural control: "elk populations can be naturally controlled by periodic severe winter weather and native predators." And just as independent scientists were scratching their heads over the new policy, wondering how predators gone from the park for four decades could keep elk numbers down, Gates, a Park Service employee, proved that the animals were there. On December 27 he took his historic photograph. With mysterious serendipity, the wolves had returned.[50]

Through the next year the number of wolf signs increased, and as they did, so too did the joy of park authorities. "The wolf may be staging a comeback within Yellowstone National Park," the park announced in November 1968. "Wolves Returning to Yellowstone," ran the headline of a story in the Wyoming Travel Commission's *Travel Log* in that year. "One thing people come to Yellowstone to see is the animals. This will give them a rare chance to look for, and see and hear wolves," Anderson told the magazine. "Besides," the magazine added, "the wolf would assist in restoring natural controls on those elk and buffalo herds which stay within the Park all year long."[51]

To heighten public awareness of this remarkable comeback, park biologists established a wolf situation map, on which sightings were identified by colored markers. They kept a pin-up board, containing photographs of wolflike animals and plaster molds of impressions that were said to be wolf prints. Through these efforts the world became aware that the wolf was not only back but multiplying. In 1968 Cole reported there were "at least six wolves." By 1971, the number had grown to "possibly fifteen different animals," and "possibly two pairs of wolves" were reproducing![52]

To independent scientists and other backcountry experts, however, the idea of a reproducing population of wolves in Yellowstone seemed too much like immaculate conception. Wolves, they knew, could not hide well. They lived in packs and often hunted in the open. They had prints as big as small plates and they left their mark on carrion.

At that time Merlin Shoesmith, now a biologist with the Canadian Wildlife Service, was working on his Ph.D. on the elk of the Mirror Plateau. For three years he had lived in this remote section of the backcountry and never saw a sign of a wolf. Yet just as he completed his studies, the wolf revival had begun, and the wolves, he was told, spent much of their time on the Mirror Plateau! Had he been blind?[53]

Shoesmith was not alone in his wonder. The Craighead brothers, who had tramped every section of the backcountry for nine years looking for bears, never saw a sign of a wolf. Neither had any of those who worked for them. Jay Sumner, one of the brothers' assistants, had spent several winters in the backcountry preparing his dissertation on grizzly hibernation. "If there had been wolves in the park before 1967," he told me, "I would have seen some sign of them." Joe Gaab, who had spent two decades and more in the backcountry on elk studies and law enforcement, never saw a sign of a wolf. He could not understand how, after living with an elk herd for months at a time, he had never seen one if they had been there. Surely, in those twenty years, a wolf should have killed an elk![54]

Jonas, still a seasonal ranger at the time Gates took his picture, wondered why, having spent years studying the beaver throughout the park, he never saw a sign of a wolf, for the beaver was one of the wolf's favorite prey animals. Many old-time maintenance personnel who spent much of their lives working on trails in the backcountry had never seen a wolf.[55]

Nor did the historical record suggest such a continuous presence of wolves. Thorough examination of the historical record in park archives by National Forest Service biologist John Weaver turned up no convincing evidence. "I very definitely think," Weaver told me, "that from 1940 to about 1968 there weren't any wolf packs in Yellowstone and it is highly unlikely there were any wolves at all."[56]

Yet after "a search of Yellowstone's files," biologist Cole was able to turn up reports suggesting that wolves had enjoyed a continuous presence in Yellowstone during the preceding forty years. Apparently a breeding population of wolves had lived in the park all along. They had just never been seen by those who had spent their lives in the backcountry.[57]

He had found, Cole insisted, records of the presence of wolves that earlier historians had missed. And, like an archaeologist digging up bones, he was able to find a few more old wolf reports each year. In 1968, he reported, there had been twenty-four reports of wolves between 1930 and 1967; by 1969 the number of reports for the same period had grown to forty-two; by 1971 the number was eighty-three.[58]

In fact Yellowstone files contained almost no reports of wolves after 1940, and park authorities, finding so little evidence that wolves had lived in Yellowstone continuously, had manufactured it. To produce this "evidence," Cole interviewed retired park employees, asking them to search their memories to see if they could remember ever seeing a wolf. If, yes, someone did seem to remember seeing something — perhaps a coyote — many years ago, which might have been a wolf, their testimony was retroactively inserted into the historical record.[59]

One of those who found himself a respondent in this poll was biologist Kittams. "In the fall of 1952," he told me recently, "I came upon a large canid footprint in the Hayden Valley. Wondering whether it might be a wolf, I showed it to Adolph Murie. He said, 'Nope, wouldn't make a pup.' But when I saw Glen Cole at the research meeting [in April 1968] in Grand Canyon, he tried to persuade me I had seen a wolf print. It made me kind of mad. He acted as though he could make me say whatever he wanted me to. But I am a stubborn sort, and wouldn't give in."[60]

Nevertheless, Kittams's "sighting" was inserted by Cole into the "historical record," and later to establish the revision of Yellowstone history as official, in 1978 Cole inserted his new numbers into the Congressional Record during Senate hearings on the Endangered Species Act.[61]

If the wolves were not remnants of a native population, where did they come from? Some thought they were not there at all, that the "reappearance" was a case of mass suggestion, like the scare, a few years previously, that black panthers were living in the Appalachians. If a myth like Yeti or Bigfoot could persist, they thought, so too could the myth that wolves had remained in Yellowstone unseen.[62]

And yet most biologists who examined the evidence came away convinced that wolves had indeed returned. Weaver's study, the most exhaustive done on the evidence, confirmed their judgment. He classified 81 of 401 wolf sightings after 1967 as "probable," and remained convinced that wolves had appeared in the park at that time.[63]

Others wondered if the wolves had drifted down from Canada, not an impossibility. Wolves often traveled hundreds of miles. But in this case it seemed unlikely. Wolves at this time were scarce in the southern Canadian Rockies (having been decimated by an anti-rabies campaign in the 1950s), and their movements were being carefully monitored by Canadian wildlife biologists. To migrate from Canada, moreover, they would have had to negotiate 300 miles of populated Montana territory, avoiding compound 1080 stations, federal predator-control agents, trappers, bounty hunters, and hostile ranchers.

Still others wondered if the wolf presence was a practical joke, if a private citizen hadn't released captive wolves in Yellowstone simply for the fun of it. But this explanation too seemed unlikely. Captives, as Eastman had explained to Anderson, were expensive. Would someone really want to release animals costing thousands of dollars simply for the anonymous pleasure of puzzling a few scientists? And would such animals, accustomed to people, remain so elusive? "Mine would have run up to the first child it saw and licked his face," Eastman told me.[64]

But if wolves had returned after a forty-year absence, not on their own or with private help, only one possibility remained: they were secretly planted by the Park Service. To many wolf biologists this remained the most likely possibility. The government had motive and opportunity. At that time wolves could easily and legally be trapped in Canada and transported without documentation across the international boundary. And an agency that could deny for decades that it had eliminated wolves in Yellowstone seemed perfectly capable of putting a few back without telling anyone.[65]

During fall 1967, Jonas told me, "all the rangers were talking about a transplant. We all believed a decision had been made. Then, when we came back the next summer and the wolves were there, we put two and two together. But we thought it was kind of funny when park authorities denied they had anything to do with it. We just naturally thought that they had done it and refused to talk about it."[66]

In spring 1968, Dr. Nicholas Novakowski, then a biologist with the Canadian Wildlife Service, attended a secret National Park Service meeting on bears held in Grand Canyon National Park. There, he remembers, he met Glen Cole.

"Cole definitely gave me the impression that wolves had either just been released in Yellowstone, or were about to be released," Novakowski told me.

In June, Harry Reynolds, Jr., now a biologist with Alaska's Department of Fish and Game, was a graduate student (and seasonal ranger) working summers on the Craighead's bear study, when he was asked to drive Superintendent Anderson to the airport in Billings, Montana. "During the drive I remember distinctly his telling me they were bringing wolves back into the park," he told me.

Jay Ward, a professional photographer in Cody, Wyoming, remembers taking pictures of bison in the Lamar Valley in early spring 1969.

"While I was there this man came up to me," Ward told me. "As we were chatting, this fellow said, 'You should have been here last year when we released some wolves. You could have gotten some good pictures.' As we parted we introduced ourselves. The fellow introduced himself as Glen Cole."

Cole and Anderson strenuously denied these stories and the official position of the Park Service remained that the wolves were remnants of a population that had been there all along.[67]

"No wolves have been introduced into Yellowstone National Park," Robert M. Linn, Chief Scientist of the Park Service, wrote to Dr. Raymond F. Dasmann, President of the Wildlife Society, in 1971. Relying on the word of Yellowstone managers, Linn told Dasmann that "Park records indicate that a residual population has been able to hang on since the days of wolf control. . . . There is an intriguing possibility that the wolves may be descendants of the supposedly extinct gray wolf. You may be assured, therefore, that we have no intention of contaminating this population with other genetic strains.''[68]

Linn's answer did not convince many in the scientific community, but the mystery remained uninvestigated. The Park Service was apparently content to say that, however they came, the wolves were there.

However they came, they did not stay. By 1972 they were seen with increasing frequency outside of the park. One was reportedly shot in the Sunlight Basin east of Yellowstone in 1971. In the same year a rancher near Sheridan, Wyoming shot two wolves that had killed a calf on his ranch; a third escaped. But the majority seemed to be heading northwest. Between 1972 and 1975 they were seen in the Gallatin, just west of the park, in Montana, then in the Gravelly and Centennial Mountains farther west, and then along the Bitterroot in the extreme western part of Montana.[69]

By 1974 they were gone from Yellowstone. Despite approximately 1,800 hours of search by plane, park authorities found only one sign of a "wolf-like canid" after 1974. Between 1975 and 1977, Weaver spent twelve months in the field looking for wolves. He covered 2,700 kilometers on foot, ski, and snowshoe; he put baits and scents at various places and placed preset cameras that would be triggered by any animal at the bait; he flew thirty hours looking for wolves; he recorded wolf howls and broadcast them across the backcountry day and night, hoping for an answer; he carried a parabolic microphone in case the answer ever came.[70]

The answer never came. He was forced to conclude there was not "a viable wolf population in the park."[71]

Where did the wolves go? If they were remnants of the native population, such an exodus would have been inconceivable. Wolves that had survived on the plateau for millennia would not suddenly leave their sanctuary to run a gauntlet of poison and mayhem in search of a new and uncertain home. But if they were transplants, such an odyssey would be entirely understandable: they were going home. Wolves, the experts tell us, can be successfully transplanted only when conditions are perfect; otherwise they will return to their original range. And conditions, according to John Weaver, were not perfect in Yellowstone.

"Wherever you find wolves," he told me recently, "you find beaver." And beaver, he said, are scarce in Yellowstone.[72]

Weaver explained that "beaver might not be essential to restoration of the wolf in Yellowstone if a transplant were done carefully and correctly. But if other conditions were not perfect and proper care were not taken in relocation, the absence of this animal could be critical."

Weaver's colleague on the Wolf Recovery Team, Dr. Bart O'Gara, agreed. "The prospects of wolves reestablishing themselves in the park are poor," O'Gara told me. "Such populations cannot live without a source of small mammals — beaver, rabbit for instance — and range conditions in Yellowstone do not encourage many of these species to live there."[73]

The story of Yellowstone had come full circle. Without wolves, elk had destroyed the habitat of beaver. When beaver disappeared, wolves could not return.

The Park Service, however, was reluctant to admit that this natural treasure had been lost. Rather, they argued, wolves continued to live in the park in small numbers. Indeed, only recent human interference in the ecosystem, they argued, had prevented predators from multiplying. For it was the elk reductions of the 1960s, according to Assistant Secretary Reed, that was to blame for the small numbers of predators in the park. "As a result of having maintained this herbivore population at an artificially low level because of the mistaken belief concerning overpopulation indicators," he wrote in a letter to the president of the National Parks and Conservation Association in 1973, "we may have been depriving predator populations of just that margin of elk numbers needed to sustain them at, or bring them up to, normal levels. This may be the cause, for example, for both timber wolf and cougar populations remaining at such low levels even though it is known that these predators do exist at levels probably capable of repopulating the park to natural population levels."[74]

Within Yellowstone, authorities took steps to ensure that all informational literature confirmed this view of a continuing population of predators. In 1972 Karen Craighead Haynam, John Craighead's daughter, submitted a manuscript for a book on the large mammals of Yellowstone, to be published by the Yellowstone

Library and Museum Association. The park's Chief Naturalist, Robert Dunmire, rejected it, however. The book, he suggested, was "erroneous and misleading" because she had falsely mentioned "the eradication of wolves from the park." Instead he wanted Mrs. Haynam to include wolves in her checklist of mammals.[75]

"I will do so if you can prove they exist in the park," she responded.

Dunmire showed her recent photographs of wolves. But they aroused her suspicions. "All were taken near the road," she told me, "and all showed the animals running away. There was a telephone line in the foreground. It looked to me like the pictures were taken just after the wolves had been released."[76]

Mrs. Haynam refused to put wolves into her book and it was rejected. She published it at her own expense. Six years later (in 1978), she requested that the Association sell her book. The new Chief Naturalist, Alan Mebane, demurred for the same reasons for which his predecessor had refused to publish it. Only when, a few months later, Weaver's study — showing conclusively that wolves were no longer in the park — had been completed, did he agree to sell the book.

Yet throughout the 1970s and into the 1980s, all other literature available in Yellowstone — both official and popular — continued to claim the wolf's presence. The 1974 *Final Environmental Statement for the Yellowstone Master Plan* stated that "predators such as the wolf and mountain lion . . . are here in low numbers." And although nine years had passed since the last wolf sighting, the *1983 Natural Resources Management Plan* (an official document used by the Park Service in planning), claimed that "occasional sightings of wolf-like canids are still being made." Aubrey Haines's 1977 history of the park listed the wolf as one of the original species remaining. So too did the *National Parkways Guide to Yellowstone National Park* and the 1982 report, *Population Status of Large Mammal Species in Yellowstone National Park*, published by the biologist's office. The official wildlife checklist given to visitors today reports that wolves are "occasionally seen in the park . . . all year." A children's book, *Animal Friends of Yellowstone,* sold at visitors' centers throughout the park, told its young readers, "A few wolves are left and are protected here in Yellowstone, but they are shy and seldom seen."[77]

Similarly, cougars enjoyed a dramatic and mysterious comeback. Shortly after wolves returned, park authorities began to report sightings of these cats, unseen for forty years. "From 1970 through October 1978," reported Houston, "105 reliable reports of cougar have been made within or *immediately adjacent to* the park" (italics added).[78]

Indeed, in many parks of the Rocky Mountain West the cougar had made a comeback. In the Wind River Range in Wyoming the rebound had been dramatic. And even in the mountains just north of the park there had been a clear resurgence. But because Park Service reports were based on unscientific testimony and were almost never accompanied by physical evidence — scat, signs on carrion, dens — independent scientists remained skeptical.

To settle the matter, Assistant Secretary Reed, in 1973, ordered Yellowstone authorities — with the help of Dr. Hornocker — to conduct a "detailed survey this winter of the cougar population." But park biologists balked at the plan. To find the lions required dogs, which conflicted with the policy of managing Yellowstone as a natural area. The study was not done.[79]

Claims of a cougar revival continued, though. Data on "sightings," park biologists insisted, were accumulating. And although protesting that, not having had time to "analyze the data" — they could not make this information available to the public, they did not shrink from drawing conclusions from it. Their reports conferred on the lion an official presence in the park. The usual visitors' guides still listed him, as did the official wildlife checklist. The *1983 Natural Resources Management Plan* mentioned the "presence of a resident population," and optimistically opined that "cougar numbers appeared to have increased in recent years."[80]

Meanwhile others, waiting for scientific confirmation, could not hide their doubt. "I know of no evidence," Dr. Hornocker told me recently, "to suggest the mountain lion is making a comeback in Yellowstone."[81]

In 1973 as the last wolves were leaving, the Fish and Wildlife Service put the Northern Rocky Mountain wolf — *Canis lupus irremotis* — on the list of endangered species. Unfortunately the animal was almost certainly extinct already, and the justification for not saying so rested on the evidence that some wolves had been seen in Yellowstone Park and vicinity.

"Although *irremotus* has been considered extinct in the U.S.," wrote the Wildlife Service in explaining its action, "wolves in the former range of *irremotus* have recently been reported from Yellowstone National Park. . . .[82]

"The best judgment," the report continued, "is that they represent members of *irremotus* descended from those few individuals in the back country that escaped persecution in the thirties, forties and fifties."[83]

With *irremotis* enshrined as an endangered species, the public was unlikely to notice that the rescue party had arrived decades too late. Nor, with the irony so characteristic of this sad history, were they likely to notice that those charged with its resurrection were those who had earlier been the agents of its destruction.

For it was the Fish and Wildlife Service, child of the Biological Survey and parent of the Division of Predator and Rodent Control, which was responsible for the resurrection of this subspecies; and any plan of restoration would succeed only if it received support from the Park Service, which controlled the vital areas (Yellowstone and Glacier National Parks) where recovery could take place.

In 1975 the Fish and Wildlife Service charged a Wolf Recovery Team to bring the wolves back. This team designated three areas for wolf recovery, each location chosen for its remoteness from large farming areas or human population centers: the Glacier–Bob Marshall Wilderness area in north-central Montana, the Selway–Bitterroot wilderness in central Idaho, and Yellowstone National Park. It also devised a system to compensate farmers for losses of livestock to wolves. And yet, despite the sensitivity of the Recovery Team to agricultural interests, farm lobbies continued to resist all efforts at revival.[84]

This resistance was entirely predictable; it was also surmountable. A similar plan had been successful in reviving the Minnesota timber wolf, despite that state's even stronger farm lobby. But to overcome resistance by farm interests, the Wolf Recovery Team needed strong support from the Park Service, which they never received.[85]

The second coming of wolves to Yellowstone had done its damage. It permitted the Park Service to insist that the species was still present and that revival, though desirable, was unnecessary. Although "fewer wolves occur in the park than originally," Cole observed in 1975, a "remnant of the original population" still inhabited the park. Thus while admitting (in the words of Harold J. Estey, Chief Park Ranger in 1976), that "it may one day be decided to augment the remnant population with transplants," it would not commit itself to a timetable for doing so.[86]

In fact, once Yellowstone authorities had adopted natural regulation, the reason for wanting wolves was diminished. For if limitations in food supply and not predation controlled elk and bison numbers, wolves and cats were no longer required by management. With no significant ecological function, they would not be missed. The Park Service dragged its feet on restoring wolves. In 1982, Superintendent Townsley reassured Wyoming's Governor Hershler that he was "against the reintroduction of wolves in the park"; pointing out that such an undertaking "would have many technical problems and would be very expensive."[87]

Chief Ranger Thomas Hobbs reported recently that wolf recovery remained "definitely on the back burner." And lacking strong Park Service support, the Wolf Recovery Team made little progress. For ten years its besieged members wrote and rewrote drafts of a recovery plan, and each time they were sent back to revise it, in the face of organized opposition by stockmen.[88]

To accommodate this resistance, writers of the most recent draft no longer insisted that wolves be restored to Yellowstone. Instead, they conceded that wolf recovery would be complete when the animals were restored to two of the three recovery areas. It would therefore be possible, according to the new plan, to satisfy the requirements of recovery stipulated in the Endangered Species Act *without* bringing one wolf back to Yellowstone.[89]

"I think we have made that policy very clear," Assistant Secretary of the Interior Arnett said at a Senate hearing in 1983, "that there was going to be no introduction of the wolf into Yellowstone at this time." Arnett was supported by Russell Dickenson, Director of the Park Service. Said Mr. Dickenson at the hearings, "There is no active proposal [to bring back wolves] afoot and none that we can foresee in the immediate future."[90]

Without wolves the park had become just the place early managers had hoped for: a giant game farm, a breeding ground for the victims of hunters, a place safe enough to be a good neighbor to sheep.

Yet return of the wolves remained over the rainbow. Never

having admitted that it had destroyed these animals, the Park Service could not say loudly that they must return. Having made no real census of mountain lions, it could not call for their restoration. Touting the ecosystem as essentially intact and claiming that predators had no essential part in it, the Service saw no emergency in their absence.

A cloud hung over the scientific and historical record. Wolf biologists remained deeply curious about wolves' origin and frustrated that the matter had not been fully investigated. For although they all strongly favored restoring wolves to the park, they feared that what had happened was like Watergate — the coverup had been worse than the crime. If there had been a secret transplant, then the Park Service, having inserted into the scholarly literature the story that *Canis lupus irremotis* was alive and well, had intentionally confused our knowledge of the natural and genetic history of these animals, making it appear that an extinct species had remained far longer than it had.

These were the feelings that prompted a career Park Service officer whom I shall call Peter to talk with me. As a young ranger Peter was working in Yellowstone at the time the wolves reappeared. He told me, "It was an orchestrated operation from the beginning."

"Before any wolves were brought in we were put to work revising the record, concocting evidence that wolves had been there all along, so that when they were reintroduced the Service would be ready with an account of how the animals were part of a remnant population."

"They had me doing things like preparing plaster casts of wolf prints to show the press," he continued. "At first I did not know why they were asking me to do this. Then after wolves appeared, people began winking at each other. It was as though they were in on a terrific secret. I got very curious. Finally one day I just asked Superintendent Anderson directly whether wolves had been planted. He told me they had. We stood there laughing like little boys who shared a naughty secret. Yet I also felt like a Boy Scout. Bringing the wolf back was our good deed. It was only later when I began to reflect on it and began to see accounts of this 'revival' appear in the scientific literature, that it seemed it was done all wrong."

11

Rendezvous at Death Gulch

SUMMER 1982 was another bad one for grizzly bears. In the previous year, seventeen had died, fifteen of them killed by man. Now seventeen more had died; worse still, six of these had been "removed" by bear managers and researchers themselves.[1]

By August, the authorities were greatly disturbed. That spring Dr. Richard Knight, Team Leader of the Interagency Grizzly Bear Study Team, appointed by the Assistant Secretary of the Interior to study the grizzly, had reported that the Yellowstone population had not, as had been supposed for a decade, more than 350 grizzlies, but fewer than 200. That summer a mathematician with the Battelle Memorial Institute, Dr. Lee Eberhardt, commissioned by Knight to analyze the health of the population, had more discouraging news. The grizzly population was declining. The number of adult females, in particular, was dwindling so rapidly that the future of the entire population was endangered. "The presently available evidence," he reported to Roland Wauer, Chairman of the Interagency Grizzly Bear Steering Committee, "indicates that the Yellowstone grizzly population most likely cannot sustain its present level unless adult female survival rates improve."[2]

Wauer, urged by Knight, sent a memorandum to members of his committee. "The survival of grizzly bears within the greater Yellowstone ecosystem," he wrote, "is dependent upon reversing the current population decline." His message quickly leaked to the press, and within two weeks the *New York Times* had reported that the Yellowstone grizzly was "imperiled."[3]

And although it passed unnoticed, black bears too were dwindling. They were seldom seen in the park any more; they did not beg along the road or come into campgrounds. Only the rare and lucky visitor saw one at all. Yet for more than a decade, authorities had insisted that they were more numerous than grizzlies. A "fact sheet" made available by the park biologist's office in summer 1982 had claimed "all our information indicates that about as many black bears inhabit the park now, approximately 650, as when the park was established." But Knight's research had found blacks to be less numerous than grizzlies and declining as well.[4]

For the Park Service, these revelations were cause for an agonizing reappraisal of bear management; for the nation they provoked disillusion. For a century authorities had been telling the world — and themselves — that the grizzly and black-bear population in Yellowstone were healthy and stable or even growing. Now they began to suspect, and had to admit, the worst: the bear was headed for oblivion. For the first time in history the entire country contemplated a truth that a few years earlier had seemed inconceivable: Yellowstone Park without bears.[5]

What had gone wrong?

In the first week of April 1968, a closed meeting of Park Service research scientists and management biologists was held at the Horace M. Albright Training Center in Grand Canyon National Park, Arizona. Comprising about forty officials, together with a few spouses, it was only the third time in Park Service history that all biologists had been together at the same time. Starker Leopold was there with his wife. So too was Leopold's student, Mary Meagher, and his assistant, Dr. Robert Linn. Glen Cole too was there from Yellowstone, along with William Barmore.[6]

As they assembled for their group picture, it seemed to the participants that a park renaissance had begun. And although the event was entirely unnoticed by press and public — to this day — it was indeed to profoundly affect the future of our national parks.[7]

Convened by Leopold, now the new Chief Scientist of the Park Service, the agenda was to implement the recommendations of the Leopold Report, calling for government control of science and a more ecological approach to park management.

Leopold opened the discussion by explaining, among other things, the new concept of mission-oriented research. One of their most important functions, he suggested to his fellow scientists, "is that of service to the superintendents, and to park administration in general. The NPS research program," he continued, "is intended to be mission-oriented. It is not science for science's sake."[8]

Linn followed, initiating panel discussions on "The Ecosystem Concept and the National Parks." He urged each scientist "to develop a schematic of the ecosystem(s) that depicts your parks."[9]

But the participants also had a more pressing and more specific interest than the abstract niceties of ecosystems schemas: something called the "bear problem." And for Cole and his Yellowstone colleagues, this problem had special significance.[10]

"In the Yellowstone," wrote a visitor in 1909, "bears are as the autumn leaves," and indeed, since Yellowstone Park was founded, bears and Yellowstone had seemed inseparable. For three generations of Americans, coming to the park was nearly synonymous with coming to see bears. For these travelers, it was the black bear — *Ursus americanus*, the impish, appealing, but destructive and sometimes dangerous clown who prowled the campgrounds and begged along the roads, blocking traffic and causing "bear-jams" — which provided many a vacation's thrills.[11]

Few ever saw the grizzly — *Ursus arctos*, or silvertip, as westerners call him. But he remained in Yellowstone as a remnant of a species that once roamed throughout western North America. Much larger than the black, ranging up to 1,200 pounds, wilder and more unpredictable, this bear, easily identified by its humped shoulders, long hind feet, broad muzzle, and long gray guard hairs, usually stayed away from human beings. But his presence was felt. It instilled in the hearts of backcountry hikers the most authentic of wilderness emotions: awe. Knowing the silvertip was there, they knew they were in truly untamed country. The grizzly, with its reputation for ferocity, its intolerance of human beings, its need for great range, had rightly become a symbol of wild America.[12]

When white explorers came to the Yellowstone plateau in the 1860s and 1870s, they saw few black bears, but they found many grizzlies. Yet no one knew how many of these animals were natives of the region and how many had been pushed into the mountains, along with the elk, by settlers who appropriated the river valleys. Evidence was plentiful that the grizzly was originally a plains animal, living on the rich riparian vegetation and the carrion, such as dead buffalo, drowned in rivers or driven off cliffs by Indians. Certainly that is where the first white explorers found them.[13]

"Today we passed on the Stard. side the remains of a vast many mangled carcases of Buffalow," wrote Meriwether Lewis in spring 1805, as he and William Clark poled up the Missouri through Montana, "which had been driven over a precipice of 120 feet by the Indians and perished . . . they created a horrid stench." Yet, these explorers recorded, sometimes such mass carnage occurred without the help of Indians, when the bison would be caught in the current as they crossed, leaving thousands of dead bodies to attract bears. Along the Great Falls of the Missouri the explorers said that both buffalo and grizzlies were so abundant and the latter so "troublesome that I [Lewis] do not think it prudent to send one man alone on an errand of any kind."[14]

By the time white men were exploring the Yellowstone plateau, grizzlies had already been chased off the plain. Settlers had taken prime grizzly habitat, killed the buffalo on which the bears had fed, and caught the spawning trout and salmon that once had been such an important source of food.

But whether bears were refugees that fled to the high country or whether they, and their ancestors, had been there all along, once the park became a sanctuary, they throve. For Yellowstone brought people who fed the bears, and people produced garbage, which these opportunistic scavengers consumed. The white man, having destroyed the bear's original ecosystem, replaced it with another. Garbage made up to the bear for the loss of spawning fish and drowned buffalo.[15]

Theodore Roosevelt, on visiting the park in 1903, wrote,

The effect of protection upon bear life in Yellowstone has been one of the phenomena of natural history. Not only have they grown to realize that they are safe, but, being natural scavengers and foul feeders, they have come to recognize the garbage heaps of the hotels as their special sources of food supply. Throughout the summer months they come to all the hotels in numbers, usually appearing in the late afternoon or evening, and they have become as indifferent to the presence of men as the deer themselves — some of them very much more indifferent. They have now taken their place among the recognized sights of the Park, and the tourists are nearly as much interested in them as in the geysers.[16]

Bears and human beings had developed a special symbiotic relationship, and by all accounts (as we saw in Chapter 3), thanks to this relationship, they began to multiply: more people produced more garbage produced more bears. Creation of the Park Service in 1916 accelerated this process. The new bureaucracy, eager to attract tourists, saw them as a major attraction. They set up bleachers around a feeding area, where visitors could sit to watch the grizzlies come to eat while listening to ranger-naturalists explain their behavior.[17]

Black bears, being smaller, were unable to compete with the grizzlies at the dumps, and so they roamed the roads and campgrounds, searching for scraps left on tables and in the garbage cans. These bears became the welcome, wild camp followers of the park.

But this policy succeeded too well. As visitation to the park increased, so too did the amount of trash and garbage. And as their source of food grew, so too did the number of bears. Censuses taken at the dumps showed that the number of grizzlies increased from 40 in 1920 to 260 in 1930. Sometimes as many as a hundred grizzlies could be seen at one time at the Trout Creek dump, near Hayden Valley. And as the bear population grew, so did the "bear problem."[18]

The problem, in fact, would have been better described as a "black-bear problem." For it was this smaller and tamer species that did most damage to property and people. Between 1959 and 1965, black bears were involved in 298 cases of personal injury and 1,738 cases of property damage, but during the same time grizzlies accounted for seventeen injuries to people and thirty-nine cases of damage to property. Grizzlies, congregating at backcountry dumps for the most part kept out of harm's way.[19]

The greatest troublemaker was the black, but the grizzlies got the headlines. Larger and more fearsome, their few encounters with people were far more dramatic and terrifying. Unlike blacks, they occasionally killed people, which had a way of concentrating the public mind, causing many to exaggerate the danger they posed.

In the first ninety-five years of national park history, three people had been killed by bears (all, as it happened, in Yellowstone). The last death had occurred in 1942. On the average, however, each year around a half-dozen people died in the park from every conceivable cause: drownings, falls, traffic accidents. People were also gored by buffalo or savaged by moose, and they occasionally died from these mishaps.[20]

Park authorities were aware that the grizzly had never deserved the bad press it received, and policy from the beginning was aimed as much at protecting this bear from the public as vice versa. The first person known to be killed by a bear in Yellowstone was a young man, out for a walk with his wife in 1907. He chased a grizzly cub up a tree and began poking it with an umbrella. The sow appeared and tore out the man's breastbone and a lung with one swipe. Although pressure was put on the chief ranger (or scout, as he was then called) to kill the bear, he refused to do so. The bear, he said, had the right to protect her young. That was forest justice.[21]

But as time went on, the Park Service was less willing and less able to invoke forest justice. The growing number of people and bears led to more and more "bear incidents," and increased the threat of a major human calamity. To cope with this danger they issued, in 1960, new regulations, closing the landfills to visitors and attempting to educate people about the dangers of feeding bears. Plans were laid (but not implemented) to replace the dumps with incinerators.[22]

And indeed, this effort seemed successful. While the dumps continued to act as a magnet, keeping the more dangerous grizzly away from people, greater sanitation in the campgrounds lessened problems with blacks. In the 1950s approximately one in 26,000 visitors was injured by a bear, during the first seven years of the 1960s the figure was one in 49,000, a drop of 50 percent. Yet as the rate went down, unease over it went up.[23]

Paradoxically, during a time of growing environmental awareness, American tolerance for bears was diminishing. We had become a more litigious people. If, as had happened, Sequoia National Park could be sued by the family of someone killed by lightning there, the bear was a walking liability. He was to Park Service lawyers what icy doorsteps were to homeowners. A bear injury often resulted in a tort claim against the park.[24]

As park visitation grew during the 1960s, so too did official anxiety over human–bear encounters. Although the relative number of bear incidents was in decline, the *potential* for disaster seemed to be increasing. For studies completed just shortly before the

Grand Canyon meeting revealed something managers found disturbing.

Since 1959, the noted naturalist twins, John and Frank Craighead, had been conducting the first scientific study of the grizzly ever undertaken in the park. Their research, most of it privately financed, had begun in 1959. By tagging and radio-tracking individual bears, techniques they were the first in the world to develop, they had accumulated by 1967 data that even today provide the most complete information available on the grizzly's population, fertility, and behavior.[25]

In July 1967, at the request of the Park Service, the Craigheads submitted a report on the population and status of the grizzly. Their population, the Craigheads reported, was *growing*. Between 1959 and 1966, according to their census, grizzly numbers had increased from 154 to 202 — an average of six bears a year.[26]

A month earlier, two graduate research assistants from Colorado State University, Victor Barnes and Olin Bray, had completed a two-year study on the black bear in Yellowstone. Shooting elk and placing the carcasses as bait in the best black-bear habitat they could find, they estimated the number of black bears in these areas to be about 5.2 per square mile. Cole took that figure — a density figure based on observations of the bears *attracted to bait* in the best habitat in the park — and applied it to the entire park. Multiplying 5.2 by the number of square miles in the park, he extrapolated a black-bear population of 650. Although this conclusion was completely unscientific and based on a study that was not even intended to be a population estimate, it had shock value. A black-bear population of 650 (compared with a 1925 census of 200) was exceedingly high.[27]

If, as these studies revealed, both blacks and grizzlies were increasing, then Yellowstone's bear problem would only get worse. And while officials were considering the implications of this growth, events confirmed their worst fears.

In August 1967, two months after the Barnes and Bray study had been completed, a month after the Craigheads had submitted their report, and just eight months before the meeting in Grand Canyon, two young women, in separate incidents, were fatally mauled by grizzlies in Glacier Park on the same day. Coincidentally, these tragedies occurred in the same month in which Cole had been appointed Supervisory Biologist of Glacier, Yellowstone, and Grand Teton parks. He had been handed a hot potato.[28]

He and others raised the bear problem at Grand Canyon. Clearly, something had to be done. The public, the participants agreed, had the right to visit our national parks without risk of being mauled by a bear. They had to find some way to control this species. And, they thought, they had found a solution that would be not only good management, but good science as well. The secret for controlling the bears would be found in the new science of ecology.

Garbage, government scientists believed, was not natural. In fact, it was an example of "artificial feeding of wildlife," which the Leopold Report had explicitly proscribed. Now evidence was on hand that feeding the bears had resulted in an artificially high number of bears — far above the natural carrying capacity of the range.

"I would consider as a goal of the management of natural areas of the National Park System," Linn wrote a month after the Grand Canyon meeting, "to maintain, or re-create when necessary and to the extent possible, those ecological conditions that would currently prevail were it not for the advent of post-Columbian man and his cultural impact." The Park Service, therefore, should not "manage the resources to augment the grizzly population. . . . Why should there be continued growth of a population? This would only lead to overpopulation. Again, the objective is to maintain or restore natural ecosystem relationships."[29]

Government scientists believed, then, that the bear problem had been the result of failure to manage the park according to good ecological principles. To manage the bears scientifically required weaning them of garbage so that, rather than concentrating around the dumps and campgrounds, they would disperse throughout the park, and rather than continuing to grow to artificially high numbers the population would drop to the "natural carrying capacity" of the range. Thus the ecosystems approach, by thinning out the bears, would cure the park manager's headache. The bear problem would disappear, naturally.[30]

In their 1967 report, however, the Craigheads had suggested that a program which might reduce the number of grizzlies was fraught with difficulty. Although substantial populations still lived in Canada and parts of Alaska, the Yellowstone bear was perhaps the largest remnant still living in the lower forty-eight states. Perhaps as few as five hundred remained south of Canada. Recognizing this scarcity, the Committee on Rare and Endangered Wildlife Species in spring 1967 officially listed the grizzly as an endangered species. The government, by law, must now take steps to *increase* the numbers of this species.[31]

The Yellowstone grizzly population, the Craigheads warned, moreover, was extremely fragile. They observed in their 1967 report that "annual recruitment is only slightly in excess of annual mortality; therefore the growth of the population is very slow." Just a slight increase in the death rate or drop in the birth rate could lead to a decline.[32]

Management of the grizzly, the Craigheads urged, should be directed at increasing the number of bears. They concluded, "Yellowstone Park may now be supporting a near-stabilized population of grizzlies, but on the basis of biological information obtained in our investigation, we believe that a higher population could be maintained, and probably should be in order to afford maximum population security."[33]

Increasing the number of bears required, they thought, that the Park Service modify its plan to close the dumps. For these pits, however unsightly, concentrated the bears in the backcountry, in the center of the park, where they were safe from human beings, and people were safe from them.

Grizzly bears, the Craigheads said, "have been feeding at refuse dumps in Yellowstone Park since the late 1800s. We have been unable to detect any adverse biological effect on the bears from this habit. . . ." Indeed, over the past century they had become dependent on garbage as an important source of food. But if the dumps were closed, the bears, deprived of a traditional source, would be forced to wander looking for food, bringing them in contact with human beings, endangering both people and bears.[34]

"Since the refuse dumps within the park attract and hold grizzlies for extended periods of time," they wrote, "they reduce outward movement in summer and fall and by so doing lower the mortality." If the closure is abrupt, they continued, "the result most certainly will be increased grizzly incidents in campgrounds, accelerated dispersal of bears to areas outside the park, and greater concentrations of grizzlies at the public dumps in Gardiner and West Yellowstone, where food will still be available but where adequate protection will not. The net result could be tragic personal injury, costly damages, and a drastic reduction in number of grizzlies."[35]

Instead of quick closure, therefore, the Craigheads suggested a controlled, gradual closing of some dumps with simultaneous careful study to monitor the effect on the bears. During the transition, and longer if necessary, supplemental carrion feeding in remote areas, or continued operation of some dumps, might be used to help the bears. They recognized that this suggestion ran counter to the mandate of the Leopold Report to minimize "observable artificiality."

They said in their 1967 report,

> We recognize the artificiality of this [maintenance of the dumps] as a management technique. However, any purposeful management of wildlife populations or their habitats can be considered "unnatural." Moreover, within Yellowstone some of the natural population-regulating processes have been so altered since the establishment of the park that these are not now effective. Since there is no possibility of these being wholly restored and since management must do the job, artificiality becomes inevitable. Maintenance or establishment of the natural situation, although a commendable ideal to work towards in national parks, and we fully endorse this concept, does nevertheless have limitations that must be recognized.[36]

Following the Grand Canyon meeting, however, the Craigheads' advice was rejected. "It seems to me to be unscientific," Linn wrote in May 1968 to Joseph Linduska, supervisor of the Craigheads' study, "to recommend augmentation of population numbers when the entire idea is to present a natural population."[37]

Eating garbage, Cole insisted, even at pits in the backcountry, caused bears to become addicted to human food and to lose their fear of people. Concentrations of bears at the dump, he emphasized, resulted in higher cub mortality as adult boars killed cubs they encountered at the dumps.[38]

Unfortunately little or no evidence had been gathered for these claims, as the Craigheads were quick to point out. Concentration of bears at landfills, they wrote, was no more unnatural than similar congregations which Lewis and Clark had found along the Missouri or which continue today at salmon streams in Alaska. Nor did bears that ate garbage in the backcountry associate this food with human beings. And crowding at the dumps did not lead to greater cub mortality. Rather, they found that the bears developed a social hierarchy at the dump sites that enhanced survival of cubs.[39]

The Park Service agreed with the Craigheads, however, that closing dumps would cause the bears to disperse, requiring many of them to be killed as they invaded campgrounds or left the park. The Final Environmental Assessment of Yellowstone's Master Plan conceded that "removal of artificial food sources for bears from inside the park may cause them to range outside the park [where, incidentally, hunting of grizzlies was permitted] until natural food habits are restored." Were grizzlies numerous enough to withstand this inevitable carnage? The Craigheads answered no, and the Park Service answered yes. Whether the plan would work, it seemed, depended on how many bears lived in Yellowstone.[40]

The underlying issue, then, was numbers. But how do you count bear? They are not easily found. If, walking through the woods, you see five bears, have you really seen five different bears? The only way to avoid duplicate counting is to do what the Craigheads (and only the Craigheads) did: capture a bear and collar it. But when do you know that you have collared them all? You never do. If you somehow managed to put collars on so many bears that after continued search you found none without collars, you might assume that you had found them all, but you could not be certain. And by the time you had gone that far, you would have no way of knowing if the first bears you counted were still alive.

Where do you go to find a bear? To collar bears, the Craigheads went to the dumps. Then, by conducting backcountry censuses, observing the ratio of dump (collared) bears to those without collars, they concluded that the bears that came to the dumps amounted to 75 percent of Yellowstone's grizzlies.[41]

Here the Interior Department parted company with the Craigheads. They said the backcountry held a much greater population. The Craigheads had, it said, fed only garbage-dump data into their computer. "If you put garbage-dump data into the computer," Cole said, "you'll get garbage-dump data out." By killing large numbers of these bears, therefore, the Park Service insisted it would not be endangering the population. It would only be making room for the backcountry population of unspoiled bears the Craigheads had never found.[42]

Besides, as with the elk, Park Service scientists thought that biological compensatory mechanisms would help. Weaning the bears from garbage, they hypothesized, would improve their diet and increase their fertility. Although more would be killed, more would be born to replace them.[43]

Unfortunately, no one had sound evidence that this backcountry population existed. The Park Service claim was based on the Barnes and Bray study of black bears, in which researchers observed, but did not tag, twenty-seven grizzlies in the backcountry, only one of which had a Craighead collar. Unable to rule out duplicate counting, they could, of course, have seen one bear twenty-seven times; but on the strength of this evidence, the park argued that dump bears were only a small percentage of the population.[44]

But, weak as the Barnes and Bray data were, they gave the Park Service justification for rejecting the Craigheads' advice. For without knowing it the Craigheads had violated a cardinal bureaucratic rule. Unaware of Starker Leopold's advice that science existed solely to serve the superintendent, they had challenged the chain of command. Linn explained to John Craighead in 1969 that "management recommendations must reflect the policies that have been established in regard to whatever it is that's being managed. . . . Recommendations that are offered to an agency should fall within the parameters set by policies, because if they do not fall within such parameters, the administrators of the organization will find it difficult or impossible to accept the recommendations."[45]

But the Craigheads had not listened. From the government's point of view they were out of control. As Deputy Assistant Secretary of the Interior Curtis Bohlen wrote to Senator Robert P. Griffin, explaining the Park Service's grievance with the Craighead brothers, "When scientists attempt to extend their products of research into the realm of policy and management decision making, this goes beyond the normal prerogatives of a scientific endeavor."[46]

Thus, within weeks of the meeting in Grand Canyon, the Park Service rejected the advice. It would wean the bears from garbage immediately, transforming them into "wild, free-ranging populations." In May, the park began to phase out the garbage dumps, to "bearproof" waste containers, to increase efforts to educate visitors on sanitation and the danger of bears, and to remove all "nuisance" bears. A year later the Natural Sciences Advisory Committee of the National Park Service, chaired by Starker Leopold, reviewed the issue and declined to recommend reversal of the new program.[47]

This policy, as Starker Leopold and Durward Allen, both on the National Parks System Advisory Board, described it later, "was to be one that protects the people from the bears; protects the bears from the people; and protects the National Park Service from tort cases in the event of mishap."[48]

It would also kill many bears. And, regardless of whose numbers were right, systematically killing a species officially designated as endangered was illegal. In March 1969, however, the Interior Department removed the grizzly from the endangered-species list, thus sanctioning any future killing. The Department explained in making its decision that "the grizzly bear is not now threatened with extinction in the United States."[49]

Now all the wheels were greased for the Park Service to take on the bears. Their future, like that of the elk and other animals, would now be decided by the new policy of ecosystems management. And the major effect of this new science, for Yellowstone, would be that the killing of elk, of which there were too many, would end, and that the killing of bears, of which there were too few, would begin.

In summer 1968, authorities greatly reduced the amount of refuse taken to the Trout Creek dump. At the end of the 1969 season they closed the Rabbit Creek dump, near Old Faithful, for good, and in the following fall shut down Trout Creek entirely. Forcibly deprived of garbage, the bears were now on their own.[50]

As the dumps closed, the grizzlies began to wander. More came into campgrounds. More people were injured. The grizzly, cut off from his usual source of food, came into populated areas in numbers, and this bear, which had been a relatively minor nuisance compared with the black, now became, for the first time in park history, a real menace. According to the Lake Bear Logs kept by District Rangers in the Fishing Bridge area, the kind of bear activity was dramatically reversed. During summer 1967, twenty-five of the thirty-one bear incidents or management actions in which a species was identified involved black bears. By 1968, immediately after the food at the Trout Creek dump had been reduced, seventy-eight of ninety-one incidents or actions involved grizzlies. In June 1972, near Old Faithful, a grizzly fatally mauled a camper named Harry Walker, the first such fatality in thirty years. Just as the Craigheads had predicted, the new management was turning the silvertip into a menace.[51]

In response, the park began to kill more bears. The number of "control actions," according to a report of the National Academy of Science, rose from an average of thirteen a year in 1967 and earlier to 63.3 a year between 1968 and 1970. The number of grizzlies reported killed by control actions, according to these official figures, rose from an average of three a year before 1967 to nine a year between 1968 and 1970, and the annual man-caused mortality rose from an average of 18.9 bears per year to 31.5. Between 1970 and 1971 alone, 101 bears, according to the NAS, were "removed" from the Yellowstone population. In all, according to the NAS, 189 grizzlies were reported to have been killed between 1968 and 1973.[52]

Even by official estimates this was a bloody period, but we have good reason to believe that not all bear kills were reported. According to Harry Reynolds, a district ranger in Yellowstone during the late 1960s, "Cole, with or without the approbation of Superintendent Anderson, endorsed or initiated a Park practice of not making records of all 'controlled' or otherwise deceased grizzlies." Consequently, according to Reynolds, many bear kills went unrecorded, especially those caused by an accidental overdose of tranquilizer. "By this time," he said, in a letter to the National Academy of Sciences, "bears were not infrequently killed in this way by the use of drugs in inexpert hands."[53]

Reynold's story has been confirmed by many working with the park at this time. Dick Randall, with the Fish and Wildlife Service in Grand Teton, remembers happening on bear removals that he later discovered had not been recorded. Several seasonal rangers whom I interviewed testified seeing bears dropped from helicopters. Others tell of finding burial pits filled with bear carcasses in the backcountry. James June, a biologist for the state of Wyoming, has told me of finding piles of bear carcasses near Turbid Lake on two occasions in the late 1960s, when he was doing field research on geese.[54]

According to Ben Morris, one of the helicopter pilots who transported bears during this period, grizzlies would be hauled to the backcountry three times. "If they came back a fourth time, they were killed. But so far as I could tell, if a black bear caused trouble they just killed it. The park VIPs felt they couldn't waste helicopter time and money hauling black bears."[55]

Usually, Morris said, they would take a grizzly to a steep divide on the park border that sloped out of the park. After the bear was revived from the tranquilizer, they would chase it downhill, out of the park. He added, "Many of the bears were killed by accidental overdoses of tranquilizer." The problem, he thought, was inexperienced park personnel. "But the people working with the drug called it a bad batch of tranquilizer." Morris recalled a time in 1971 when three grizzlies died in two days from tranquilizers, and he suspected there were more. He remembered also another pilot accidentally dropping a grizzly, killing it.

As Morris's testimony implied, grizzly mortalities were probably less than fully recorded because so many occurred out of Yellowstone. Bears released at the boundary could, at this time, be legally hunted in Montana; and individuals caught at the West Yellowstone or Cooke City dumps might be destroyed by state officials. Neither of these types of mortalities would show up on park records. Yet, as bears continued to leave the park, Montana officials became desperate to find places to send them.[56]

"We must have a place to put the bear," a Montana official complained at a management meeting in spring 1972. "We have to know once and for all where we can put them."[57]

Perhaps, the conferees suggested, they could be sent to the Cabinet Mountains, or the Bob Marshall Wilderness, or the Tobacco Roots, the Anaconda-Pintlar, Lander's Fork, Fleecer-High Rye, the Lincoln Backcountry. And wherever they went their fates would show up on no official tally of Park Service "management actions."[58]

Yellowstone archives have almost no material on bear management between 1968 and 1973. In 1967, according to park archivist Tim Manns, Superintendent Anderson discontinued the long-established practice of requiring monthly reports from district rangers. Surprisingly, even the original "bear logs" kept by the rangers are nowhere to be found. Granted access to park bear files under the Freedom of Information Act, I could find no original bear incident reports. All that were there, it seems, were logs that were neatly typed and, apparently, edited.[59]

Yellowstone authorities, moreover, had reason to be secretive. A year before the Trout Creek dump was closed, Congress passed the Environmental Policy Act. This law required that no major federal action be taken until an environmental impact statement (EIS) had been completed. But no such review was begun until 1974. "Suddenly, in the early seventies," one former senior Park Service official explained to me, "just as the Park Service was in the midst of killing bears, they found what they were doing was in violation of the EPA. They had to cover it up."[60]

But they had other reasons for secrecy. In spring 1972, in response to environmentalists' objections to poisoning predators on public lands, President Nixon signed Executive Order 11643, prohibiting use of chemical toxicants on animals on federal lands. But at that time, the medicine cabinet of Yellowstone authorities was filled with varied toxicants that suddenly became subject to drug enforcement. Many of these poisons were used on bears: M-99 Etorphone, M-285 Cyprenorphine, M-50-50 Dyprenorphone, Succinylcholine Chloride (Succostrin), Phencyclidine Hydrochloride (Sernylen, known on the street as "angel dust"), and Alpha Chloralose. Park authorities were also shooting bears with drugs considered experimental at the time — such as "Bayer 1470" (or Rampon). Whether this use was a violation of law or not, it certainly would not meet with public approval. The best thing was to use it but keep silent.[61]

In May 1971, President Nixon appointed Nathaniel Reed as the new Assistant Secretary of the Interior for Fish, Wildlife and Parks. Reed, scion of a prominent family and benefactor of many conservation groups, quickly endorsed the official Yellowstone bear policy. In the same year the Craigheads' research permit came up for renewal. The park agreed to renew it only if they would sign an agreement not to speak to the public or publish without first obtaining Park Service permission. The Craigheads, seeing this requirement as an infringement of academic freedom, and even the First Amendment, refused. Their twelve-year study came to an end.[62]

Immediately after their departure, the Park Service began to remove the collars from bears the Craigheads had studied. They did so, Superintendent Anderson explained, in preparation for the park's centennial. Collars were considered unsightly and every effort was being made to spruce up the bears for the celebration.[63]

With the collars gone, the numbers dispute between the park and the Craigheads could not be settled and no one could gather fertility or longevity figures; no further information on the bears' movements and population dynamics would be available. Yet Cole showed little interest in more research. "Additional research on bears in Yellowstone," he wrote Linn in 1972, "is of lower priority than having adequate funds and men to carry out the present grizzly management program."[64]

Three years into the new era of ecosystems management, science was moving relentlessly in the wrong direction. Now all population statistics were kept by Cole. These consisted of reported sightings by "qualified observers." Yet this system invited double counting and hopeless inaccuracy. Examining these reports, I found that fully a third of the observations did not identify the sex of the bear. Yet if these "qualified observers" could not sex a bear, how could they distinguish separate individuals? As you would expect, this system of reporting produced numbers that, if believed, were reassuring to those worried about the bears.[65]

"Monitored observations of grizzlies in the backcountry of Yellowstone National Park," Reed wrote to Senator Philip A. Hart in 1973, "indicate a very healthy, viable population with an increase this year over last in the number of cubs produced."[66]

"If we are guilty," Reed wrote Congressman James A. Haley, "let the charge read that we are guilty of freeing the grizzly bear from the trap of man's garbage, restoring them to their natural distribution and behavior, and significantly reducing bear injuries to man."[67]

The grizzly bear dispute, meanwhile, had become a public controversy. Closure of the dumps, John Craighead told a meeting held in Mammoth in 1972, was reducing diet for the bear, lowering the reproductive rate, and increasing movement "resulting in increased deaths." The population of grizzlies was "rapidly declining." Five different computer projections, he told his audience, all predicted that the Yellowstone grizzly would become extinct if present trends continued. "The earliest extermination date is 1988," he said. "The latest extermination date is the year 2000."[68]

Lewis Regenstein of Fund for Animals and Stephen Seater, a biologist working for Defenders of Wildlife, joined the controversy on the side of the Craigheads. Regenstein wrote in 1973, "The management program in Yellowstone is an extermination campaign designed, in part, to make the Park 'safe' for campers."[69]

Seater, who remained close to events in Yellowstone during this time, was equally outspoken. "It is sad, but unfortunately true," he wrote to the *Wall Street Journal* in February 1974, "that the National Park Service [NPS] and the Assistant Secretary of the Interior, Nathaniel P. Reed, are guilty of numerous violations of federal law and regulations in their frantic attempts to conceal the truth about the program in Yellowstone."[70]

But Seater paid a price for his opinions, for they made officials at Defenders of Wildlife uncomfortable. "I'm sorry to tell you this," he wrote Regenstein in 1973, "but I'd appreciate having my name removed from the grizzly bear article we coauthored. . . . Miss Harris [Executive Director of Defenders] feels the subject matter is too controversial for Defenders of Wildlife and that it would be best not to have our name associated with it."[71]

Seater's caution, however, apparently came too late. Shortly afterward Defenders fired him — according to Seater, on pressure from Reed. "Secretary Reed and NPS have done everything in their power to silence those who disagree with their management of grizzly bears," he later wrote to the *Wall Street Journal*. "In my case, Mr. Reed lobbied one or more influential Directors of Defenders of Wildlife to 'quiet me down.' . . . He ultimately prevailed and I was fired without so much as a day's notice."[72]

Although Reed denied Seater's charges, the latter's departure from Defenders did not quiet the controversy. Seater was promptly hired by the Fund for Animals, where, together with Regenstein, he continued his role as gadfly of the government. Reacting to this continuing pressure, the Interior Department created, in 1973, the Interagency Grizzly Bear Study Team (IGBST), composed of personnel from the Park Service, the Forest Service, the Fish and Game departments of Montana, Idaho, and Wyoming, and the U.S. Fish and Wildlife Service, and directed by a Park Service employee, Dr. Richard Knight. The team in turn was supervised by a Grizzly Bear Steering Committee, coordinated through Interior Department offices in Washington.[73]

And in February 1973, Interior Secretary Rogers C. B. Morton wrote to Dr. Philip Handler, President of the National Academy of Sciences, requesting that the Academy conduct "an overview investigation of the events and scientific data concerning this continuing controversy," hoping that, once and for all, the matter could be laid to rest.[74]

The Academy in turn created a Committee on the Yellowstone Grizzlies, which in 1974 issued its findings. The committee's report was a nearly complete confirmation of the Craigheads' conclusions. It described Cole's population estimates as having "little if any meaning." It suggested that he exaggerated the number of back-country bears, agreeing with the Craigheads that the dump bears were between 65 and 75 percent of the population. It criticized lax record keeping by the Park Service of both control actions and injury rates, suggesting that both black-bear and grizzly kills may have been higher than reported. It suggested that "uncontrolled sources of bias" in the figures on bear-caused human injuries "make any comparison of injury rates valueless." It asserted that man-caused mortality rates of grizzlies were too high, that by 1974 these were averaging more than thirty a year, and that the population could not, in the long run, endure a man-caused mortality rate greater than ten.[75]

On only one major point did the committee criticize the Craig-heads. The brothers' population projections, the committee averred, had been in error because they did not account for the probable effect of "compensatory mechanisms."

"The rigid characteristics of the model of Craighead et al.," the committee wrote, "seem, a priori, to be unrealistic, because no compensatory mechanisms are built into the model." Presumably, the committee argued, as the population declined a mechanism of self-regulation would be activated, stimulating the birth rate and thus gradually compensating for the high mortality. But the Craig-head projections, "in which a given reproductive parameter is low-ered, will obviously lead to extinction if compensatory mechanisms are not built into the model," the committee explained. Yet, its members were convinced, this result was unlikely. They concluded, "There is little question that increased adult mortality will be offset by compensatory increases in juvenile recruitment."[76]

The Craigheads were wrong, in short, because their model pre-dicted eventual extinction of the grizzly, and this outcome, ac-cording to the committee, thanks to the miraculous biological device of population regulation, could not happen. Once again, as with natural regulation of elk, biology would cover for the mistakes of management.

Included in the committee's recommendations were these: First, "we urge the creation of a nongovernmental coordinating body . . . directed toward the well-being of Yellowstone grizzlies. It must be chaired by a respected neutral individual." Second, repeating the recommendations of the Robbins Committee a decade earlier, "the National Park Service and the U.S. Forest Service [should] pursue a policy of supporting and encouraging independent research on Yellowstone grizzlies. The freedom of scientists to conduct research throughout the Yellowstone ecosystem is imperative if the data essential to successful management of Yellowstone grizzlies are to be obtained."[77]

These recommendations of course conflicted with the policy of government control of scientific work in the park. Putting a grizzly study effort under the supervision of an independent scientist, wrote David A. Watts, Assistant Solicitor for Interior, "would be an improper and unauthorized delegation of authority." Instead of an independent effort, therefore, all research was placed in the hands of the IGBST.[78]

Like the Craigheads, however, Knight and his colleagues were not permitted to do serious research. Park authorities continued to base population estimates on unscientific sightings by "qualified observers." Knight was given an office next to Cole and Meagher in the administration building in Mammoth, and it was made plain to him that he was there to support management. Those working with the team who objected to the arrangement were prohibited from working in the park. Whistle-blowers quickly learned that they paid a price for speaking out.[79]

One such whistle-blower was Robert B. Finley, Jr., now a researcher at the University of Colorado Museum. "I was involved in grizzly bear research 'planning,' from 1972 to 1974, as Chief of the Section of Wildlife Ecology on Public Lands of the Denver Wildlife Research Center," wrote Finley,

> when the Interagency Grizzly Bear Study Team was established by direction of Nathaniel Reed, then Assistant Secretary of Interior, supposedly for the purpose of obtaining factual data to resolve the disagreement. . . . It quickly became apparent that the team leader, Dick Knight, was expected to run the team in a way that gave support to the policies of Yellowstone National Park. Views of the other team members were overruled or ignored. [Team members] were unable to get access to original data on which the National

Park Service claims were based. . . . Dissenting agency biologists
on the study team were excluded from working in the Park. . . .
My efforts to promote conditions for a sound, honest research effort
convinced me that the Department of the Interior was more interested in suppressing controversy than resolving scientific disagree-
ment or saving grizzly bears.[80]

Meanwhile, the Fund for Animals, believing that only law could
protect the grizzly from the Park Service, carried on a campaign
urging that the grizzly be declared, once again, an endangered
species. Officials of the state of Montana, however, which permitted, and wanted to continue, hunting of the grizzly, objected
to federal protection. The idea of endangered status also disturbed
Yellowstone authorities. "None of us believe that there is scientific
evidence or facts indicating that the grizzly is threatened with extinction . . ." wrote Superintendent Anderson in March 1975.
"Plainly it [the grizzly] does not need help in the Yellowstone
area."[81]

Nevertheless a compromise was reached. In 1975 the bear was
declared a "threatened species," a lesser status than "endangered,"
permitting Montana to hunt the grizzly in the Bob Marshall Wilderness, south of Glacier Park, while protecting the species in
Yellowstone.[82]

Once the animal was federally protected, Yellowstone authorities could no longer prohibit tagging and collaring. Knight was at
last permitted to do what he called "real research." And according
to the Endangered Species Act, the Fish and Wildlife Service was
required to identify and study "critical habitat" and to formulate
a recovery plan for the species. Presumably also, the new classification would curtail the aggressive management actions that continued to take such a toll. For, as Dr. Russell W. Peterson, then
Chairman of the Council on Environmental Quality, reminded the
Secretary of the Interior at the time, the law "clearly restricts the
use of regulated taking [that is, killing] to the 'extraordinary case'
where population pressures *cannot* be otherwise relieved." And
because the Yellowstone population showed no signs of such pressure, no case could be made for killing.[83]

Now that he was permitted to tag and trap, however, Knight
could find few bears. By 1980, after five years of searching and a
little more than a decade since viewers could count more than a

hundred grizzlies at the Rabbit Creek dump in an evening, the IGBT had found a total of forty-six animals![84]

Indeed, nearly all the bears that had been alive when the dumps were closed had disappeared. By 1980 Knight had found only six bears more than twelve years old. "That means," one Park Service official confided to me, "that nearly every bear that was alive when the Craigheads worked in the park was killed. One entire generation of bears was eliminated."[85]

John Craighead told me, confirming these findings, that "as late as 1971, twenty-three adult female grizzlies which we had tattooed on the lower lip were still alive. The IGBST, since beginning work in 1974, has been able to find only two of these. This indicates massive predation between 1971 and 1974."[86]

A similar holocaust, Knight discovered, had engulfed black bears. In 1965 Barnes and Bray had made 899 sightings of black bears in one season along the roads of Yellowstone (not ruling out duplication). By 1982 the IGBST, after ten years of flying over the park looking for bears, could report only 391 sightings — an average of 39 (duplicated) sightings a year![87]

Devastated as they had been by the aggressive management of the early 1970s, yet the bears' plight continued to worsen. The population of both grizzlies and blacks clearly was declining. Whereas in 1974 the IGBST saw an average of 1.58 black bears and 2.5 grizzlies on every observation flight, by 1980 the ratio had dropped drastically, to .22 for blacks and 1.16 for grizzlies.[88]

Contrary to the sanguine assumption that compensatory mechanisms would lead to an increase in bear fertility, the team found that fewer cubs were being born. The average life expectancy was declining. Whereas the life expectancy of bears studied by the Craigheads had been more than twenty-five years, now it seemed to be little beyond twelve. "Few bears," the 1980 report observed, "are reaching the older age classes."[89]

A growing imbalance appeared in the sex ratio, a dangerously low number of sows relative to boars. "It is evident," the team stated, "that females are being produced and recruited into the population at a lower rate than males and that they are dying at a higher rate than males." Larry Roop, a biologist on the IGBST, told me that this change was probably a result of control actions. Sows, protective of their cubs, were often seen as nuisance bears, and were therefore more likely to be removed from the population.[90]

"Removals" by authorities continued to be a principal source of bear mortality. Deprived of the dumps, their hunger made many of the bears mean and less afraid of people. Often invading human settlements, they became victims of control actions. Trevor S. Povah, president of the park concession, Hamilton Stores, wrote in 1975 that "we talked to two different groups who had been back in [the Hellroaring backcountry of Yellowstone] for several weeks who stated that they had seen helicopters coming over and dropping off bears . . . they had observed three or four dead bears. . . ."[91]

More bears wandered farther outside of the park, to be killed by state officials, shepherds, ranchers, hunters, and poachers. More maulings occurred outside the park. "Grizzly bears," Roop told me in 1982, "are wandering farther each year. We are finding them in the northern part of the Beartooth Mountains, also near Bozeman — seventy-five air miles from the park. These are places where grizzlies haven't been seen in a hundred years." Indeed, IGBST statistics confirmed Roop's observation. Between 1978 and 1981, the average male grizzly range more than doubled, from 152 to 318 square miles.[92]

The pattern continued to repeat: To keep bears from garbage, the authorities would "sanitize" a community by closing its dump; the bears would then disperse, boldly invading campgrounds and other towns, where, declared to be problems, they would be removed. Grizzlies beat a trail from one town to another, following their noses to find a pit still open. Leaving Rabbit and Trout Creek in 1971, they fled to West Yellowstone; when that landfill was fenced they moved on to Cooke City; after that dump was moved, fenced, and locked, they flocked to Gardiner and back to West Yellowstone.[93]

As they wandered, the number killed outside the park rose dramatically. John Craighead, using Park Service estimates of man-caused mortalities (excluding deaths due to legal hunting, which ended when the grizzly was declared a threatened species in 1975), recently calculated that the number of grizzlies killed outside the park in the decade since the closing of the dumps was two and a half times what it was during the decade prior to the closing of the dumps.[94]

Yet, federal authorities insisted, they were not responsible for the diaspora. "A great many . . . bears are probably killed or captured outside the park," Reed's successor, Bob Herbst, acknowledged in a letter to Senator Richard Stone, "after being lured from within the park by various techniques."[95]

Mr. Herbst's imaginative reply was understandable, however, for he had received strong advice from the Fish and Wildlife Service to tell the public as little as possible about events in Yellowstone. "Due to the sensitive nature of this situation," the Associate Director for Fish and Wildlife Resources wrote to him a week earlier, "FWS does not think we should go into details of this at this point."[96]

Meanwhile, the architects of the government bear policy continued, for the public, to display fresh optimism.

Cole announced at a meeting of biologists in Corvallis, Oregon in 1975 that "a more natural grizzly population was restored as evidenced by spaced distributions of natural foods in summer and a reduced need to control bears to protect humans. . . . My conclusions are that the management program . . . has made measurable progress toward accomplishing its objectives."[97]

"All our information indicates that about as many black bears inhabit the park now as when the park was established," Yellowstone biologists continued to insist (up through 1982), adding that "an estimated 350 [grizzlies] inhabit the Yellowstone ecosystem."[98]

In a paper by Meagher and Jerry Phillips, a Yellowstone resource-management specialist, it was concluded that "the bear management program within Yellowstone National Park during the 1970s had achieved the goal of restoring populations of grizzly and black bears to natural foraging conditions. Concurrently the management program reduced bear-caused injuries to humans in developed areas."[99]

Likewise, it was stated on the bear fact sheet used to train naturalists in 1982 that "the results [of Yellowstone's bear management] thus far have been encouraging — for the first time in almost a century these truly magnificent creatures are living relatively wild and undisturbed lives."[100]

Starker Leopold wrote in 1980, "Current indications are that the wild populations are doing very well indeed."[101]

Former Yellowstone historian Schullery wrote in 1980, "Healthy wild populations of blacks and grizzlies now exist."[102]

Clearly, to Park Service officials, the operation was a success even as the patient was dying. From the time of the Grand Canyon meeting, bear policy had been guided by two goals: protection of bears and protection of people. The new ecosystems management, they had assumed, would accomplish both objectives: Bears, restored to natural foraging conditions, would leave visitors alone. Now it was turning out that these policies, though reasonably effective in protecting people from bears in the park, was failing to protect bears from people. Closing the dumps led to fewer bear maulings in the park because it led to fewer bears. But measuring success by a declining trend in maulings, their bear program would be a complete success when they had eliminated all bears: zero bears, zero bear mortality; zero bears, zero maulings.

National Park Service bear policy had confused management goals (visitors' safety and protection) with scientific objectives (protecting a natural population of bears). And, though bear policy was turning out to be successful management, it was terrible ecology. Protecting people by eliminating a species was hardly enlightened science.

Bureaucratic boundaries also limited responsibility. Park policy was to protect the bear *in the park*, and to reduce human injuries caused by bears *in the park*. But the bear problem had shifted to the surrounding areas, and what happened outside was no longer Park Service responsibility. From the standpoint of park managers at this time, therefore, if their policy led to exportation of the bear problem, it seemed successful.[103]

Despite their public optimism, however, the horrible implications of Knight's discoveries began to dawn on Park Service scientists. Even principal architects of the new policy, such as Leopold and Allen, began to sense that things had not worked out as they had hoped. In 1977, as members of the National Parks Advisory Board, they wrote, "It would appear that progress in bear management is far from satisfactory. We seem to have grossly underestimated the problem. In the past ten years grizzlies have killed more people in the parks than in the previous century. . . . The task ahead is still enormous."[104]

Some senior officials and scientists, living in Washington and seldom on the scene in Yellowstone, began to wonder if park authorities had been telling them the truth. Was this a case, like Vietnam, in which men in the field convinced those in Washington there was "light at the end of the tunnel" by keeping them in the dark? If so, then Starker Leopold's mission-oriented research, wherein scientists served the superintendents, had turned out to be a disaster. The chickens of government biology had come to roost on the wreckage of a bear program.

All along, these officials had believed they had been fighting for a scientific bear policy. But now they wondered. Had they been seeking an ideal of enlightened environmental science, while, in Yellowstone, unknown to them, park managers were aiming to eliminate the bears? Had they been fighting for a policy that all along had been run by and for managers who saw extermination as the only way to solve the bear problem?[105]

"When Dick Knight told me, in 1980, that in five years he had been able to find only forty-six bears," one senior official told me, "I suddenly realized we had been had. . . . The Park Service got what it wanted," the official continued. "They wanted to get rid of the bear, and they succeeded."

Disillusionment spread, as a generation of dedicated professionals quietly tried to forget "this morass," as one put it. "It will continue to haunt me," one participant in the Grand Canyon meeting wrote me recently, "till death does us part."

There is a place in the Yellowstone backcountry near Cache Creek known as Death Gulch. At the bottom of the gulch is a spring that emits a poisonous gas. On windless days the gas saturates the gulch. Anything that visits on these days dies.

One of the first white men to see this place, in 1897, reported finding "a large group of recumbent bears; the nearest to us was lying with his nose between his paws, facing us, and so exactly like a huge dog asleep that it did not seem possible it was the sleep of death."[106]

This place is now marked on maps as Wahb Springs, named after the bear in Ernest Thompson Seton's 1899 story, *Biography of a Grizzly*. This is where, Seton tells us, bears go to die when they know their time has come: "and feeling, as far as a Bear can feel, that he is fallen, defeated, dethroned at last, that he is driven from his ancient range by a Bear too strong for him to face, he turned up the west fork, and the lot was drawn. . . . But as he climbed . . . the west wind brought the odor of Death Gulch, that fearful little valley where everything was dead. He turned aside into the little gulch. . . . The deadly vapors entered in . . . he calmly lay down on the rocky, herbless floor and as gently went to sleep, as he did that day in his mother's arms by the Graybull, long ago."[107]

What was the poison that killed the bears at Wahb Springs? No one knew. In August 1983, I paid a visit to the springs with Wayne Hamilton, a park geologist, to identify this source of death. Wahb Springs, we learned that day, could kill in many ways. A vent at the bottom of the gulch poured forth enormous quantities of hydrogen sulfide, carbonyl sulfide, and a variety of gases known as the mercaptans. All were extremely lethal.

As it happened, we were at the gulch in the same week as Congressional hearings on the grizzly were held in Cody, Wyoming, only a few dozen miles away. Dr. Christopher Servheen, Coordinator of the Grizzly Bear Recovery Program, had told Senators Simpson and Chaffee at the meeting that "there has been a significant downward trend in the number of females with cubs since 1959. . . . This trend indicates that we must be very concerned about the future of the grizzly bear."[108]

The downward trend, of course, did not start until 1968, but the point that Servheen was making had great significance. A government spokesman was publicly admitting for the first time that despite eight years of protecting grizzlies as a threatened species, government policies were still leading to their extinction in Yellowstone.

Indeed, leakage of the Wauer Memorandum in fall 1982 had quickly changed official attitudes about the bear. Nearly overnight, spokesmen had been publicly transformed from optimists to pessimists. Talking about the decline of the bears had become acceptable for bureaucrats.

For the first time in seven years there were renewed public demands to do something. Government agencies and environmental groups rediscovered the problem. The IGBST was reorganized. Interior allotted hundreds of thousands of extra dollars for bear management, and Congress was considering further appropriations. Yellowstone authorities were at work on a new management plan. Subcommittees on population analysis and supplemental feeding had been created. John Craighead, for the first time in a decade, had been asked by officials to participate in a review of policy. Enormous media attention — including a CBS special — was being focused on the problem. Supervisors of the national forests surrounding the park now gave grizzly protection their highest priority. The Audubon Society threw its great influence behind grizzly recovery. A new environmental group, the Greater Yellowstone Coalition, dedicated, in part, to saving the grizzly, was being organized.

Surely now, some hoped, the bear might be saved.

"In my opinion," wrote Hornocker to the editor of the *Wildlifer* a month after the Wauer Memorandum appeared in the press, "the Craighead brothers have been completely vindicated. To me, this is a victory for the grizzlies because now, presumably, the Craigheads' recommendations will be followed."[109]

Hornocker's optimism, however, rested on his belief that we would learn from the past. But what had been learned? There were, after all, many ways, as Hamilton and I discovered that day at Death Gulch, for bears to die in Yellowstone.

12

The Grizzly and the Juggernaut

DURING spring 1984, a sow grizzly and her three cubs walked across the ice of Yellowstone Lake to Frank Island, two miles from shore. While they were there, the ice bridge melted. Stranded on the island, they quickly exhausted their food supply. After subsisting for a while on fish remains, small rodents, vegetation, one elk carcass, and ants, they began to starve.

Rangers spotted the four bears, but authorities decided against a rescue. Instead, they decided, according to a later report, "to let the situation develop naturally."[1]

Naturally the bears were dying. By July, emaciated, the sow, which should have weighed at least 250 pounds, was down to 150. Her cubs, normally thirty-five to forty pounds, had less weight to lose. Two weighed only twenty pounds and one only ten.

At last park authorities decided to move the bears to the mainland, but for the smallest cub it was too late. Weakened by starvation, it died after it was shot with tranquilizer.[2]

Although extreme, the experience of these four animals symbolized the dilemma of Yellowstone bears in the 1980s. For whatever the bureaucrats might do, the fate of the bears remained unchanged. They continued to dwindle in numbers, and it became increasingly evident that they were suffering from lack of food.

Estimating that at least thirty-six adult females were among the population, Knight and Eberhardt, using a mathematical technique known as regression analysis, computed a total population of fewer than two hundred bears in 1982. But this population, they saw, was declining at the rate of about four bears a year. They observed,

"There are various grounds to believe that the population may be declining, and good evidence that the number of adult breeding females is small. The prospect of extirpation thus has to be considered." Although they looked, they could find no sign of the expected compensatory mechanisms: "Long-term stability at such levels requires some sort of inherent regulatory mechanism. Unfortunately, the nature of that mechanism is as yet unknown."[3]

Grim as the official numbers were, though, they may have been overly optimistic. For although the Interagency Grizzly Bear Study Team was estimating a population of about the same size as the one the Craigheads had found in 1967, they were actually finding half as many bears. Whereas the Craigheads had trapped an average of 6.8 adult sows a year, the IGBST was trapping only 3.5. Whereas the Craigheads had identified (but not trapped) an additional 15 a year, authorities now were able to find only 7.4. Could it be that the actual population was half the official estimate? At least, as John Craighead explained to me, "based on these parameters, the official estimates must be seriously questioned."[4]

But whatever their true numbers were, there was little doubt that the nutritional levels were poor and getting worse. The average weight of the animals had dropped drastically. Whereas the spring weight of adult males during the dump era was 543 pounds, by 1982 it was 408. The spring weight of five-year-old females had fallen forty-five pounds in eight years. Mean litter sizes had dropped from 2.2 to 1.9. Maturation of females slowed from five to seven years, thus shortening their reproductive life. The age structure continued to be skewed toward young bears, suggesting a shorter average life span.[5]

Clearly, Knight and Eberhardt commented, the bears missed the dumps. "The population has not recovered from the reductions following closure of garbage dumps in 1970 and 1971, and may continue to decline," they wrote. "These dumps had provided a steady seasonal food supply for bears since the 1920s and undoubtedly exerted a major influence on the distribution and dynamics of the population." The Craigheads, of course, agreed. John Craighead told a meeting of the Audubon Society in 1983 that "the size of the Yellowstone bear population has declined as a direct result of the dump closure."[6]

Officials at last began to wonder: What was the "natural" carrying capacity of the range? "Can the area sustain a grizzly population?" asked the Wildlife Committee of the National Parks Advisory Board in April 1985. "This critical question demands a far more basic holistic view than researchers and managers have given to date. It would appear that the recovery ship has been launched . . . while it has not been established where or if the actual destination port actually exists."

Perhaps no one could ever know how many grizzlies lived in Yellowstone in primeval times, how far they once roamed in search of food, how much they had depended on concentrations of carrion and spawning fish for survival. Perhaps we would never discover all the ways in which Yellowstone's range had declined at human hands. Still, no attempt had been made to find out. Although some studies had measured bears' use of habitat, no systematic work had been done on carrying capacity. Neither historical nor archaeological nor anthropological research had been undertaken to determine the original range of the animals or their relations with the Indians. No one had tried to measure present carrying capacity. And even though required to do so by law, the Fish and Wildlife Service had not yet designated critical grizzly habitat and showed no signs of planning to do so.[7]

"The only way to determine the carrying capacity for the Yellowstone ecosystem," John Craighead said, "is to do a study of the habitat, comparing it with other areas such as Glacier and Alaska, where conditions for bears are known to be good." Indeed, Craighead himself had developed a way to do just this type of study. Called ecosystem habitat analysis, it used remote-sensing multispectral scanners on the LANDSAT satellite to chart zones of vegetation, thus producing a detailed map of the habitat. He had used this system effectively to study the habitat of the western arctic caribou herd in Alaska, as well as for the bear in Montana's Lincoln/Scapegoat Wilderness, among other places. He urged both the IBGST and the Forest Service to do the same for the Yellowstone bear. Yet so far neither agency has taken his suggestion.[8]

And so the question — how many grizzlies can live in Yellowstone in natural conditions — remained unanswered. Seventeen years into the "new" bear policy and no one had tested the premise on which it rested. No one knew whether Yellowstone could support a viable population of grizzlies. Instead, that there was sufficient habitat remained an article of faith among those who wanted to believe the government, and this belief, repeated often by journalists, became firmly fixed as part of the conventional wisdom about Yellowstone bears. The grizzly suffered from neither habitat deterioration nor lack of food, writer Thomas McNamee commented in 1984, in a typical defense of the official policy, explaining, "the let-burn wildfire policy is improving both."

But the growing volume of data collected by the IGBST suggested another view: If bears were losing weight they could not have enough food. Knight and Eberhardt wrote that "the major functional unknown in assessing the prospects of a local extinction is the probable impact of environmental fluctuations. Results of the study thus far have shown that at least one such effect does occur in the form of reduced food supplies in dry years."[9]

Indeed, all the researchers knew that Yellowstone did not offer bears as much palatable vegetation as Glacier Park or Alaska did, and there were reasons to believe, as we have seen, that this habitat was deteriorating. A century of misguided fire-suppression policies and failure of "natural-burn" remedies surely affected the bears. The park was (as we shall see in Chapter 13) constructing a major new development (Grant Village) on the site of five prime grizzly fishing streams. In a few years, Knight told me, the pine-bark beetle, which was already widespread throughout the park, would begin to infest the whitebark pine, whose nut was the principal fall grizzly food. When that happened, he said, the effect on the grizzly would be "devastating."[10]

Some began to suspect that the bears were suffering from competition with the burgeoning population of elk. Researcher Kay wrote in 1984, "Grizzly bears are primarily herbivores that favor succulent vegetation and berries; mesic or riparian habitats at low elevations are critical for bears" — the kinds of places that grizzlies had given up with the settling of the West. Now much of the riparian habitat remaining in Yellowstone was, Kay suggested, being destroyed by the elk. "Overgrazing by an unnaturally high population of elk in the Park," continued Kay, "has greatly reduced berry production of shrubs and completely altered willow and aspen habitats so that they presently have almost no food or cover value for the grizzlies."[11]

Knight, too, began to suspect that perhaps the "natural" carrying capacity was far lower than most had thought. He told me,

The garbage dumps without doubt sustained a larger population of grizzlies than the habitat could support, and there is no question that there are far fewer bears than there were in Yellowstone in 1970. The bears may have reached an equilibrium or they may still be declining — we cannot say for sure. But if they have reached equilibrium, it is at a level far lower than 1970. It may simply be the fact that Yellowstone cannot support the population of grizzlies we had hoped, and this is a consequence that the Endangered Species Act and Grizzly Recovery Plan did not cover or anticipate.[12]

To cope with the problem of nutrition as well as to provide a magnet that might keep the bears in the park longer, the Craigheads continued to urge that feeding centers be established for the bears in remote locations of the Yellowstone backcountry. Such a program of supplemental feeding need not have meant reopening the dumps. It might have entailed, as the brothers had first suggested in 1967, killing a few elk and leaving their remains in a central remote area of the park, to encourage grizzlies to stay there and to give them the protein necessary to raise their reproductive rate and to endure hibernation.

"Man cannot take habitat from the bears," John Craighead explained to me, "dotting it with towns and crisscrossing it with roads and then expect the animal to survive in a land we have helped to impoverish. If we take something from the bear, we must give him something in return."[13]

Concentrating bears in this way, Frank Craighead urged in 1984, "is a natural phenomenon and one that we might again encourage in Yellowstone." To emphasize their naturalness, the brothers called them "ecocenters."[14]

John Craighead explained to the Audubon Society in Casper, Wyoming that

> ecocenters are areas that seasonally attract and hold large aggregations of bears for prolonged periods. They represent large dependable sources of high-calorie food. They should not be confused with rich or extensive sources of pine nuts, berries or forbs that attract relatively small numbers of bears for short periods of time. Here the food is much more limited and dispersed. Bears establish traditional movements to ecocenters and feed on the high nutrient food for extended periods of time: two to three months. Ecocenters can be natural or man-made; the bears don't differentiate. At these

sites they exhibit behavioral traits that have evolved over long periods of evolutionary time. . . . We can speculate that such ecocenters were present in the West before the arrival of the white man. Bear aggregations may have occurred, for instance, at sites where large numbers of bison drowned or were winter-killed at the Great Falls of the Missouri or along its upper reaches. They were no doubt attracted to the Indian piskum or buffalo jumps in and around Yellowstone National Park and elsewhere. Evidence suggests they concentrated at spawning sites of salmon, steelhead or Dolly Varden trout in the Upper Clearwater and Salmon rivers of Idaho.[15]

Such ecocenters, John Craighead concluded, "have many positive values for the bear populations associated with them. They seem to confer a margin of safety that is absent in ecosystems where neither natural or man-made ecocenters exist."[16]

Gradually more of those working closely with the bear began to see the value of the ecocenters. In 1981, the late John Townsley, then superintendent of Yellowstone, an outspoken and powerful voice in the Park Service, suggested supplemental feeding for the grizzly. A grassroots organization established in 1983, the Campaign for Yellowstone Bears, circulated a petition recommending, in part, that if man-caused mortality could not be eliminated then there should be an "attempt to reestablish an ecocenter by providing a dependable food source for the grizzly bear near the geographic center of the ecosystem."[17]

Knight too began to suspect that such a step might be desirable. "If we want the bears," he told me in 1985, "we must consider the options. An option is supplemental feeding." But, he added, "feeding grizzlies is also an admission — an admission that present policies are not working."[18]

It was an admission that many found too painful to make. Government officials and leading environmentalists remained steadfastly opposed to supplemental feeding. They did not want to interfere with nature.

"If we are not going to have wild bear," Durward Allen told me, "we are going to have no bears at all."[19]

"We cannot play God," Meagher told me.[20]

"We don't want to make Yellowstone a zoo," Wauer and Servheen said.[21]

Supplemental feeding doesn't fit with the "national park concept," said Yellowstone bear manager Gary Brown, who added, "We don't believe it is necessary, and it's not part of the natural process we are attempting to allow to take place."[22]

Human refuse, as well as elk killed by rangers, these people agreed, did not belong in wilderness. To make a plea for garbage or supplemental feeding, therefore, was unscientific and an indication that one lacked sound ecological values.

Yet this question had nothing to do with science. It was, one might say, a matter of taste.

The question What is natural? had piqued debate for millennia. It was a metaphysical issue that involved, not facts about the world, but how a culture saw itself in relation to the world. And though the natural was defined by contrasting it with the unnatural, seldom in history had human beings been thought to be unnatural. Primitive societies, such as the American Indian or the early Greek (cultures that the German poet Schiller called "naive"), tended to make no distinction between man and nature. Completely un-self-conscious, they saw the world, including themselves, as one. Rather, alienation from nature occurred only after a culture had matured, and its people began to live in cities, cut off from the country.[23]

In the fourth century B.C., the Greeks, struggling to define their new urban culture, were preoccupied with the question of what was natural and what was conventional. Itinerant teachers who called themselves sophists (or "teachers of wisdom") roamed the Peloponnesus, arguing that what was man-made was conventional and unnatural. But most subsequent Western philosophy was a reaction against this relativism. The philosopher Plato devoted his entire career to an attempt to refute sophism. Nearly everything about humanity, Plato argued, was natural, including the political organization of the state.[24]

Similarly, Aristotle suggested that the natural was anything that possessed the source of its own movement. By his definition not only was man natural, but, by implication, garbage as well, so long as it was composed of or derived from living organisms. The Romans, following the Greeks, believed that anything born was natural. (Our word *nature* comes from the Roman verb *nasci*, meaning "to be born.") Early Catholic church and natural law tradition contrasted the natural with the supernatural: everything on earth that could be discovered by reason was natural, including the laws of society. By the sixteenth century, physical scientists, or natural philosophers, as they were then called, supposed that all things governed by the law of cause and effect were natural. Eighteenth-century philosophers such as Hume and Rousseau argued that anything derived from sentiment (as opposed to reason), was natural.[25]

Not until the nineteenth century did Western societies, increasingly urban, revert to the views of the Sophists and begin to see themselves as entirely alien from nature. At the same time Westerners changed their attitudes toward garbage. "Since human beings have inhabited the earth," wrote historian Martin V. Melosi in his book *Garbage in the Cities*, "they have generated, produced, manufactured, excreted, secreted, discarded, and otherwise disposed of all manner of waste." Yet, Melosi said, only when people began to live in crowded conditions did they come to see their waste as a problem. Nomadic peoples simply left their waste behind them. In primitive cities such as ancient Troy, waste was "left on the floors of homes or simply thrown into the streets."[26]

Believing garbage to be artificial was, therefore, not a scientific hypothesis but a cultural bias. And because philosophers had failed to resolve the question after two thousand years of debate, it was unlikely that bear managers would settle it before the future of the Yellowstone grizzlies was sealed.

Scientific bear management, therefore, would have ignored the word "natural" altogether. If, as urged in the Leopold Report, the purpose of our national parks was to re-create a vignette of primitive America, then the first task of science was to discover what primitive America was like and to learn humanity's part in its early ecology. But as we have seen, this investigating was not done. Although the grizzly's future depended on its resolution, the unanswerable cultural question about the naturalness of garbage remained at the center of the dispute and the answerable scientific questions went unasked.

In this way the grizzly controversy became an ideological dispute. Bureaucrats and environmentalists, sharing a metaphysical ideal, joined in common cause to defend bear policy against the Craigheads and others who urged supplemental feeding. How one stood on the question of garbage became the litmus test of environmentalism.

"The principal argument against supplemental feeding is philosophical," as writer Thomas McNamee put it, "[and] the source of the philosophical obstacle [is] the Leopold Report." Indeed Leopold, as a leading bureaucrat and environmentalist, embodied the new order. Relinquishing their self-appointed post as watchdog, environmentalists formed, with the Park Service, an unholy alliance.[27]

By the 1980s, many of those officials who, while in Interior, had been architects of bear policy, now occupied influential positions in prominent environmental organizations. Besides Leopold himself (who remained an important member of several environmental groups), they included Reed (Chairman of Audubon's Executive Committee, member of the Boone and Crockett Club's Conservation Committee, and a director of the Nature Conservancy); Amos Eno, Reed's assistant at Interior, now on the Audubon staff; Allen (on Audubon's board and adviser to Boone and Crockett); Sigurd Olson (former member of the National Parks Advisory Board and honorary President of the Wilderness Society); Herbst (President of Trout Unlimited); Bohlen (World Wildlife Fund), and former Secretary of Interior Andrus (Audubon).

Nearly the entire early bear bureaucracy, in fact, was safely ensconced atop the masthead of the mainline environmental groups. Endowed by these offices with unassailable ecological authority, some continued to defend policy and others simply refused to criticize. In this way a delicate balance was upset. The watchdogs had joined with the wolves against the sheep. The government, no longer facing resistance from its traditional critics, had no real opposition. And the public, depending on the environmental movement to keep it informed, was instead kept in the dark.

In India, near the town of Puri in Orissa, was a temple to the Hindu god Vishnu. Each year in religious celebration, the faithful dragged a huge wagon carrying an effigy of this god, which they called "Juggernaut," through the streets of the town and to a temple on the outskirts. Occasionally, slipping in the mud or sand, people fell under the wheels and were crushed. Sometimes zealots threw themselves under the wheels as a sacrifice. But these deaths did not stop the procession. The Juggernaut continued forward, no matter who got in its way.[28]

From 1968 on, Park Service bear policy had been a Juggernaut, a religious symbol pushed by zealots, rolling inexorably, crushing those who tried to stop it. The Craigheads, paying a price for principle, had been only its first victims.

Hornocker wrote in 1982 that "the Craigheads stuck to what they believed in and they paid a very dear price. Their research was stopped. A large segment of the profession ostracized them. They were criticized by many. . . . They have been men of principle, and because of this the rewards of their profession often have been denied them. This is a sorry state of affairs."[29]

Banned from research, they sometimes found themselves followed when entering Yellowstone. Government scientists ignored their research and seldom cited their publications in official bibliographies. The Park Service prohibited visitors' centers from selling Frank Craighead's *Track of the Grizzly* and tried to prevent concessionaires from carrying the book as well. The brothers became, as John Craighead said to me, "nonpersons."[30]

"The Craigheads never knew what hit them," a former senior Park Service scientist told me. "They thought the issue was science. But it was not. The issue was management. The Park Service is a large, powerful government bureaucracy. It is a Juggernaut. It goes where it wants. Anyone who gets in its way gets crushed. It wanted to close the dumps and it closed them. So the Craigheads got crushed."

Along the way, others did too: environmentalists Seater and Regenstein, ranger Reynolds, wildlife official Finley, university professors Pengelly, Peek, Jonas, and many others. Once critics, these and others were dubbed, in Park Service parlance, "not credible." Writers, no matter what their qualifications, if they criticized policy were branded yellow journalists. To venture open criticism was to risk ruination of one's career.

As time passed the Juggernaut picked up momentum. In fall 1984 Kay submitted his article on deterioration of bear habitat in Yellowstone to *BioScience,* a publication of the Institute of Biological Sciences. Eager to have his work reviewed by disinterested scholars, he asked the editor that it not be refereed by anyone associated with the grizzly controversy. Instead, the editor submitted the work to Meagher and Servheen. Both recommended against publication and it was rejected.

Kay later asked me, "Does this mean that university scholars must have permission of the Interior Department before they may publish?"[31]

The period following publication of the Wauer memorandum in fall 1982, therefore, had been a time of false hope for those such as Hornocker who had expected to see reversal of policy. The course of the Juggernaut had been set. Sharing an aversion to intervention, both government officials and environmentalists responded with a flurry of activity that, giving the appearance of change, ensured that everything would remain the same.

As the bearer of bad news, Wauer was reassigned to the Great Smoky Mountains National Park. Bear management was put into the hands of higher-level bureaucrats. The Steering Committee was replaced by the Interagency Grizzly Bear Committee (IGBC), chaired by the regional director of the Fish and Wildlife Service and including as members, among others, the regional director of the Park Service as well.[32]

Working closely with Yellowstone bear managers to deflect criticism from government grizzly policy, the National Audubon Society lobbied hard for congressional support of Interior bear management. Audubon President Russell Peterson wrote in 1983 that "Audubon's policy is to actively encourage the IGBC and to build public support for the recovery plan."[33]

Amos Eno, the Society's Director of Wildlife Programs in 1984, wrote that "the National Audubon Society strongly supports the Interagency Grizzly Bear Committee as the appropriate forum for addressing the management problems of the grizzlies in the Yellowstone Ecosystem."[34]

Audubon successfully brought in more funds for continuing policy. In 1984, according to Polly Plaza, Rocky Mountain Regional Representative of Audubon, the Society's efforts resulted in adding "$1.2 million to Fish and Wildlife Service and Forest Service budgets [for grizzly management] for Fiscal 1984." By the end of that fiscal year, thanks to this support, the Departments of Interior and Agriculture had spent $2,759,000 on the bear.[35]

This National Audubon Society involvement perplexed and disturbed some observers. "It makes me sick," Regenstein commented, "to see people like Nat Reed, who helped to exterminate grizzlies in the seventies, now posing as their protector. And no one says anything! Don't people have any memory?"

As influential member of the board of Audubon, Reed had moved from administrator of Interior bear policy to conservationist watchdog over this policy without missing a step. The bear problem, he now discovered, had become the "Knight problem."[36]

As Knight had found so few bears, the steering committee had asked him to embark on a program of "saturation trapping" with the hopes of finding more. But Knight objected, feeling that intensive trapping would only hurt the bears. "I was coming up with the wrong numbers," Knight explained to me. They want me to continue with research until I find more grizzlies. They wanted me to do this because as long as they can say, 'We don't know enough,' they can postpone making any management decisions. But I can't find any more grizzlies, because there aren't any more."[37]

As with the Craigheads, it was once again time to blame the messenger. "Dear Nat," Meagher wrote Reed in August 1982. "As for the Knight problem . . . Dick is obviously committed to one objective, a paycheck. . . . When it comes to research design, he simply goes the line of least resistance."[38]

As Audubon had embraced the grizzly issue as one of its major concerns, even though it too was opposed to saturation trapping, the criticism of Knight from Reed's former subordinate Meagher was disturbing. If his research, bringing such bad news, were not sound, they worried, he was a weak link in the government's recovery efforts.

Reed sent copies of Meagher's letter to some members of the board of directors of Audubon, including Dr. Russell W. Peterson, Leopold, and Allen, marked "Confidential — Your Eyes Only."

"The Knight situation remains clouded," Reed told his Audubon colleagues. ". . . I urge the senior staff to concoct some stratagems for our consideration. For instance: . . . Meet with Chris Servheen and state our problems. We could invite him to Milwaukee [site of the September board meeting], or staff could meet him in Washington. Servheen could be the key man. . . . My point is that when *Audubon* magazine's article ["Breath-Holding in Grizzly Country" by Thomas McNamee] is circulated we may lose a valuable moment unless preliminary work is organized and accomplished."[39]

As Reed had noted, Servheen turned out to be the key man. As Recovery Plan Coordinator, he had become the most prominent spokesman for government grizzly policy. And as manager Servheen's star ascended, biologist Knight's waned. Knight had become another victim of the Juggernaut. Efforts were made to replace him, and he and his assistants failed to get expected promotions. Meanwhile, Servheen attempted to gain more control of research data that might prove embarrassing. According to the minutes of the August 11, 1983, Grizzly Bear Research Subcommittee: "Servheen said that considering that this raw data was fairly sensitive, it would be most useful and most secure to have the data analysis under Servheen's control."[40]

Security was important because, as in 1968, the issue was data. Because the data contained little good news, leaks to the press proved embarrassing to management. Shortly after the Wauer memorandum appeared, therefore, authorities embarked on a program of damage control: they created two new committees.

In January 1983, the IGBST formed "the ad hoc subcommittee on population analysis." This effort, one of its members explained to me, was "a quick and dirty way to get a handle on grizzly numbers."

The committee quickly concluded that grizzlies were not declining. It announced in a May press release that "no significant trend is indicated for 1974–1980."[41]

This conclusion was in error, of course — a fact that officials discovered shortly after it was released to the press. The report was calculated, Knight pointed out later, "on the wrong data." But not until December was it publicly conceded by the IGBC.

"Population Trend should be changed from 'No significant trend is indicated for 1974–1980' " stated the *erratum,* "to 'A significant downward trend is indicated for 1974–1980.' "[42]

A month later, the Department of the Interior created another "ad hoc" committee, this one to consider supplemental feeding. It was to be chaired by Knight, and John Craighead was asked to join as well — the first such invitation he had received in a decade.[43]

The committee's report confirmed that since dump closure the grizzly had suffered from lack of food. Spring weights of five-year-old males, for instance, had fallen from 500 pounds in 1971 to 312 by 1982; females of the same age fell from 271 to 225. Adult weights had also dropped dramatically. "The nutritional level of the population may be lower at present," the committee observed, "than when bears were feeding in garbage dumps. . . . It seems apparent that those captured during the 1959–1970 period averaged larger and heavier than those captured during the 1975–1982 period."[44]

Mean litter size, the committee also said, was shrinking while the reproductive life of females continued to decline. "The differences in population characteristics indicate that the Yellowstone grizzly population was on a higher nutritional level when it was being supplementally fed than afterwards."[45]

The committee also recognized the value of ecocenters in concentrating bears in the park. Annual man-caused bear mortalities had risen, it said, from 4.18 a year in the dump era to 7.8 today, mostly because bears were wandering more. "Supplemental feeding concentrates a population in time and space and this can be highly protective."[46]

Nevertheless the committee refused to recommend supplemental feeding. Killing elk to feed bear, the committee explained, "would mean cutting the present herd approximately in half," and would make the elk "more secretive than at present, seriously damaging their value to park visitors. . . ."[47]

The report concluded, "The population should be able to sustain itself at the present level of nutrition using natural foods in the environment if human-induced mortality is limited. Therefore, supplemental feeding is not necessary from a nutritional standpoint for the Yellowstone population."[48]

Carefully adding two and two, the committee had reached the sum of three. While conceding all the arguments in favor of supplemental feeding, it had nevertheless refused to endorse the obvious conclusion.

"The supplemental feeding committee was a farce," one official who was close to the committee deliberations told me. A majority, including Meagher and Brown from Yellowstone, had long been on record as opposed to supplemental feeding, and stated their opposition early in the proceedings.[49]

John Craighead, disturbed by the appearance that he had been co-opted, agreed. He told me, "The committee did not consider supplemental feeding seriously. I had the choice of quitting or staying on with the hope that the report would at least contain some meaningful data. I chose the latter. If the public knew bears were stressed nutritionally and were declining, I had hoped it would see the value of supplemental feeding, no matter what the committee recommended."[50]

The public, however, had little opportunity to make up its own mind. Only the conclusion of the report made the newspapers, not the supporting data. And the conclusion was quickly embraced by

environmentalists. The committee, wrote Hank Fischer, representative of Defenders of Wildlife, "concluded that supplemental
feeding is not necessary nutritionally. In short, grizzlies are not
starving to death."[51]

"John Craighead signed on to the IGBC ad hoc committee report," Eno said: "The supplemental feeding question is now twelve
years after the fact. The dumps are now closed and the Yellowstone
grizzly population is now a long way toward adjusting to that fact."[52]

With the bothersome issues of population decline, starvation,
and supplemental feeding now settled, bear policy was once again
safe from its critics. Continuing the policies of "protecting against,
removing, or compensating for the activities of man," Yellowstone
authorities responded to the new bear crisis by closing portions of
the backcountry to human travel. And although this move would
reduce somewhat the possibility of further encounters, it was not
clear just how much it would help the bear. For in its search for
sustenance the grizzly spent less time in the park.[53]

Instead, more wandered out to be killed near human settlements.
Black-bear hunters in Wyoming baited bears to their blinds and
sometimes killed grizzlies "by mistake." The Forest Service continued its practice of leasing land for sheep grazing around the
edges of the park, where shepherds sometimes shot grizzlies. People who lived in border communities also occasionally shot hungry
bears visiting town dumps. Poachers killed bears that wandered
out of the park, so long as a carcass could bring $15,000 when sold
in parts. (The pancreas, for instance, is prized in the Orient as an
aphrodisiac.) Grizzly habitat around the park was being nibbled
to death by geothermal development, oil and gas exploration, construction of new ski resorts and summer homes, grazing, and increased backcountry use by hikers and sportsmen.[54]

As grizzlies wandered into these places where people killed them,
bear managers and environmentalists more and more frantically
sought ways of keeping man and beast apart. Somehow, they thought,
they must stem the tide of mayhem surrounding the park.

Management continued its aggressive control actions and accelerated efforts to "sanitize" the gateway communities and national
forests surrounding the park. Various agencies took steps to reduce
human predation and to protect wildlife habitat along park boundaries. The Fish and Wildlife Service lobbied with some success for

stiffer legal penalties for poachers and better sanitation in border communities. Wyoming prohibited black-bear hunting in its Jackson district and bearbaiting in the Cody district.[55]

The Audubon Society threatened to sue Park County, Montana, if it did not fence the Gardiner dump. Claiming that poachers were a major threat to the bears, the Society, in a widely publicized press release in fall 1982, also offered a $10,000 reward for information leading to the arrest and conviction of anyone who killed a grizzly illegally. Calling poaching "a scandalous situation," Reed, in announcing the bounty, laid blame for the bear's fate squarely on these illegal hunters. When he left office in 1976, he said, "we thought we could see recovery in the grizzly population. One thing we did not take into consideration was hunting." Now, he added, "it may be too late to save the Yellowstone grizzlies, but I think we ought to make one hell of a try."[56]

The Wilderness Society lobbied for passage of the National Parks Preservation Act, designed to give more protection to lands surrounding national parks. The Northwest Section of the Wildlife Society, whose president, Fred Bunnell, had been a member of the ad hoc Committee for Population Analysis, announced that it "supports and encourages the continued efforts of the Interagency Grizzly Bear Committee."[57]

A new organization calling itself the Greater Yellowstone Coalition (GYC), established in 1983, called for protection of the "greater Yellowstone ecosystem" (an area that included both the park and the surrounding national forests) in part to save the bear. "We believe the top priority should be to maintain and preserve wildlife habitat," GYC president Rick Reese, a former ranger, told the *New York Times*.[58]

The plight of the grizzly soon became, for environmentalists, a symbolic issue for preservation of wilderness. Protecting habitat was essential, they rightly declared, for the bear's survival. Yet as the bear's distress accrued value as a political argument for preserving and perhaps expanding wilderness, the grizzly became a hostage in the war over land use. "We just hope the grizzly can hold out," one member of the board of directors of the Audubon Society told me, "until we can secure the habitat."[59]

But it was not clear that the grizzlies could hold out. Important as efforts to reduce human predation and to protect wildlife habitat along park boundaries might have been, these actions alone would not save the bears. For although overlooked by those who sought to place all blame on poachers and hunters, the greatest threat to the grizzly lay elsewhere.

"Now, 9:00 P.M. 8.21.84 I sit at my employee cabin, Lake Station," wrote Frank Phillips, Chief of the National Park Service's Exhibit Production, and his wife Elizabeth, to the Cody (Wyoming) *Enterprise*. "We heard the report of a large rifle in the Bridge Bay Campground area. Almost every evening large rifle shots are heard in bear habitat. What is going on here? . . ."[60]

"Thursday 8.23.84," they continued, "a yearling grizzly strolled by an employee campground near Fishing Bridge. The bear was digging roots and eating dandelions as he ambled by. He also sniffed at some laundry hanging on a clothesline, before moving on. Someone called the rangers and within minutes, rangers Gary Brown, Tim Blank, John Osgood, Patricia Murphy, and three other unknown personnel arrived on the scene armed with rifle and shotguns to track the innocent bears down. Is this bear management? . . . If you are serious about the welfare of bears and your national constituency, you will impound all bear traps and ranger firearms. Then, observe the natural return of bears to Yellowstone Park."

Others too began to wonder. The Wildlife Committee of the National Park System Advisory Board reported, "From 1980 through September 1984, there have been 41 known human-caused grizzly deaths in the Yellowstone region. . . . In examining these categories of 'causes,' one of the largest ones is 'controlled action' [13 mortalities]. When this source is coupled with research mortalities [6] the most prominent cause of recent mortalities seems to be at the hands of resource managers."[61]

More than a decade after the dumps were closed, managers were still the biggest killer, and the bears showed no signs of adjusting to the loss of ecocenters.

By 1983 Cooke City, under pressure from authorities and the Audubon Society, had bearproofed its dumpsters and installed a chain-link fence around its landfill. Once the area was "sanitized," the bears moved on. Some traveled to Gardiner. One of these was sow number 10, a habitué of the Cooke City dump, and her three cubs. In May 1983, she was caught in a snare at the Gardiner dump and transferred to British Columbia, which permitted hunting of grizzlies. Her cubs were released inside Yellowstone. Less than a month later, two of her cubs returned to the dump, where they were captured again. One, caught by the hind foot, became tangled in barbed wire as well. Suffering a broken foot, it was destroyed by authorities. The other was again released. One month later it too was killed by bear managers.[62]

Similar steps were taken to sanitize West Yellowstone. This town, having had a bear problem ever since the Rabbit Creek dump near Old Faithful was shut down, had finally solved the problem — or so authorities thought — by trucking the garbage to Ennis, seventy miles away. But once the dump was gone, bears began to come into town and campgrounds more often. One of these was a boar, number 15, nicknamed Dum-Dum.

Dum-Dum was a model bear. Captured at least fifteen times by researchers, they called him trap-happy: quite willing to be caught in exchange for food. Often he came to feed at the dumpster at the West Yellowstone airport. There authorities managed, by the end of summer 1982, to wean Dum-Dum of this bad habit by mixing ammonia with the garbage. The next spring, having lost 173 pounds over hibernation, Dum-Dum did not return to the dumpster. Instead, in the early morning hours on June 25, he visited the Rainbow Point campground near West Yellowstone. Ripping open a tent, he dragged a young man named William May from his sleeping bag and began to eat him. When May's body was found the next day, around 200 feet from the tent site, approximately seventy pounds of the victim had been consumed or lost as fluid.[63]

The tragedies and removals continued throughout the summer. By August sow number 38, with two cubs, moved on to sheep rangeland in the Two Top Mountain area of Idaho just west of the park, killing at least four sheep. Trying to drive them back into the park, authorities fired a propane cannon at them every two minutes, sounded a whistle gun, fired rifles into the air, and flashed car headlights. Finally trapped, they were transported by helicopter to Mount Washburn in the park. A year later she was killed by an overdose of tranquilizer.[64]

A few weeks later another sow with cubs wandered into West Yellowstone, where they were trapped. The cubs were transported into the park. But as they were lifted by helicopter, the cable holding their trap detached and fell to the ground, killing one cub.

The second cub, only a few months old, was released alone in the park.[65]

The sow was sent to Professor Jonkel in Missoula for behavioral research, while authorities made plans to kill her. Outraged, the residents of West Yellowstone pleaded for her life. What, they wondered, had this bear done wrong? The sow, although guilty of having wandered through a campground, had caused no trouble. "Everybody in West Yellowstone is hotter'n hell about it," said Elizabeth Bailey, owner of the Airport Cafe. "They just kill the mother and dump the cubs. The cubs aren't big enough to fend for themselves."[66]

"We want a stay of execution until they can decide if the bears belong to the people or to the god-dang bureaucracy," said Walker Cross, a local motel owner. After persistent opposition, the sow's life was spared. She was sent to a zoo in Great Bend, Kansas.[67]

When the 1983 season had ended, six more grizzlies were dead, including three females. All were victims of management-control actions.[68]

By 1984 it seemed Yellowstone was caught in a time warp. History was moving in tight circles, the same events repeating themselves. Reminiscent of the period immediately after dump closure, grizzlies seemed to be everywhere. They came into the campgrounds and numbers were seen along the road. More people were mauled, and once again tragedy struck. On August 1, the body of Brigitta Fredenhagen, a young Swiss woman backpacking alone near White Lake in the park, was found near her campsite, eaten by a bear. Only forty pounds of her flesh remained. In early August a grizzly entered the Grant Village campground at night, ripping open the tent of a sleeping twelve-year-old boy, mauling him. Also in August, a grizzly attacked a seasonal park naturalist and her husband while the couple were walking in the Hayden Valley. There were many other, less well publicized, close encounters. Trail crews working in the backcountry sometimes found themselves chased up a tree by an angry bear.[69]

Their aggressiveness made the bears vulnerable to management control. In July, a young male died within two weeks of being transported by park managers, perhaps of an aftereffect of the drug. In August, bear number 93, the last surviving cub of number

10, was captured in the Indian Creek campground in Yellowstone and transported to a zoo. In the same month a seven-year-old male, number 88, was captured at Fishing Bridge and killed by authorities. In September, authorities accidentally killed thirteen-year-old sow number 38 with a drug overdose. In October, a yearling male was trapped in a snare by the neck in West Yellowstone and died. Shortly afterward, another young male was trapped near the community of Lake in Yellowstone and sent to a zoo.[70]

By the end of the summer another human life had been lost and ten more bears had died (six while in the custody of management), four of them females.[71]

Yet, as in the 1970s, some bear managers, searching for a silver lining to this dark cloud, saw the bear activity as a good sign. Shoshone National Forest Supervisor Steven Mealey told a September 1984 meeting of environmentalists and government officials at the Brinkerhof retreat near Jackson, that "things are better for bears than ever before." In Yellowstone, Mealey suggested, where vegetation is scarce and elk prolific, the bear, in the past predominantly vegetarian, has become a major carnivore. "The grizzly is now back to eating elk," he said. "In 1974 succulent green plants were the main diet supplemented in the spring by meat. Now meat has become a large diet source. . . ."[72]

"If we don't have more bears now," he concluded, "we soon will have."[73]

If his idea was correct, then in one fell swoop Mealey had solved both elk and bear problems. Now the bear, once a starving vegetarian, was on his way to becoming a fat carnivore at the expense of the overabundant elk. But he was not. Although grizzlies did eat winter-killed elk in the spring and bulls weakened by the rut in fall, their overall use of ungulates was very small. Between 1977 and 1981, scat analysis found that elk comprised only 5.4 percent of the average yearly grizzly diet. And though during some falls when pine nuts (the principal food) had been scarce, use of elk had risen above the average for this time of year, use of carrion had more dramatically *declined* in the spring. In 1979, 1981, and 1982, Knight and his colleagues found "very little spring consumption of ungulates by grizzly bears."[74]

John Craighead's examination of elk and bison carcasses confirmed this drop in spring use of elk. He wrote recently that during the 1970s the carcass utilization patterns began to change. Use by bears declined until, during the 1980s, relatively few carcasses were being scavenged, even when severe winters caused heavy die-backs in the bison and elk herds. During 1971 and 1972, 36 of 41 elk (87.8 percent) and 6 of 7 bison (85.7 percent) carcasses examined had been utilized by grizzly bears. In 1980 and 1981, none of 9 elk and 9 bison carcasses had been used and in 1982, none of 83 elk carcasses and only 3 of 35 bison carcasses (8.6 percent) had been used by bears.[75]

Indeed, most bear managers conceded, the bears were neither increasing in numbers nor enjoying a more robust diet. They were more visible, and were encountering people more frequently, Servheen explained to the Casper (Wyoming) *Star-Tribune,* because they were wandering more than usual in search of food. It had been another bad year for "natural foods," Servheen reported, especially for the whitebark pine nut. "It's a bad fall for food," Roop told Buzzy Hassrick of the *Star-Tribune.*[76]

Nevertheless, as the maulings and removals continued, both the public and the authorities grew more worried. Following the death of William May the conviction had been growing among many that the grizzly was too aggressive.

Bear number 15 (which killed May) "was one of the most docile bears I ever knew about," said Bart Schleyer, a member of the IGBST who had trapped him many times. What had caused this change in behavior? "Bears don't do what bear number 15 did, or we would have to kill them all," commented University of Montana grizzly researcher Professor Charles Jonkel.[77]

Indeed, said Jonkel, when bears kill, they usually do so from surprise. They do not, he suggested, kill people to eat. It could be that "the long series of captures, druggings, and disturbances could have affected his behavior." The bear had been tranquilized many times with Sernylan, whose street name is angel dust. Another drug commonly used on grizzlies, M-99 Etorphine, was up to 10,000 times more powerful than morphine. Could either of these substances affect bears as they affected people? Jonkel in fact doubted this thesis, but wanted to study the bear in order, at least, to rule it out. But instead authorities quickly killed the bear. This act, Jonkel thought, was irresponsible. "If the authorities get sued over this, it would not surprise me," he said.[78]

Others began to suspect that hunger might be encouraging the grizzly to prey on people. "A grizzly that is having difficulty in normal feeding," writes biologist Stephen Herrero, in his recent *Bear Attacks,* "either because of old age or some other reason, may try to prey on people."[79]

Whether angel dust, M-99, or hunger was the culprit, some authorities were growing suspicious that too much trapping and tranquilizing hurt the bear and perhaps made him more aggressive. Hank Fabich, a Montana game warden and veteran of ten years of trapping bears, told me in 1982 that he believed trapping was the greatest source of man-caused bear mortality:

> Every time you catch a bear you risk killing it. Sometimes you overdose it. Other times the bear goes berserk and breaks a leg or tears out its claws or teeth in the cage. I have found a couple with useless limbs because they had been hobbled by park authorities and released with the hobbles on. How do you think bears without teeth or claws or good legs can make it? They can't hunt well. So they go into human settlements looking for easy food. They are hungry and mean — they have a reason not to like us. Then they are killed by administrative removals because they are nuisance bears. Others are never listed as being killed by humans. They just die of starvation. I've seen bears — young bears that should have been perfectly healthy — dying in their sleep, in hibernation. Why did they die? Well, some were missing teeth, others claws. One we found with a broken foot. I love the bear and I hate to see this happening. But you can't convince me that we are not doing this to the bear.[80]

Rather than restrict control actions or research, however, officials sought more direct ways of countering the bear's apparent boldness. The state of Montana pressed for the right to hunt the animal. Surely hunters, its spokesmen reasoned, would make the bear afraid of man! "I think we could solve our grizzly bear problem if we could hunt two or three a year," remarked Don Bianchi, public information officer for the Montana State Department of Fish, Wildlife and Parks.[81]

The IGBC, however, had other ideas. Perhaps, its members believed, the bear could be taught to get along with people. They would embark on a program of behavior modification! At a closed meeting during fall 1984, therefore, they decided to begin a series of "aversion" experiments. If bears could be made to associate bad experiences with people, it was theorized, they would avoid such encounters. To find out if this could be done, they asked the Wyoming Department of Game and Fish to experiment by attacking the bear with nonlethal weapons: rubber bullets, electrical shocks, and various types of repellents. By 1985 the program was off and running. "What we are looking for is a non-lethal means that will keep bears out of camps and away from people," reported Dale Strickland, assistant chief game warden and supervisor of the program. Soon the IGBC received the support of environmentalists. "Research on aversive conditioning of bears," stated a 1985 Greater Yellowstone Coalition position statement, "should continue."[82]

Despite attempts to keep human being and bear apart, wildlife managers were driving them together. Despite a desire not to interfere with nature, they were trying to change the bear's behavior, using helicopters, angel dust, propane cannons, rubber bullets, and electric shock. Despite reverence for ecology, they were trucking tons of garbage more than a hundred miles (one way) to Livingston or Ennis to be incinerated, consuming thousands of gallons of nonrenewable fossil fuels (more than $70,000 worth a year), and contributing to air pollution.[83]

In short, despite a desire not to play God, they were playing God.

Yet no one wanted to be first to admit the need to feed the bears. For the Park Service, reopening the dumps or shooting elk would be to admit that a system of official government biology had produced years of deception; that hundreds of bears had died in vain. To environmentalists, it would be a confession that a beautiful ideal did not work, and, worse, was responsible for nearly eliminating a species; that prominent individuals, still influential in the Park Service and the environmental movement, were dreadfully mistaken.

Many still did not accept, or did not want to accept, Knight's numbers. Some were timid, not wanting to be the first to favor a radical departure from policy. Others suggested that supplemental feeding would not keep bears in the park a significant length of time, or, they said, it was too late to reverse the migration. Still others argued that favoring one animal over another — killing elk to save bears — could destabilize both populations.[84]

They did not see that park policy was killing bears by driving them into human settlements. They did not realize that no matter how strong the efforts to protect the bears outside might be, if policy remained unchanged and grizzlies continued to wander, they would continue to die.

It was much easier to chase red herrings. Confusing symptoms with cause, they continued to seek the source of decline outside the park rather than in it.

And so the Juggernaut rolled on. As an alliance (with a growing budget) of game officials from three states, land and wildlife managers from the Departments of Interior and Agriculture, biologists from within and outside of government, and influential environmental organizations, it had picked up too much momentum to stop.

And as bears continued to lose weight, as they wandered and were killed, as bear managers followed them with helicopter, snare, and dart gun, as officials continued to tell the public that the policy was working while warning that the bear was in trouble, as blame for the animal's demise continued to be placed on everyone but those who ran the park, a mystery remained.

The architects of this policy — both government officials and environmentalists — were neither evil nor mean-spirited. They were all decent people, well intentioned and devoted to wildlife conservation. It seemed nearly inconceivable, therefore, that the spectacle we have witnessed — decline of the range, disappearance of species, decimation of bears, in short, destruction of Yellowstone — could happen when so many people were dedicated to its preservation.

Yet a simple fact remained. Since the Park Service had assumed authority in 1916, not a year had passed that officials did not kill an animal in Yellowstone. Wolf, cougar, lynx, bobcat, wolverine, fox, marten, fisher, pelican, coyote, elk, bison, antelope, beaver, ground squirrel, mole, rat, mouse — and now black and grizzly bear — had, at one time or another, fallen victim to a program of removal. And each had been dispatched in the name of an environmental ideal.

How could the National Park Service — an agency with a billion-dollar budget whose mandate, after all, was "to conserve the scenery and the natural and historic objects and wildlife" within our parks — permit Yellowstone to decline?

How could environmentalism, the movement of Thoreau, Muir, Aldo Leopold, and Rachel Carson, dedicated to the highest ideals of wildlife preservation, remain silent as it witnessed the slow decay of our natural heritage? How could any group dedicated to conservation actually become an unwitting partner with the government in promoting practices that were bringing about this decline?

It is the answer to these questions that we shall explore next.

The Rangers

I was to learn later in life that we tend to meet any
new situation by reorganizing . . . and a wonderful
method it can be for creating the illusion of progress
while producing inefficiency and demoralization.

— Petronius (died A.D. 66)

13

Grant Village and the Politics of Tourism

ALONG the south shore of Thumb Bay on Yellowstone Lake, in a clearing cut in a lodgepole forest, stand the first buildings of the intended model wilderness development known as Grant Village. Now half built, it will be, when completed, a summer community of 3,000 people, with 700 motel units, several restaurants, marina, dormitory, and trailer park for concession employees, housing for Park Service staff, information center, overnight rest areas, employee "staffeteria," recreation area, audiovisual center, pub, dancing area, laundromat, playground, Young Adult Conservation Corps facilities, fishing tackle store, gift shop, filling station, post office, maintenance area, several parking lots, and sewage- and water-treatment plants. The entire project, when completed, will cost more than $60 million.[1]

The village, under development for nearly fifty years yet languishing for decades, is now gathering momentum. Since 1981 millions have been spent on construction, as new buildings spring up amidst relics of earlier construction. A gigantic concrete breakwater and marina, designed for motor launches and houseboats, more than twenty-five years old and never used, juts into the bay, its foundation subverted by erosion. A complex of square prefabricated plywood multistory motel units stand by themselves in a raw clearing. A sewage-treatment plant, two decades old, lies next to the lake shore. The accommodations center, eaveless and entirely covered with shingles, towers next to the recently constructed but more conservatively designed Hamilton Store. The visitors' center, built in an earlier era, sits by itself, facing apparently in no

particular direction. The din of jackhammers, draglines, bulldoz-
ers, cement trucks, and other construction equipment reverberates
through the clearing.

Called a "wilderness threshold community" by the Park Service,
something is vaguely disturbing about Grant Village. The line of
trees demarking the clearing is too abrupt, a boundary obviously
man-made. The dark canopy of surrounding lodgepole looms like
a huge fence, making visitors feel cordoned off from the natural
beauty they came to see, and focusing their attention inward on
the clearing. Yet the nearly windowless architecture of the buildings — a curious mixture of Cape Cod and Star Wars — gives a
sterile and futuristic impression. They are too large, too plain, too
square to fit within the irregular lines of a natural setting. The
perspective of the open spaces is askew and unsettling to the eye.
The whole arrangement lacks focus. Something is missing.

The something that is missing is visual evidence of why Grant
Village is there at all. It has no apparent reason to be. Nearly every
human settlement since time began has been built around a natural
feature that announces its intent — a cave for protection, a harbor
for commerce, a mountaintop for defense, a river for communication. These structures express our ancestors' efforts to blend with
nature. Even in Yellowstone other communities have followed this
rule. West Thumb and Old Faithful lie next to the hot springs and
geysers many come to see. Fishing Bridge sits at the outlet to
Yellowstone Lake, a natural junction near the center of the park
and a perfect spot for fishing. The village of Lake lies on a point
of land that affords perhaps the best view anywhere of the Absaroka Mountains. Tower Junction is built where the Bannocks
once forded the Yellowstone and where later Jack Baronett built
a bridge.[2]

But no such natural purpose is apparent at Grant. Rather it
seems contrived, as though it were plunked down in that lodgepole
forest for no compelling reason. The effect is of something strangely
urban: like an inner-city project in the heart of primitive America,
a wilderness ghetto.

Grant Village seems artificial because it was intended to be so.
The location was chosen precisely because it was thought to contain
little of natural value. It is meant to be a place where visitors can
stay without harming the environment, where humanity and nature

can be kept apart. This artificial community was put there, in short, to implement the policy of natural regulation.

Even beyond the irony of its stated purpose, however, the village's reason for being remains obscure. During its fifty-year gestation, at every stage of planning, various government agencies and officials have tried to stop the project. Yet, like a schoolchild who flunks every test but is passed along to the next grade, Grant Village, despite consistent opposition over the decades, has continued to grow. The Fish and Wildlife Service, realizing that construction would badly hurt grizzly and trout, urged that the Park Service reconsider the need for it. Park Service accountants and planners argued that it would lose money and was not needed. Archaeologists pointed out that it would destroy important prehistorical sites. Many Park Service biologists privately opposed it. Rangers objected that it put the Service in the motel business, conflicting with their mission to preserve the park. The federal Office of Management and Budget denied the Interior Department funds for the project. Several presidents — including both Carter and Reagan — tried to stop it. For twenty years park concessionaires refused to build there. Seven times, in fact, Grant Village had been stopped. Yet today it rolls along unopposed.

Grant Village is the project that would not die. Yet to most people in the Park Service its resilience remains a mystery. A high-ranking Park Service officer asked me, "Why has Grant Village gone ahead when everyone seems against it? Where is the public outcry? Who are its friends?" Why, where, who, indeed.

Who built Grant Village?

Shortly after World War II, according to ranger lore, backpackers hiking in a remote part of Rocky Mountain National Park were startled when a naked lady ran across the trail in front of them and disappeared into a thicket of alders. Skinny-dippers are not unusual in our parks, and this would have seemed just another of those moments when someone who thought herself alone was surprised to find she was not. But this was not one of those moments.[3]

Not long after, another group of hikers ran into this same apparition, skipping across the trail in the altogether. Soon another group saw her, then another, and another. The story spread. Just as accounts of Bigfoot and Yeti seized the popular imagination, so too did this mysterious nature woman of Rocky Mountain. The local papers did stories on her; the wire services picked them up. Shortly the entire country knew the tale. She was quickly dubbed "Eve" by the press, and visitors flocked to the park in hopes of gaining a glimpse of her. The game went on until a young man arrived in Estes Park from Chicago, doffed his clothes, and headed for the high country. A voice had told him to trek west and play Adam to the elusive Eve.

With the arrival of Adam the jig was up for Eve. She was, it turned out, in the backcountry for a reason. The local rangers had staged the whole thing as a stunt to promote visitation to the park. The flimflam had worked, but now, with a naked man searching for her, and who knows how many more self-appointed Tarzans or Adams or voyeurs on their way to the trailheads, it was time to call her in from the cold.

Although rangers at Yellowstone were neither as imaginative nor as bold as those at Rocky Mountain, they shared a problem. By the end of the war the Park Service was in trouble. Visitation to national parks had dropped dramatically during the war (at Yellowstone, falling from 581,761 in 1941 to 64,144 in 1943). As visitation dropped, so too did revenues, and park facilities fell into disrepair. Yet although visitation rose again in 1946, popularity did not prime the money pump. In the meantime Congress had lost interest in the park system. Appropriations for the Park Service were a million dollars less in 1955 than in 1940, even though visitation had more than tripled during the same time.[4]

The problem was that the Park Service needed better friends in Congress. It had just lost a powerful political constituency and had not yet found a new one.

For seventy-five years the park system had been the beneficiary of an unusual political alliance between railroads and preservationists. The railroads saw national parks as good for business, attracting patrons West by train. They began early to promote the national-park idea and helped to lobby for bills creating Yellowstone and other parks, pushing them through Congress. Yellowstone in particular (as we shall see in a later chapter) was a child of the railroads. And after its creation, Yellowstone Park became the magnet attracting patrons on the Northern Pacific (which served the gateway community of Gardiner, Montana); the Burlington (which brought Yellowstone visitors to Cody, Wyoming); the Union

Pacific (to West Yellowstone); and the Milwaukee (to the Gallatin gateway, Montana). The first Yellowstone concessionaires and hotels were funded with railroad money.[5]

To preservationists, an alliance with the railroads made sense. These corporations were powerful allies, and yet their interests in parks were relatively benign. Visitors coming to parks by train did little damage to the natural scene. They relied on public transportation to get to the park, and once there, explored on foot or horseback. Ecologically speaking, railroad visitors were harmless, or so it seemed, and politically they were a necessity. For without them the preservationists had no constituency. They needed to convert people to their cause by showing them the grandeur of the unspoiled West. "Even the scenery habit in its most artificial forms," wrote John Muir in 1898, "mixed with spectacles, silliness and kodaks; its devotees arrayed more gorgeously than scarlet tanagers, frightening the wild game with red umbrellas — even this is encouraging, and may well be regarded as a hopeful sign of the times."[6]

The railroads were at that time the only way to get the public to the parks, and so as the national-park idea grew in the minds of both preservationists and trainmen, this alliance grew as well. A coalition of these interests successfully lobbied the National Park Service bill through Congress in 1916, convincing the Service's first director, Stephen Mather, that the new service could not survive without the political support of both. Consequently he set a course with a twofold agenda, one preservation, the other tourism. "Our national parks," Mather wrote in 1915, "are practically lying fallow, and only await proper development to bring them into their own."[7]

Mather and his successors were aware that these agendas might sometimes be inconsistent; indeed, that the twin goals of preservation and providing for the "benefit and enjoyment of the people" were inherent strains on the institutional mission soon became a cliché in the Park Service, and remain so today. Great numbers of visitors, park managers soon realized, posed a threat to nature.

The problem, as they saw it, was that tens of thousands of visitors, picking flowers, pocketing arrowheads, purloining discarded elk antlers, trampling grass, defecating in the backcountry, scaring waterfowl, scarring fragile thermal areas, being gored by buffalo, getting mauled by bears, catching fish, could destroy the parks.

And the solution to this problem, as they saw it, was law enforce-ment and social engineering.[8]

By building the rangers as a police force dedicated to protecting parks from the people, and by placing hotel accommodations and other visitor facilities away from the fragile areas, they could min-imize the potential conflict. In this way they could eat their cake and have it: they could promote tourism and then take steps to see that people did as little damage as possible.[9]

Indeed, the invasion of our national parks by millions of people had to be taken as a real threat. But this invasion was not the only consequence of promoting tourism. It was what the Park Service did in its attempts to attract people, not what the people did them-selves, that most endangered wildlife.

Growing as a ministry of tourism, the Park Service became a bureaucracy less capable of satisfying its mandate for preservation. Rangers became a force of policemen rather than wildlife scientists or historians, and most wildlife-management policies came to be designed, not to preserve parks as pristine, but to attract visitors. It was these policies that hurt the animals. Visitors did not kill wolves, coyotes, or cougars; rather, as we have seen, they were victims of an ignorant Park Service policy designed to protect pop-ular animals at the expense of those thought to be less popular. Nor did visitors ruin the range; rather Park Service policies — designed to promote the popular elk and antelope — did the dam-age. Yet that the conflict between promotion and preservation might take this course, that it might affect wildlife management policies and thus might be a problem that better law enforcement or building codes could not solve, occurred to no one. Instead, the potential conflict seemed manageable. Relying on support from the alliance between railroads and preservationists did not need to be a Faustian bargain. Rather, the new service had a constituency it could grow on.[10]

As the parks grew, however, the interest of the railroads waned. In Yellowstone, fewer people arrived by rail each year, the number falling from 86 percent of visitors in 1915 to fewer than 3 percent by 1940. The era of the car had begun.[11]

Sagebrushers, these early visitors were called, enjoying the new freedom to travel in their Franklins, Fords, and Hudsons, carrying pup tents and camping out-of-doors. And their numbers grew like Topsy. The first car entered Yellowstone in 1915; four years later the number was 10,000. By the end of World War II the railroads got the message. They had no future in the national-park business.[12]

The Northern Pacific discontinued service to Gardiner in 1948; the Burlington ended service in 1956; the Union Pacific and the Milwaukee in 1961. And the passing of the railroads was a crisis for the Park Service. For the revolution of the automobile was not just technological, but social and political as well. Visitors who came to the parks by train were well-heeled members of the upper classes.[13]

These people — the Roosevelts and Rockefellers — were powerful allies who had many friends in Congress. But with the auto arriving, Yellowstone was becoming the destination for the average citizen. The car was democratizing the parks. By the early 1950s, Eisenhower was building the interstate highway system; Dinah Shore was telling us to see the U.S.A. in our Chevrolet; the national parks were beginning to be deluged by an entirely new kind of clientele.

But the political clout of the Park Service seemed to be nearly nil. Clearly it was a new ball game. The political potential of the new visitors — who were, after all, the entire American people — was nearly limitless, but to be realized it had to be cultivated. The Park Service would need to woo and organize its newfound friends. It had to begin all over again to build a new constituency. And by 1955, it had formulated a plan to do so. The plan was called Mission 66.

The brainchild of Conrad L. (Connie) Wirth, Director of the Park Service, Mission 66 was designed to prime the pump of federal dollars by packaging Park Service demands in a way that would capture public imagination and impress on Congress how many millions of voters such appropriations would serve.

"As I pondered our dilemma, I asked myself," Wirth wrote, " 'What would I want to hear from the Park Service if I were a member of Congress?' " Wirth, a landscape architect by training and city planner by profession, decided that they would be most impressed by an ambitious, long-range building plan with a short, "provocative" title. Mission 66 had the right ring to it.[14]

Conceived in 1955, it was a ten-year construction program aimed at increasing and refurbishing facilities for visitors in time to accommodate the eighty million people expected to visit national parks by 1966, the date of the Park Service's golden anniversary.[15]

It was also intended to capture the patronage of the motorist and the imagination of the environmentalist. "The rising curve of park travel," wrote Lon Garrison, Chairman of the Mission 66 Steering Committee, "coincided beautifully with the number of new automobiles, but would this continue?" To ensure that it would, the Park Service put high on its list of priorities construction of roads, campgrounds, trailer parks, filling stations, and roadside rest areas. Yet this building was to be done in the name of conservation. "Appropriate development of facilities," wrote Garrison, "such as roads or trails actually could be viewed as a conservation and protection measure, as it tended to channel and restrict use."[16]

They had found a way to revive the old alliance. The American motorist had replaced the railroad baron, but otherwise the coalition of travel promoters and environmental activists remained intact. Mission 66 was off and running. The American Automobile Association sponsored the kickoff dinner in Washington, D.C. in February 1955.[17]

Yellowstone was to be the showcase of Mission 66. Garrison was sent there as Superintendent in 1956, explicitly, he said, "to make it work." Plans called for $70 million of construction, mostly for roads and bridges. They also called for nearly doubling the capacity for overnight accommodations. In the quaint language of Yellowstone planners, the number of "pillows" in the park were to be increased from 8,500 to 14,500.[18]

To increase the number of accommodations for visitors, Mission 66 planners decided to create two new "villages" in the park. The first would be at Canyon, next to the old and elegant hotel built with railroad money in 1906. The second would be built along the south shore of Thumb Bay, on Yellowstone Lake, to be called Grant Village.[19]

Putatively named after Ulysses S. Grant, the president of the United States who signed the Yellowstone Park Bill into law in 1872, it also happened to honor Wirth's patron, U. S. Grant III, a city planner and executive officer of the National Capital Park and Planning Commission. Yet even before baptism Grant Village had a fitful history. Conceived in 1936 as a replacement for the small complex of visitor facilities at West Thumb, it died in embryo when World War II broke out. Surveyed in 1947, the project was again aborted in the early 1950s for lack of funds. Now it was to be a centerpiece of Mission 66.[20]

There was, however, a problem. The government did not offer overnight accommodations in the park. Rather, private concessionaires did, and they did not want to build Grant Village.

For the first seventy-five years of Yellowstone history, railroads had kept the hotels afloat, providing low-interest loans for a business of marginal profitability. But by the 1950s the railroads were pulling out. The Yellowstone Park Company, a family business, was already heavily in debt when the Park Service asked them to invest millions to build hotels at Canyon and Grant and spend millions more to upgrade facilities throughout the park.[21]

To avoid losing their concessionaire's license, however, the Nichols family, owners of the YP Company, as it was called, borrowed the money to build the Canyon Village. Completed in 1957, it was a financial disaster. Each night the old and elegant Canyon Hotel filled to overflowing while the newer village went nearly empty. The Nichols were going broke. In desperation, they closed the hotel, selling the hulk for salvage in an attempt to force people to patronize the newer accommodations. During winter 1959–1960, the old hotel burned to the ground under mysterious circumstances.[22]

That was the last straw for the Nichols. They felt that the Park Service, in an attempt to attract visitors, had coerced them into an unprofitable venture. Now it wanted them to build Grant Village, which, they suspected, would be an even greater disaster. They refused to build. For the third time, the Grant Village project seemed dead.[23]

The Park Service, however, was not to be denied. The village seemed to have become an idée fixe in the bureaucratic imagination. Garrison wrote, "The creation of Grant Village was one of the 'horizon' events from the beginning. . . . The whole thrust at that point in time was to serve more visitors." With this goal, the Service proceeded to build a village for which, without the YP Company's cooperation, it had no foreseeable tenant. Hiring Ted Wirth, Connie Wirth's son, as architect, it clear-cut the forest, bulldozed roads, strung telephone lines, buried water and sewer lines, put up a ranger station and visitors' center, built a campground, and constructed a huge marina. The Park Service, it seemed, was still determined to put Grant Village on the map, even if it was to be a ghost town.[24]

But, in fact, the Park Service had a tenant in mind. The Nichols family, owners of the YP Company, decided to sell out and the Park Service found the buyer it thought to be perfect. His name was Laurance Rockefeller, and they asked him to purchase the YP Company. The asking price was $5 million.[25]

Rockefeller epitomized the kind of tenant the Park Service was looking for. He was a generous donor to countless environmental groups, a philanthropist whose family had given the people of the United States three national parks. Yet he was also a businessman who, through his company, Rockresorts, Inc. and operating foundation, the Jackson Hole Preserve, Inc., had made a business of operating concessions in, and developing large tracts of land around, national parks. As someone who had walked the tightrope of development and conservation, he was a perfect match for the Park Service. He embodied the coalition the Service sought: an alliance of environmental and business interests.[26]

Rockefeller decided not to buy, however. The run-down condition of facilities and the Park Service's demand that the concessionaire build Grant Village made the $5-million price too expensive. Instead, shortly thereafter the YP Company was sold to the General Hosts Corporation. But this company too refused to build the Village.[27]

Designed as a showcase to promote visitation, Grant Village had to be judged a failure. Three concessionaires had told them it was a bad idea. The best thing, surely, would be to scrap it. But $7 million had already been spent preparing the ground there, and writing the project off would be to admit that this huge sum had been wasted. Rather than face such embarrassment, it was easier simply to spend more. The Park Service, therefore, continued to build the infrastructure at Grant, and pushed ahead with an ambitious roadbuilding program. It constructed a costly bypass around West Thumb and a four-lane thruway, complete with cloverleaf, at Old Faithful. It planned another bypass at Fishing Bridge, and built another major highway — known as the John D. Rockefeller, Jr., Memorial Parkway — connecting Yellowstone with Grand Teton.[28]

All this construction, unfortunately, was done with little thought about the environment. The West Thumb bypass damaged an important grizzly stream; the parkway ran through critical grizzly habitat. The marina, designed for large cabin cruisers and house-boats, changed lake currents, causing the beach to erode. It obliterated an important archaeological site. The sewage plant, built within a hundred feet of Yellowstone Lake — one of the purest lakes in the world — and a scant mile from the intake for drinking water, was *designed* to pour up to 300,000 gallons of treated sewage into the lake every day. Lacking denitrification tanks or settlement ponds, it achieved only the stage that sanitation engineers call "primary treatment" of sewage; the liquid, after passing through a clarifier, was piped directly into the lake.[29]

But this was an era in which environmental protection was considered an extreme political position, and the Park Service, unfortunately, was no more enlightened about these niceties than the rest of us. And it was probably no worse than what happened later in Glacier National Park, where a sewage-treatment plant was built in a flood plain, risking contamination for miles downstream each time MacDonald Creek left its banks.[30]

In Grant Village, however, sloppy planning escaped notice because the village remained unused. It was a town without buildings, without people, a town no one except the Park Service wanted, a town developed for tourism that the tourist industry had thrice rejected. Yet by the end of Mission 66, the Park Service, in its persistence, had destroyed another part of the park. It seemed a gratuitous waste. "We were still destroying wilderness," Garrison recalled ruefully; "Grant Village was a normal outcome of this growth pattern."[31]

At its conclusion, Mission 66 seemed an anachronism. The American love affair with the automobile was diminishing, and the uneasy alliance between preservationists and motorists was breaking up. Increasingly, environmentalists lobbied against exploitation of national parks by purely commercial interests. They also called for removing everything artificial, including the car. Indeed, Yosemite banned cars from part of the valley in 1970. At this time Grant Village, conceived as a way to woo the motorist, epitomized a condition that Edward Abbey, in 1968, called "Industrial Tourism."[32]

"No more cars in our national parks," he wrote in his *Desert Solitaire*. "Let the people walk. Or ride horses, bicycles, mules, wild pigs — anything — but keep the automobiles and the motorcycles and all their motorized relatives out." Abbey's voice seemed to ring clearest in these heady days of environmental awakening. Clearly, Grant Village was an idea whose time had passed. The Park Service could no longer promote both tourism and protection, nor could it maintain, as its constituency, an alliance between environmentalists and those in the tourist industry.[33]

Mission 66 and the spending boom it represented had come to an end. The tide of environmental awareness had turned against it. Motoring was out and protection was in. For the fourth time it seemed Grant Village had died.

By 1968 the era of natural regulation had begun and the new strictures required minimizing the influence of people in the park. The pendulum had swung, and the emphasis was not, as in Mission 66, on attracting as many people as possible to Yellowstone, but rather on protecting the park from the people. But this swing presented a problem. How, in the context of natural regulation, would future development in the park proceed?

To answer this question the Park Service formed a blue-ribbon committee of Park Service officials and leading environmentalists (including the prominent conservationist and writer, Sigurd Olson). The document written by this committee, the Master Plan, appeared, as we saw, in 1973 and was to affect Yellowstone for the next twelve years.[34]

With publication of the Master Plan, preservation for the first time was put ahead of promotion. The committee wrote, "All planning for public use of national parks must give priority to the preservation and maintenance of the natural values for which each park was established." To that end, and to conform to the new mandate that Yellowstone was to be managed as a natural area, the Master Plan was designed to separate people from fragile natural resources.[35]

To protect the park from the people, they urged that it be zoned into various "visitor-use corridors and enclaves." They endorsed the policy of closing garbage dumps to "eliminate artificial relationships between bears and man"; they urged that no further development be done at Canyon, that overnight accommodations at West Thumb and Old Faithful be abolished to protect the "fragile thermal zone," and that the latter be converted to a "scenic day-use area." They recommended that all development at Fishing

Bridge — including the campground, cabins, and recreational vehicle (RV) park — be dismantled "to facilitate restoration of critical wildlife habitats."[36]

In fashioning this design to protect the park, the architects of the Master Plan declared war on the automobile. They stated, "An important objective is to lure the 'scenic drivers' from their automobiles." To this end they urged that a highway system be built that would skirt Yellowstone and Grand Teton, diverting through traffic away from the parks altogether while encouraging park visitors to leave their cars at the gates and do their sightseeing by public transport. The number of permissible "pillows" in Yellowstone would be reduced from 14,500 to 8,500, and the gateway towns would become the "primary visitor hubs" providing the bulk of facilities. There would be no further development of the park road system, and instead, as an alternative to the private automobile, they would build an "optional [public] transit system to tie into the accommodations centers on the periphery of the park." They envisioned and indeed studied carefully the possibility of a monorail system throughout the park.[37]

By these means Yellowstone would be a "sanitized" nature preserve insulated from human beings and protected from the automobile, a place where visitors would sleep in enclaves at night and travel through the park via monorail or bus, viewing animals through glass during the day. For Grant Village this new scheme would seem to mean the end of the road. Conceived as a magnet to attract people and cars into the park, it was clearly out of step with the very idea of environmental management.

Instead, however, Grant Village became the center of the Master Plan. By an apparent contradiction, it was now to be a "wilderness threshold community," a staging area where people might be encouraged to jump out of their cars and into the backcountry. "If we truly want to get the visitor off the road and into the park we must create people-oriented or social spaces to accommodate him," the committee mused. Grant Village should be such a place, "a tightly knit development creatively interspersed with plazas or green spaces." For these reasons, the committee concluded, "Grant Village will become a major development, containing several classes of accommodations."[38]

Thanks to the Master Plan this project, pushed for twenty years as a way to boost tourism in Yellowstone, became, without missing a step, the heart of a new, environmentally inspired vision of social engineering. Designed as a tourist magnet, it became part of a plan to protect the park from the people. Built with little regard for nature and already nibbling away at the remaining wilderness, it was now to proceed to "protect the resource."

Turning Yellowstone into a wilderness island with Grant Village as the port of entry was a grand, attractive idea. But like other grand ideas — such as those brought forth in the Leopold Report — its appeal rested more on eloquence than fact. In fact, many of the Master Plan recommendations were either ineffective or actually harmful.

That visitors would voluntarily leave their beloved cars at the park gates was hopelessly utopian. Both Old Faithful and Fishing Bridge had buildings that were candidates for the National Register of Historic Places; some, such as the Inn at Old Faithful, were true architectual treasures, and it was not clear that they could be removed without violating the National Historic Preservation Act.[39]

Both of these older communities, too, were built around natural scenic attractions and had won the affection of millions of people, but Grant Village, then nothing more than a mosquito-infested clearing in a lodgepole forest, was a purely artificial location with no natural features to beckon the visitor. For these reasons it was not at all certain that the public would either accept dismantling of the older communities, or desire to patronize the new one.

Building a new highway system around the park, rather than preserving the wilderness, would destroy tens of thousands of acres in the pristine forests that ring Yellowstone, opening even vaster areas to the automobile. And clearly, encouraging people to get out of their cars and into the woods was just the wrong thing to do.

As long as they stayed in their cars, slept in motels, and walked on the boardwalks, visitors did little damage. But there is practically no limit to the harm that can be done by backpackers in the high country. Thousands of hikers turn trails into ruts and greatly accelerate soil erosion. In their clothes they carry seeds of exotic plants into pristine country. They strip tree branches in collecting firewood. They camp under or around critical bird nesting sites. They startle and sometimes molest game. They clean their fish or wash their dishes in streams, leaving their garbage to decay in the water. And, increasingly, they spread intestinal diseases.

Visited by thousands of people a year, popular spots in the backcountry — and the nearby streams — become hopelessly polluted by backpackers who defecate on the ground and fail to cover their leavings. This refuse is blown or washed into the watershed, where it incubates bacteria that are in turn spread from stream to stream by wild animals and by other hikers. Many such diseases, such as giardiasis, have actually reached epidemic proportions in many parts of the Rocky Mountain backcountry, and have spread into the water supply of sizable towns, such as Red Lodge and Bozeman, both in Montana near Yellowstone.[40]

In the final analysis, however, the justification for Grant Village rested, not on its role as a wilderness threshold community, but on the supposition by the Master Plan committee that to protect the fragile geothermal resources at Old Faithful and West Thumb, and to protect the fragile biological resource at Fishing Bridge, visitors' accommodations must be concentrated at Grant. But how did they know that this trade-off would benefit the environment? How did they know what damage the buildings at Old Faithful did to the geysers and hot pools? How did they know that Fishing Bridge was more critical as wildlife habitat than Grant itself?

In truth, they did not know. No studies had been done by the Master Plan committee at any of the three locations. The community at Old Faithful had become an eyesore to be sure, thanks to the thruway bypass and proliferation of honkytonk concessions, yet it was not clear how the buildings were affecting the geothermal system. In fact, it seemed that Park Service incompetence and mother nature, more than overnight accommodations, threatened the area. Recently, for instance, engineers, failing to consult a geologist, installed a water system six feet deep to prevent it from freezing, and were astounded when they opened the valves that only steam escaped. The ground, hot enough to bring geysers to boil, did the same to the water line.[41]

For the grizzly bear, knowledge of habitat was critical. If Grant was as important for the bear as Fishing Bridge, then shifting services would not benefit him. Many biologists, in fact, strongly suspected that the location at Grant was at least as important as that at Fishing Bridge. One of these was Professor Jonas, who wrote to the park superintendent to voice his worry:

It is my understanding that one of the reasons for eliminating all or parts of the Fishing Bridge complex has to do with designation of this area as "prime" grizzly bear habitat. I hope there is good, solid scientific data to confirm that claim. Too often I have seen biological reasons given for various actions when the real reasons are economic, social or political. I suspect more bears might be there or sighted there because of the presence of the humans rather than anything innate about the natural habitat. . . . If the Fishing Bridge complex is reduced or eliminated, I assume replacement visitor facilities will be constructed elsewhere. If that "elsewhere" is Grant Village, I suggest we might be moving one problem from one place to another at taxpayers' expense. Grant Village area is as good (probably better) a habitat for grizzlies as the Fishing Bridge system.[42]

John Craighead concurred with Jonas: "Fishing Bridge is critical habitat to the grizzly," he told me recently, "because there is human food there." Indeed, both Jonas and Craighead suggested that human fishing activities in the vicinity were the principal reason why bears began to congregate there. "When a fisherman caught a sucker, he just threw it on the beach. That kind of thing brought the grizzlies," Jonas told me.[43]

"There were few problems with grizzlies at Fishing Bridge," John Craighead told me, "until the Fish and Wildlife Service resumed its fish-trapping studies there." In fact, in 1968, the same spring in which reduction of garbage at Trout Creek dump began, fisheries biologists began catching and marking trout and suckers at a weir that stretched across Pelican Creek, a stream entering Yellowstone Lake near Fishing Bridge. And although they took considerable care to remove dead fish, they could not remove their odor. Some bears, following their noses, were attracted to the area. Occasionally, this activity forced the biologists to retreat. In 1974, for instance, the trap had to be abandoned prematurely on July 30 "because [according to a Fish and Wildlife Service report] of the presence of a sow grizzly and her yearling cub which attempted to feed there during the day while the trap was being operated."[44]

Craighead's and Jonas's suspicions were justified. The rationale for moving facilities from Fishing Bridge to Grant was a result of the Park Service grizzly management policy. For Fishing Bridge, as we saw, did not begin to have dangerous bear problems until authorities began to reduce garbage at the Trout Creek dump in

1968. Even in that year the number of bear "control actions" at Fishing Bridge was 33 percent lower than the number at Grant; but by 1971 the number of actions at Fishing Bridge was nearly *four times* as many as those at Grant.[45]

Yellowstone grizzly management had created a bear problem at Fishing Bridge, and the idea of closing it down and moving people to Grant — a place whose popularity with grizzlies was, after all, unknown — began to seem attractive. Yet five cutthroat-trout spawning streams entered the lake at Grant, and in the summer, grizzlies came to these streams to fish. Clearly it too was prime grizzly habitat. And that the bears might frequent Fishing Bridge after the dumps were closed because the people were there — and that they might follow the people to Grant — did not occur to the Master Plan architects.

Instead, they seemed to have accomplished the impossible. They had given a preservationist rationale for a major development to be placed in the midst of the prime, fragile habitat of a threatened species.

No sooner was the Master Plan published than the Park Service began to have second thoughts. As a clear statement of environmentalist antipathies to the automobile, it threatened to fracture the fragile constituency that served as the Service's advocates in Congress. The automobile and recreational-vehicle lobbies were strong and the Service could not afford to alienate them. In 1974, moreover, as the Arab oil embargo and first fuel crisis swept the country, visitation in Yellowstone dropped dramatically. Rather than continuing to increase as planners had anticipated, it had dropped below two million for the first time since 1963. In these times, declaring war on the automobile seemed to be overkill; the Park Service began to back away from the Master Plan.[46]

In 1974 the National Park Service published the Final Environmental Statement on the Master Plan. But this document, though it was written in consultation with the Wilderness Society and the Sierra Club, among other organizations, and though it reaffirmed commitment to the principle of "permitting the evolvement of indigenous ecosystems without interference by man," was in fact no environmental statement at all. Rather than conducting any studies of Grant Village's effect on the park, its authors were content to conclude that they did not know what these effects might be. "Environmental effects of the proposed wilderness threshold communities," they said, "particularly at Lake and Grant, are unknown." Surprisingly, they still urged "centering the major visitor services at Lake and Grant Village."[47]

Even though they accepted the Master Plan recommendations for building Grant Village, however, the authors of the Environmental Statement rejected the Plan's overriding goal: to restrict automobile travel. Encouraging visitors to leave their cars at the park gates, they suggested, would not work. The use of public transportation "means separating the visitor from the 'security module' that is his personal automobile." Diminishing park visitors' facilities would hurt the gateway economies: "The tourist industry in the surrounding states," they observed with crocodilian sympathy, "grosses some $60 million from the presence of Yellowstone National Park. It is difficult to estimate the loss that would occur, but a large percentage of the visitor-oriented businesses near the park gates and all the park concessions would fail." And yet they opposed making the gateway communities the "primary visitor hubs," as suggested in the Master Plan, even though this suggestion would surely help the economies of these communities. These communities (whose lifeblood is tourism), might not like the extra business: "The additional crowding and noise are objectionable to those who prefer the quiet of a small town."[48]

In the Environmental Statement, in short, while refusing to make a judgment about the environmental price, the authors endorsed construction at Grant Village, even as they rejected the principal justification for the authors of the Master Plan, discouraging automobile use in the park. As one would expect, the Yellowstone Superintendent, Jack Anderson, thought it a bad idea and refused to support it. And the new owners of the YP Company, General Hosts, like their predecessors seeing it as a bad investment, still refused to build it. Moreover, visitation in the park, leveling off at two million by 1976, seemed too low to justify such a major investment. For the fifth time it seemed the project had died.[49]

Appearances were again deceiving, however. In 1976 Anderson was replaced, and in the new Superintendent, John Townsley, Grant Village had found its champion. Townsley was attractive, forceful, a consummate politician, and a supersalesman. Having spent much of his career in Washington, he knew politics and had connections in Congress that most lobbyists would envy. These contacts gave him power far beyond his station. As a salesman, he could, one official said, "sell ice cubes to Eskimos." He was, in fact, a man hard to stop.

He ran Yellowstone with a personal style that defied the usual chain of command. His favorites, often junior officials, retained power beyond their rank, while many senior officers — not the court favorites — languished in a power vacuum. He was intensely loved or hated and often feared. No one felt indifferent about him.

He was also competitive, and enjoyed challenging the manhood of his subordinates. One of his favorite pastimes was "making the loop": driving the 142-mile circle of roads in the park on a snowmobile at speeds of up to ninety miles an hour. Often he would take a novice along and challenge the newcomer to a race to test his mettle. On one such trip he reached a place on Mount Washburn where the road was blocked by a nearly vertical snowdrift. In typical Townsley fashion, he dared his companion to cross the field. Going first, Townsley gunned his machine and made it across. His companion was not so lucky. Halfway across, his machine capsized, and although he managed to stop his fall, the snowmobile kept going. It fell hundreds of feet, and could not be retrieved until spring.[50]

With Grant Village, as with his snowmobile, Townsley went full bore and never looked back. The project became his mission. For many in the Yellowstone administration, his singular pursuit seemed beyond the call of duty. "In Grant Village," many said to me, "Townsley was building a monument to himself."[51]

Townsley soon had a plan on which to build his monument. He would revive the old environmental–recreational coalition. And in doing so, he had the help of two new studies commissioned by the Park Service.

In the first, the Yellowstone National Park Concessions Management Review, it was concluded that the YP Company, because of its failure to invest sufficiently in park accommodations and its refusal to build Grant Village, was in default of its concessions agreement with the Park Service. The writers of the report urged the Service to purchase the possessory interest of the company. If private enterprise would not build Grant Village, the study's authors said in effect, then the Park Service should.[52]

The second report, the Greater Yellowstone Cooperative Regional Transportation Study, added a valuable political argument. Wyoming, it stated, was especially dependent on Yellowstone visitation. Large numbers of tourists in the park meant more money to Wyoming, and Grant Village, near the south (Wyoming) entrance, would particularly help the state.[53]

Armed with these new studies, Townsley went about mending fences, rebuilding the coalition of economic and environmental interests that would be necessary to put the project through. He lobbied the new Park Service Director William Whalen. He made the rounds in Congress, emphasizing to legislators the need for public funds to buy the YP Company and restore the dilapidated accommodations in Yellowstone. He visited Wyoming officials, selling them on the idea that Grant Village would be a magnet pulling dollars into Wyoming. He courted the mainline environmental groups, invoking the name of the Master Plan.

Throughout he had one great advantage not available to most lobbyists: Yellowstone itself. Most of his persuasion was done by entertaining groups in Yellowstone, a background that virtually guaranteed that they would have a good time and that he would establish rapport between himself and his target. As one Park Service official told me, "Few realize the great advantage Yellowstone gives to the Park Service [for lobbying]. You have a Congressman out here, show him the beauty, let him catch fish, and he can't say no. No other federal agency can show people such a good time, because no other federal agency has Yellowstone."

One by one, supporters fell in line. Wyoming interests became enthusiastic boosters of the idea. Well coached by Townsley, they saw Old Faithful as a magnet attracting Yellowstone visitors through West Yellowstone, Montana, and Grant Village a draw for Wyoming. They became convinced that building the latter and phasing out the former was all in their interest. "If accommodations do not exist at Grant Village," Wyoming's Governor Ed Hershler responded to a critic, "tourists may choose travel routes to and from the Park that would avoid Wyoming entirely. We simply can't afford to allow West Yellowstone in Montana to become the chief overnight center for Yellowstone vacations."[54]

By 1978 Congressional support for the YP Company buyout had materialized and by summer Director Whalen gave his approval for it. More surprisingly, so too did the environmentalists. Many such groups, William Turner, Rocky Mountain Regional Director of the National Audubon Society told me recently, "follow the lead of the National Parks Conservation Association on matters to do with Yellowstone." But the NPCA, an organization founded by Mather to *promote* our national parks, was an unlikely watchdog. It decided to support the project. "We were anxious to see the buildings removed from Old Faithful," Destry Jarvis, Federal Programs Coordinator of the NPCA told me recently, "and Townsley convinced us the only way that would be done was by building Grant Village."[55]

Townsley had done his homework with other groups as well. "He came to Jackson [Wyoming] and sat down with me and a few representatives of other environmental groups," Phil Hocker, Treasurer of the National Sierra Club told me, "and pretty much convinced us Grant Village was a good idea. In fact, afterward we occasionally interceded on the government's behalf to help the project along."[56]

By September 1978 a request for $15 million for the buyout was sent to the Office of Management and Budget. In the same appropriations package, buried on page 1058 of the "Justification Material" submitted with the budget was a request for $3 million to build "100 Motel-Type Guest Units" at Grant Village. At last Grant Village had its foot in the appropriations door.[57]

The Park Service, however, had put the cart before the horse. It had asked for funds before conducting an environmental review, probably a violation of the Environmental Policy Act, which required that no project proceed without an Environmental Impact Statement, or, at the very least, a less thorough Environmental Assessment. It also violated Section 7 of the Endangered Species Act, which prohibited the Park Service from proceeding until the Fish and Wildlife Service had determined that construction would not jeopardize the future of the grizzly — a threatened species. But the Fish and Wildlife Service knew nothing about Grant Village. And, for that matter, the public knew as little.[58]

Indeed, not until June 1979, a year after funds had been requested, did the Park Service publish its Environmental Assessment of Grant Village. And only then did it discover problems with the project.[59]

Sewage was spilling. Although the treatment plant now had settlement ponds, the ponds had to be constructed on a hill above the plant, requiring all sewage to be pumped up the hill. When a power outage occurred, the pumps did not work and sewage spilled into the lake. The Assessment stated, "Raw and treated sewage will continue to be discharged to the lake during power outages. . . . Overflows in the vicinity of 700 gallons per minute are possible."[60]

Grant Village, they now realized, was prime grizzly habitat. "Five [cutthroat trout] spawning streams enter the lake in the Grant Village area; the easternmost and westernmost streams are heavily used by grizzlies." Thanks to this traffic, the document's authors wrote, "increased visitation associated with the construction of 100–200 lodging units will increase the probability of encounters between humans and bears. The likelihood that bears will be killed, injured, or transported from the area will also increase as will the possibility of humans being injured or killed."[61]

To cope with this danger, they decided, the Park Service "will consider blocking all but the easternmost and westernmost spawning streams to trout, so that bears will not be attracted to the area." Blocking these streams would destroy at least 5 percent of the gene pool of cutthroat in the lake, but, they concluded, "loss of [this] spawning area would be acceptable to park biologists."[62]

It would seem that an agency directed by law to "conserve the wildlife . . . in such manner and by such means as will leave them unimpaired for the enjoyment of future generations" might have second thoughts when confronted with these biological probabilities. Instead, they made matters worse.

According to the Environmental Policy Act, planners are required to compare the environmental effects of various so-called action alternatives with a so-called no-action alternative. That is, they must consider the possibility of doing nothing at all. For Grant Village, however, the Park Service did not do so. It was explained in the Environmental Assessment that "If the no action alternative is implemented in its entirety, no development *beyond what is already programmed* will occur at Grant Village. *Because of prior approval and programming* [italics mine], the 100–200 lodging units and utility improvement described in the Phase 1A study will be constructed." It was, the Assessment told us, too late to stop Grant Village. The project had already been approved; the die was cast.[63]

This statement unfortunately was not true. Although the Park Service had applied for funds for Grant Village nine months earlier, it would be another six months before the president would sign the appropriations bill for fiscal 1980 approving this expenditure. Funds had not yet been approved. And even after the funds were approved, the Park Service always had the option not to use them. A federal agency can ask Congress for permission to "reprogram" funds for another project. Such reprogramming is done all the time.[64]

Rather, Yellowstone had been caught in an upward ratchet. The Park Service had followed an old bureaucractic maxim: It is easier to apologize for something done than to ask permission to do it. Presenting the public with a false fait accompli ensured that no environmental niceties would thwart the agency's will.

That the timing and content of the Environmental Assessment suggested that the Park Service was breaking the law in several ways, however, went unnoticed. Nor did anyone call attention to the predicted effects on the grizzly bear. Townsley had prepared the way well. The Assessment sailed through hearings virtually unopposed. No one asked for a much more ambitious Environmental Impact Statement. The development was, after all, the Park Service wrote in approving the Assessment, "a minor Federal action with minimal environmental, social and economic impacts."[65]

If it was then a small project, however, it would not long remain so. Within weeks the Park Service drew up ambitious plans to build 700 motel units over three years. For the second stage of development (for the 1981 fiscal year) it submitted (in September 1979) a request to OMB for $14,736,000, along with a further request for a commitment to receive an additional $11.4 million for more construction in future years.[66]

Although still unknown to the public, these plans did disturb many people within the government. Scientists and planners quietly expressed their unease. Some geologists discovered that streams flowing off the nearby Pitchstone Plateau carried a high — perhaps unsafe — level of natural radiation, and they feared the same might be true of some of the streams flowing through Grant. Budget Examiners at OMB objected that building Grant put the Service in a business for which they were not qualified and invited misuse of funds. Park accountants and engineers observed that Grant Village was expensive and unneeded, and that owning motels put them in potential conflict with their mission to protect nature.[67]

One of the latter was Richard Bowser, a Park Service engineer and accommodations expert, now retired, who was charged with evaluating the project. According to Bowser's calculations, the taxpayer would be required to subsidize the stay of each visitor by more than $120 a night. Yet Bowser, like many in the Service, was convinced that they should not be in the motel business at all. "Most rangers," he told me, "have never even owned a home. They — even superintendents — have no idea of household finances, much less the motel business."[68]

The justification for constructing the village moreover, Bowser told me, was based on projected visitation, and these figures, he said, were misleading. The Park Service justified its appropriations requests to Congress by the size of park attendance, and so it was in their interest to exaggerate this number. To compute attendance, Bowser explained to me, statisticians counted the cars and multiplied by a number around 3.5. Yet such a multiplier, he said, had little statistical justification. City-traffic experts usually used a figure around 1.8. Thus Park Service claims that more than two million people visited Yellowstone a year was surely too high. Not enough people, he feared, would visit to fill all the planned accommodations. The park would be overbuilt.

These cautions, however, went unheeded. On October 3, 1979, Townsley invited construction firms to make bids for building Grant Village. Three weeks later — eighteen months after funds were first requested for Grant Village — the Fish and Wildlife Service at last reported to the Park Service its opinion of how construction of the village would affect the grizzly bear.[69]

The Fish and Wildlife Service is, of course, a branch of the Interior Department, and it answers to the Secretary of the Interior, as does the Park Service. And it was the Secretary of the Interior, Cecil Andrus, who had approved the Grant Village project. Yet under the provisions of the Endangered Species Act, the FWS was charged with stopping the development if it endangered the bear. So charged, the FWS was in an impossible position. It was charged to consider stopping a development to which its own boss was already committed. It had power to stop the project in theory, but not in fact.

And yet FWS biologists expressed fear for the grizzly. They concurred with the Environmental Assessment that the proposed community increased the "likelihood that bears will be killed," and added, "considering the cumulative impacts to the grizzly, to our knowledge these are losses the Yellowstone grizzly population cannot afford." They suggested that Park Service plans to close the streams to spawning trout was probably a violation of the Endangered Species Act, commenting that "spawning streams are important biological components in grizzly bear habitat and may constitute one of the biological elements within officially designated critical habitat that are essential to the conservation of the species." They were not convinced that closing Fishing Bridge alone justified building Grant Village. "We also believe," they said, "the project will negate many of the benefits acquired through the phaseout of facilities at Fishing Bridge and view such a 'trade-off' as an unfavorable solution to a wildlife conflict that, with development of Grant Village, will likely be duplicated rather than eliminated."[70]

Instead they urged the Park Service to reconsider: "We suggest that the Park Service review and evaluate the need for extensive development at Grant Village in light of this biological opinion. . . ."[71]

Yet in the end they reluctantly issued a "no-jeopardy" decision. Although Grant Village would damage the bear it would not, they concluded, "jeopardize [his] continued existence." Grizzlies would survive despite the depredations of Grant Village, they suggested, because Park Service bear management was otherwise so successful. Indeed, they stated, "the Yellowstone grizzly population appears to be stabilized or increasing."[72]

In issuing this decision, the government had fallen victim to its own propaganda. Having convinced the public that the grizzly bear program was working, it now based a crucial decision on this misinformation. For the "no-jeopardy" decision was based on faulty information: the grizzly population, as we know, was actually declining, not increasing; the margin of error assumed by the Fish and Wildlife Service did not exist.

Even so, in reaching its conclusion the FWS stipulated several conditions. The spawning streams should not be damned, and the community at Fishing Bridge — including the recreational vehicle park, general store, and overnight accommodations — must be removed by 1985. "The adequacy of the Section 7 consultation," the FWS said, "depends upon the commitment of the Park Service to follow through with the phaseout of facilities. . . . The benefits to the grizzly from closure of Fishing Bridge will be greatly reduced unless the entire area is eventually restored."[73]

Within a week of receiving this opinion, the Park Service also learned that neither Congress nor the president was happy with the project. The cost of purchasing the YP Company, it was discovered, would be $19.9 million, $4.9 million more than the Park Service would receive from Congress, and four times the price that Rockefeller had decided fifteen years earlier was too much. On November 8, the Senate and House Interior and Related Agencies Subcommittees of the respective appropriations committees directed the Park Service to come up with the extra money itself. Specifically, they directed the Park Service to reprogram funds already designated for other projects toward the $4.9 million more needed for the buyout. Clearly implied was that any money earmarked for Grant Village should be used for this purpose.[74]

Then on November 15 came the final blow: The Director of OMB, in his "passback" letter to Secretary Andrus, ordered the Park Service not to use public funds to build Grant Village. Even the $3 million already approved, he stipulated, should not be spent for concession facilities. "No funds," the letter said, "are provided for Grant Village . . . the Department should seek a concessionaire agreement at Yellowstone which will finance any future new construction."[75]

This clause, it seemed, was the last straw. Within three weeks the Fish and Wildlife Service had directed the Park Service to reconsider the need for Grant; both houses of Congress had ordered it to reprogram funds for the YP Company buyout, and OMB, speaking for the president, had ordered it not to build the village. Thus on November 20, Assistant Secretary of the Interior Robert Herbst drafted a letter reprogramming the Grant Village money toward the YP Company buyout. The project, it seemed for the sixth time, was dead.

A week later, however, Herbst reversed himself. He recommended to Interior Secretary Andrus that the Park Service reprogram other funds for the YP Company buyout. It should build Grant Village after all. Townsley, Herbst told me, had convinced Director Whalen of the project's importance. Whalen in turn had convinced Herbst, and Herbst persuaded Andrus. The chain of command at the Park Service, apparently, was balanced precariously on its head.[76]

Grant Village quietly proceeded. In spring 1980, the YP Company purchase was completed, and formal bids for construction of Grant were received. But the Park Service faced a problem. Told by the president (OMB) not to build the village, the Park Service could not very well advertise the fact of its defiance. How could they construct it without OMB's knowledge? The Wyoming governor's office, thanks to the "magnet" argument, and environmentalists, thanks to the magic of the Master Plan, were firmly in favor of Grant Village and could be counted on to remain silent. But the state of Montana stood to lose. How could their cooperation be ensured?

"Townsley," former Supervisor of Concessions Management for the Rocky Mountain Region, L. E. "Buddy" Surles told me, "was very smart. To line up the support of the Montana Congressmen, he was sure to see that the first phase of Grant Village construction was done by a Montana firm."[77]

With serendipity, both problems were solved at once. Yellowstone concessions managers awarded the first contract — to build the first one hundred motel units at Grant — to Kober Construction Company of Billings, Montana, on December 12, 1980, during the transition between the Carter and Reagan administrations. The formal directive to proceed with construction was made on January 21, 1981, the day after Reagan was inaugurated. By this timing it was reasonable to assume that the budget examiners at OMB who had prohibited Grant Village construction would have left with Carter. The new Reagan people would not know that the Park Service had been told not to build it.[78]

And the construction would go ahead. The Kober company began purchasing materials and collecting a work force. They planned to begin construction April 6, 1981. Then everything hit the fan.

The country was in the midst of another fuel crisis that had crippled Montana's tourist industry. The merchants of West Yellowstone, Montana, in particular, had been so severely hurt by this recession that, in 1980, they had qualified for economic disaster relief from the federal government. In January, a newspaperman in that town named Joe Cutter happened to see the announcement in the Billings newspaper that the Park Service had awarded Kober Construction a contract to build Grant Village. Why, he wondered, would the federal government spend taxpayers' money in economic disaster relief for the motel industry in Montana, at the same time spending taxpayers' money to create subsidized competition for this industry in Yellowstone? Grant Village threatened the economy of West Yellowstone. He called David Stockman, Reagan's new Director of OMB.[79]

Stockman had never heard of Grant Village. He referred Cutter to Dale Snape. Snape was the Budget Examiner in charge of Park Service appropriations. He was also a Carter man whom the Park Service had assumed had left along with other Democrats at OMB. But through a fluke he had remained, and he knew all about the project.

"In telling the Park Service not to build Grant Village," Snape told me, "we assumed that was the end of the matter. We had heard nothing further from them. Then all of a sudden Joe Cutter called and I learned they were pouring cement in Yellowstone." Snape was outraged. He said, "It was a colossal example of bureaucratic bad faith."[80]

"Unfortunately," Snape added, "the Park Service sees OMB as a bunch of bean counters who have no knowledge of the finer points of politics. Apparently they thought they could make the end run; and in fact they did. A heavily scrutinized agency like the Defense Department could never get away with a thing like that, but the Park Service has such a low profile it can do things other agencies can't."

Snape blew the whistle on Grant Village, calling the project to the attention of James Watt, the new Secretary of Interior. Watt turned the matter over to Rick Davidge, Special Assistant to the Assistant Secretary for Fish, Wildlife and Parks.

"The day I arrived in Interior," Davidge told me, "Grant Village landed on my lap. Contracts were being let and no one seemed to know what was going on. There was a great deal of alarm and concern, especially by the gateway overnight accommodations industry." To Davidge, the Grant Village construction seemed a bad idea. He told me, "Government was getting into an area of business that should be left to private enterprise." Clearly it was a violation of the very principles that Watt was trying to bring to Interior: that government should not do what private enterprise could do. Thanks to Davidge's urging, the Park Service issued a stop order on Grant Village construction on April 6, 1981. Grant Village for the seventh time was dead.[81]

The stop order, however, did not please everyone. The Kober Company had already spent $500,000 on materials and had hired sixty workers. If the Park Service did not put them right back to work, the company would sue. Montana Congressmen were embarrassed. Grant Village was hurting one set of their constituents but helping the other. Montana Senator Max Baucus, with close ties to both environmentalists and labor unions, quickly sided with the Park Service. The environmental groups stayed away from the fight. Still bewitched by the thought of removing Fishing Bridge and Old Faithful, and still strong supporters of Park Service grizzly policy, they saw it as a parochial economic dispute.[82]

But for Davidge and the Reagan administration the stop order was even more embarrassing. As Davidge had said, the Grant Village project was a violation of Reaganomics. But it was also one that Wyoming saw as in its interest, and Senator Malcolm Wallop of Wyoming, a strong Reaganite, was for it. Wallop's aide in charge of Park Service issues, Patty MacDonald, a former (seasonal) ranger herself, strongly favored the plan. Key staffers in Congress, such as Anthony Bevinetto, senior minority staff member on the Senate Energy Committee and a former assistant superintendent of Grand Teton National Park, also supported the project.[83]

This move in turn put Watt on the spot. The Secretary was from Wyoming and had especially close ties in the state. If he supported the project, he violated the principles he had been appointed to promote; yet if he stopped the project it would anger some to whom he owed political debts.

Pressure from the Wyoming delegation to resume construction mounted. Davidge suddenly found that in trying to stop Grant Village he had embarrassed his boss. He was told to reverse field.

On April 14, just one week after the stop order had been issued, Director Russell Dickenson signed the order to resume construction. It was too late to stop it, the Park Service rationalized. The threat of a lawsuit by Kober was real, and besides, as Davidge said to me, "the infrastructure was already there." Better to spend another $60 million than to admit that $10 million had been wasted. But as a gesture to Reaganomics, they would turn the project over to private enterprise. They would complete the first one hundred units using public funds but seek an agreement with private industry to build the remaining units.[84]

By fall the government had found a partner willing to run concessions in Yellowstone and build Grant Village. It was TW Services,
an arm of the Canteen Company, which in turn was a subsidiary
of a conglomerate called the TW Corporation. In 1982, this company signed a five-year contract with the Park Service, an agreement hailed by the administration as a model of Reaganomics.
With this new agreement the Park Service would get all it wanted,
but the financing would be more creative.[85]

In a typical franchise contract, the concessionaire purchases possessory interest in facilities and pays the Park Service a franchise
fee for the right to conduct business in the park. Indeed, the law
is explicit on one thing: the concessionaire must pay a franchise
fee. This money goes into the general fund of the Treasury of the
United States — that is, the Park Service does not get the money.
The idea is that the Park Service should always be required to ask
Congress for money. Only in that way do they remain answerable
to the people of the United States.[86]

In the contract between the Park Service and TW Services, however, the government remained the owner of the physical plant
and the franchise fee was waived. The company had no investment
in facilities and for the first time in history the Park Service was
made a partner of private enterprise. It was to receive 22 percent
of the company's gross revenues. This money was set aside for
capital investment in Yellowstone (such as building the rest of
Grant Village), and the Park Service used appropriated funds for
maintenance. Thus when the company made money, so too did
the Park Service. And this money did not go back to the Treasury;
rather it was spent in Yellowstone. It escaped congressional oversight.[87]

Under this contract TW Services had minimal risk. "With no
possessory interest, no inventory, no capital investments that might
depreciate, they are practically guaranteed a profit," Don Hummell, former Chairman of National Park Concessioners explained
to me. But more important, the Park Service was put in a position
of conflict of interest: it had a motive for maximizing sales of TW
Services. The temptation to expand tourism, to give the concessionaire new ways of making money, once simply a political incentive, was now an economic one as well.[88]

"You see," another concessions expert told me, "the Park Service has almost unlimited funds for maintenance, but it must go

to Congress and justify a project each time it wants to build new facilities. With this arrangement it need not do that. It can build the new facilities from the 22 percent of sales it receives from the TW Corporation, and use appropriated funds for maintenance." In other words, if it wanted to expand Grant Village, or build another village for that matter, it need never seek congressional approval again. It need not suffer OMB's scrutiny as it had in the past.

Grant Village had become a game of political football with the state of Wyoming, a game pitting ideology against special interest, where in the end special interest won. It had forced Watt — the apostle of free enterprise — to intercede to save a government spending project that the Carter administration had thought an inappropriate use of federal funds and too expensive to boot. Indeed, the agreement with TW Services was not what American voters had a right to believe Reaganomics was all about. For, while giving the Park Service a profit motive, it protected the company from the risks of a free market.

Townsley died of cancer in fall 1982 and his successor, Robert Barbee, arrived just in time to pay the piper. For six years Townsley had argued that Grant Village must be built to permit removal of Old Faithful and Fishing Bridge. The Fish and Wildlife Service had made its "no-jeopardy" decision on the grizzly bear contingent upon removal of Fishing Bridge by 1985; the same rationale had won the silence of many conservation groups. The support of the National Parks and Conservation Association had been won by promising removal of overnight accommodations at Old Faithful. Now it was time for Barbee to keep those promises.

But this part of the bargain, Barbee soon discovered, was not so simple. With their removal imminent, few wanted to see these old communities dismantled. Now that Grant Village was growing, the older communities refused to shrink.

Old Faithful had many faithful friends. The beautiful and historic inn had just been renovated at great expense and absolutely no one wanted it torn down. Many other buildings were on the National Register of Historic Buildings. "Geyser gazers," those faithful Yellowstone fans who came just to see the thermal features, were opposed to all change and began to organize against it. They pointed out that no studies had ever been done to prove that this

community threatened the geyser basin and demanded better justification for the plan.[89]

The state of Wyoming, now having lobbied successfully for Grant
Village, suddenly saw the advantages of Old Faithful. The Wyoming State Historic Preservation Office wrote to the Park Service,
urging "that NPS reconsider its position and not destroy a valuable
part of the Yellowstone heritage." Governor Hershler now decided
that Old Faithful too was a magnet for Wyoming tourism and
should not be hastily removed. He wrote the Park Service in 1983
that "drastic reduction in overnight accommodations at Old Faithful
contemplated in all of the action alternatives could have far-reaching
effects on Wyoming communities and businesses." Likewise,
Randall A. Wagner, Director of the Wyoming Travel Commission,
objected to the plan. If overnight accommodations were closed at
Old Faithful, he argued, then many motorists would "drive the
extra ten miles to West Yellowstone, Montana. Needless to say,
at that point they can no longer be considered as taxpaying contributors to the Wyoming tourism economy."[90]

Almost immediately, the Park Service began to back away from
its commitment. The inn would remain and so too would a representative number of the historic buildings, they conceded. Indeed, they discovered that they might even need new facilities
there. Dan Wenk, their landscape architect, commented that employee living space was critical at Old Faithful. He urged them to
consider building a new dormitory there.[91]

Fishing Bridge too was soon discovered to have friends. It lay
near the park's east entrance, and the merchants of Cody saw it
as a major attraction for their business and a source of tax revenues.
It contained a large park for recreational vehicles, and that industry's lobby had great influence in Congress. It had the park's first
visitor center, a building on the National Register of Historic Places
that even the Park Service did not want to demolish. And it had
a shop belonging to the park's other concessioner, Hamilton Stores,
and its President, Terry Povah, was understandably reluctant to
abandon it.[92]

These friends had influence. When Superintendent Barbee began
to draw up plans for removing facilities he was told by Director
Dickenson that he must consider the "no-action" alternative as
well. The Park Service had not made a final commitment to removing facilities there, he was told. In addition, Dickenson insisted

that Barbee provide scientific evidence that dismantling Fishing Bridge was critical to grizzly survival.[93]

To his surprise, Barbee discovered that his predecessors had never done such a study. There were a few memoranda, written between 1968 and 1972, from Cole and plant ecologist Don Despain, to Superintendent Anderson about the environmental impact of the proposed Fishing Bridge by-pass road, but these did not make a strong case for closing Fishing Bridge. "These types of vegetational units," Cole had written in 1972, referring to the habitat surrounding Fishing Bridge, "i.e., meadows and open mature lodgepole pine, are not in short supply in Yellowstone." The great ecological rationale for Grant Village, Barbee discovered, had been based on a hunch.[94]

The Park Service, therefore, now had to find a biological rationale to justify what it had already done. In January 1984, Knight submitted a habitat analysis of Fishing Bridge. Although he mentioned that the habitat surrounding Fishing Bridge was critical (which few disputed), his report also called attention to the importance of Grant. He said, "The Grant Village development has the potential for a less serious but still significant detrimental impact on the population."[95]

This was not the unqualified conclusion for which the Park Service had hoped. Calling attention to the damage Grant Village would do raised all the old questions. They would have to do it again. Landscape architect Wenk told the press in February 1984 that "biologists will reconsider information that led to the park's 1974 master plan."[96]

By November, Yellowstone authorities had published a new report which indeed showed the importance of Fishing Bridge to the grizzly (based largely on the number of bear incidents that had occurred in the area since dump closure), but which failed to mention Grant Village. But by this time the closing of Fishing Bridge had become a major controversy. On pressure from several western senators, the Park Service agreed to postpone a decision until a full Environmental Impact Statement was completed. The issue was back to square one.[97]

While waiting for these studies to be done, Director Dickenson postponed the date of the removal indefinitely. In Washington it was privately conceded that nothing would be done before 1990, and then, as one official put it, "we'll all be gone. It will be someone else's problem." Indeed, the joke then making the rounds in Interior suggested that "if we wait long enough the grizzly will be gone and then the closure will be unnecessary."

Victory has a thousand fathers, John F. Kennedy remarked (paraphrasing Count Galeazzo Ciano), but defeat is an orphan. Yet Grant Village, victorious after five decades of fitful evolution, remains, if not an orphan, then a bastard. Pushed successfully through nine presidential administrations, few admit liking it and no one will concede being the prime mover.

"Townsley did a hell of a selling job on us," Destry Jarvis said to me sadly, adding, "the accommodations at Grant Village are the worst I have seen in the National Park system — and I have seen 90 percent of them."[98]

"It is not my favorite place," Amos Eno, lobbyist for the National Audubon Society, told me.[99]

The marina, a senior official confided to me, "looks like the launch pad at Cape Canaveral."

"To call the buildings ticky-tacky," a ranger I know said, "is to compliment them."

Who then built Grant Village?

Although in construction for fifty years, no one knows. The epicenter for Grant Village planning, it seems, is always down the hall. Career civil servants hint darkly that it was a product of a plot hatched between economic special interests, congressmen, and presidents. Political appointees insist that their decisions were made at the urging of career civil servants.

Yet today without apparent advocates the village continues to grow, propelled by its own autonomous life force. Three hundred motel units have been constructed and four hundred more are planned. The environmental problems anticipated by planners have begun to surface. Sewage still spills into the lake on occasion, and grizzlies have already found new victims in the campgrounds. If maulings continue, the Park Service still intends — in defiance of the Endangered Species Act — to close the streams to spawning trout. Whether this move will stop the bears or only make them hungrier remains to be seen.[100]

Just south of the park another major project is under way. The John D. Rockefeller, Jr., Memorial Parkway, or JODR, will be, when completed, more than another road. Dedicated to "the many contributions of John D. Rockefeller, Jr. to the cause of conservation," it is to become a major "vehicular staging area," eventually to equal Grant Village in size. Although no environmental impact statement or assessment was ever done for this project, it is now off and running, justified "because the number of lodging accommodations in Yellowstone and Grand Teton will remain at roughly the present level." Yet it lies in prime grizzly habitat, near where several bears have been killed in recent years because of conflicts with hikers and hunters.[101]

Indeed, this is the final irony. Despite the master planners' intention to separate bears from people, they have actually brought these two closer together, further endangering both. For both bear and man in Yellowstone, the worst may be yet to come.

14

Gumshoes and Posy Pickers

GEORGE WRIGHT and Roger Toll left Alamogordo, New Mexico in their rented Chevrolet, turning west on the Transcontinental Broadway of America toward Arizona. Wright, then Chief of the Division of Wildlife, and Toll, Superintendent of Yellowstone National Park, were on a mission for the Park Service. Working in cooperation with the government of Mexico, they were scouting new areas for national parks along the international boundary. They had just left Big Bend, a hunk of the Chisos mountains on the Rio Grande that they were considering as a national park, and were headed for a beautiful desert on the Mexican border near the town of Why, Arizona, which later became Organ Pipe Cactus National Monument. It was a good day for driving, that February 25, 1936, when these two men met their fate near the town of Deming, on their way to Why and fifty miles south of a town that would later be known as Truth or Consequences.[1]

Part of a convoy with American and Mexican officials, Toll was driving the lead car; Wright was in the passenger seat beside him. Coming toward them was a sedan with a family of tourists from Connecticut. Some distance from them a rear tire of the Connecticut car blew. It swerved and then stabilized, apparently under control. Just as the two vehicles closed on each other, however, the Connecticut car lurched head-on into Toll and Wright. Toll was crushed by the steering column and died instantly. Wright's head hit the dashboard, sending his jawbone into his brain like a spear. He died about five hours later.

This collision in the lonely New Mexico desert had greater

than ordinary consequences, and as a tragedy it was far more than personal; it changed the course of the Park Service to this day and it may have closed the last opportunity to save Yellowstone National Park. In history and politics George Wright occupied, for a few brief years, a unique position. He had the opportunity and vision few ever do: to change a bureaucracy; to close the gap between appearance and reality in the Park Service.

Few federal agencies are esteemed as highly by the general public as the National Park Service. Since its creation in 1916 it has been thought of as the one arm of government that we can trust. Its mission of preserving our wild lands was a goal we sanctified, giving the Service a purity or purpose that was beyond reproach. The rangers, in their green and gray uniforms and Boy Scout hats, projecting an image both strong and wise, epitomized this goodness. They were the uncorruptible guardians of our precious national heritage, and they encouraged an impression of complete competence. They were Smoky the Bear, Sergeant Preston, and Ranger Rick rolled into one. The parks, we felt justified in believing, were safe entrusted to them.

Yet who are the rangers? How, if they are as good as they appear to be, could Yellowstone have suffered so sharp a decline? As we have seen, over the last seventy years nearly every conceivable mistake that could be made in wildlife management has been made by the Park Service in Yellowstone. Not a year has gone by since it assumed responsibility there when the National Park Service, an agency created to preserve our wild heritage unimpaired, did not kill an animal in the name of an environmental ideal. Today its management policies threaten the very capacity of the park to sustain life. How could such an apparently good, enlightened agency do such things?

The story of George Wright, his tragic death, and all that followed, will give us the answer.

Like many other freshly minted college graduates, Ben Thompson began to worry what he would do next. Although he had just received a degree in philosophy from Stanford, he now knew he wanted a career that had something to do with wildlife. For two summers he worked as a waiter at the Ahwahnee Lodge in Yosemite, and, at the end of summer 1929 he went to work there full time. Waiting on tables he knew would not be his life's work, but at least it allowed him to remain near the wilderness he loved until he found a vocation.[2]

Meanwhile in his spare time he began to visit the Yosemite museum to learn about the park. Behind the desk at the museum Thompson found a knowledgeable and enthusiastic young ranger two years older than he who greatly increased his knowledge and enthusiasm for wildlife. They became good friends. The ranger's name was George Wright, and although Thompson did not know it at the time, in meeting Wright he had found his calling.

Wright had joined the Service two years earlier, attending the recently established school of Field Natural History in Yosemite. A former student of forestry at Berkeley, he had come to the Park Service with a great love of nature and high hopes for a Park Service career. He also came to the Park Service with a mission. Our national parks, he knew, were wildlife museums holding an irreplaceable gene pool of original fauna; yet few in the Service were qualified to care for this treasure. The Park Service at that time was run by policemen, engineers, lawyers, and soldiers, not wildlife biologists.[3]

When Wright joined the Field Naturalist School in 1927 the Service had not one wildlife research specialist. Instead of running our parks as outdoor museums, the founders of the new Service had another plan. Attracting visitors was highest on their agenda. They must build their political constituency by bringing as many people into the parks as possible — people they hoped would later advocate an expanded park system. With this goal in mind they built the ranger corps to be experts in law enforcement, dedicated to "visitor safety and protection."[4]

"Naturally, when the system was young, its first needs . . . were protection and proper direction," Albright wrote in 1933. Indeed, the Park Service's first director, Stephen Mather, a businessman and mining engineer, and his successor Albright, a lawyer, although astute politicians, had little knowledge of or appreciation for science. Rather than populating the new agency with scholars from universities they recruited soldiers from the army. Protection became the byword: protecting wildlife from poachers and visitors; protecting ungulates from predators; protecting visitors from bears, hot pools, and themselves.[5]

In creating the Park Service as a clone of the cavalry, however, Mather and Albright stamped it for all time. The new rangers had little special education. In those days it was woods savvy and ability to handle firearms that counted most. Accordingly, they were classified by the civil service, not as professionals, but as belonging to the "C (for custodial and crafts) Series" — a general-purpose blue-collar category. They needed only a high school diploma.[6]

Any efforts to understand wildlife, therefore, occurred by accident. It was by accident that Albright, when Superintendent of Yellowstone in 1919, met Milton Skinner, a man of independent wealth who had worked at various jobs in the park since coming there as a college student in biology in 1898. Skinner so impressed Albright that the latter created a position (without pay) for the former, called "naturalist." Skinner's duties were both research and instruction for visitors in the natural history of the park. During the next two years Skinner established a museum and began a program of public education. Yet it was Skinner's work as teacher that most interested Albright, for this was clearly a direct service to the public. Finding that these duties left little time for research, Skinner left in 1922 for a position at Syracuse University.[7]

Other naturalists followed Skinner at Yellowstone and at a few other parks. Yet by the time Wright came to Yosemite, the Service had only five naturalists, and their work, like Skinner's, was primarily in education rather than research. Although most were true professionals — qualified scholars with advanced degrees and classified by the civil service as belonging to the "P" (professional) series — they were considered neither real rangers nor real scientists by the Park Service. In fact, they were outsiders, having no hope of climbing the career ladder or of becoming superintendent. Their advice, therefore, was usually ignored by park managers. "Not uncommonly," as Park Service historian Frank Brockman described their status, "they were referred to by some of their associates as 'nature fakers,' 'posy pickers,' or 'Sunday supplement scientists.' "[8]

In return, the park naturalists were horrified at the ignorance they found among rangers and their policies. Milton Skinner wrote in 1928, "Almost nothing has been written to show how unscientific, how careless, we have been in the Yellowstone National Park in the past. There have been no wholly adequate studies made and very little is positively known about the wild life in the Park and the interrelations of the various species of its wild plants and animals. Careful, minute investigations should be made, on which a

wise general policy for the care of all wild life, and for such control
as may be needed, can be based."[9]

Likewise, independent scholars pleaded the case for science in
the parks. Ecologists such as Joseph Grinnell tried for years to
persuade the Service to hire scientists. But such persuasion was
unsuccessful. As parks were multiplying, only a few noticed that
wildlife was dwindling. Instead the policies of ignorance prevailed.[10]

Some of this sad story was familiar to Wright when he joined
the Park Service immediately after graduating from the University
of California. As Grinnell's student, he entered the Service with
a deep conviction of the need to better understand nature. And
possessing considerable personal wealth, he had the means to act
on his conviction.[11]

By winter 1929 he had formulated a plan. He would take a leave
of absence from the Service, hire a team of biologists at his own
expense, tour the park system, take inventory of the wildlife and
their problems, and make recommendations on management to
the Director.

He asked Thompson, "How would you like to join the team?"

In the next summer Wright took a leave from the Service, hired
Thompson and Joseph Dixon (an assistant professor at the University of California and another protégé of Grinnell's), obtained an
office and laboratory space in the town of Berkeley, packed a car
with gear, and set off on a cross-country tour of our national parks.
So began the research that became known as the fauna series.[12]

Over the next four years they made many visits to our national
parks, returning to their laboratory to analyze results. Out of these
efforts came two volumes that Wright and his colleagues persuaded
the Park Service to publish, entitled *Fauna of the National Parks
of the United States,* and known simply as Fauna Number One and
Fauna Number Two. The first was an inventory of wildlife and its
problems, and the second a set of concrete recommendations for
management. For the Park Service this fauna series was a catalogue
of both shock and hope: shock, for it opened eyes to the dreadful
danger facing the parks; and hope, for it seemed to point the way
to reform. They became classics in the field of wildlife ecology.[13]

In Yellowstone, Wright's team discovered the elk problem, first
learned of the many animals already extirpated there, and exposed

(along with Grinnell) the Park Service's policies on control of predators. Thompson (then writing his master's thesis on the white pelican) exposed the plans Park Service personnel had to stamp out the pelican by sending rangers to stomp on eggs at the Molly Islands. Their work was, by every measure, a brilliant inventory of America's wildlife and its problems, and a clarion call to reform.[14]

This work also carried a revolutionary doctrine, a prescription for managing parks by active science that flew in the face of the rangers' reliance on passive methods of protection. Over the history of our national parks, Wright and his colleagues wrote in Fauna Number One, "the policy of noninterference with wild life became more and more deeply entrenched." Yet this policy, "far from being the magic touch which healed all wounds," was insufficient to save our parks. They wrote, "The need to supplement protection with more constructive wild-life management has become manifest with a steady increase of problems both as to number and intensity." Citing the example of the cavalry's successful breeding program to save the Yellowstone bison, they urged a new philosophy of "intensive management."[15]

The response, in fact, was terrific. In 1931 Director Albright, trying to expand the Service's educational programs, created a Division of Research and Education; and within this new agency established two years later a scientific arm to be called the Wild Life Division. Henceforth no management decisions were to be made until approved by the director of this new agency. Wright (by then only twenty-seven) was named its first director. In 1933 Albright wrote, explaining his support of Wright's efforts, "Of recent years it has become evident that ranger protection and restocking are not sufficient for the complete preservation of the wild animals."[16]

Receiving funds from Roosevelt's Civilian Conservation Corps (CCC), the Wild Life Division grew in power and influence. Wright toured the country, scouting talent, looking for wildlife biologists. He recruited people like Russ Grater, a naturalist at Grand Canyon; Victor Cahalane, a forester working for the state of Michigan; and Lowell Sumner, a biologist who had recently completed his graduate training in Berkeley.[17]

By 1935 he had formed a team of twenty-seven young and prolific scientists. In seven years, according to Sumner, the Wild Life Division produced more than a thousand reports; in the year 1936–1937 alone, it published seventy-six articles. More classics were added to the fauna series: Fauna Number 3, *Birds and Mammals of Mount McKinley National Park,* by Joseph Dixon; Fauna Number 4, *Ecology of the Coyote in the Yellowstone,* and Fauna Number 5, *The Wolves of Mount McKinley,* both by Adolph Murie.[18]

Yet as it grew, the Wild Life Division made bureaucratic enemies. In urging active management Wright not only challenged the philosophy cherished by the rangers, but also offered a prescription — emphasis on science — they were not qualified to fill. Many in the Park Service's Division of Forestry in particular felt threatened by Wright's ecological approach, which included opposition to pesticides and advocacy of prescribed burning. "I remember having furious arguments with John Coffman (Director of the Forestry Division) about fire," Wright's successor Victor Cahalane told me recently. "He had the conventional idea that any fire over one foot square was a disaster. I argued that in areas such as Yosemite Valley fire was the only way to preserve the natural vegetation."[19]

Other resistance came from the U.S. Biological Survey of the Department of Agriculture. Cahalane told me, "They thought that all biologists should belong to them."

But despite this resistance, the Division of Wildlife flourished under Wright. His grand ideas came through the muddy sea of bureaucratic intrigue still afloat. "He had vision," Thompson told me, "a way with people, and powerful friends in Washington." Cahalane told me, "He had extraordinary common sense." Sumner said, "He had the ability to pour oil on troubled waters."

Extraordinary vision, powerful friends, common sense, diplomacy — all were surely part of the power that enabled Wright to do what he did, and it made the era of the Wild Life Division a golden period. For wildlife biologists this was their Camelot. At last, it seemed, our parks would be preserved by science and not the gun, by ecologists, not policemen. Although Wright's ecological ideas were considered radical by the oldtimers in the Service, he had what was needed to put them through.

These hopes came to an end on that clear February day in New Mexico in 1936. Victor Cahalane succeeded Wright as Director of the Wild Life Division, but Cahalane was a scientist, not a political man, and could not contain the counterrevolutionary forces growing within the Department of Forestry and the Biological Survey. The waters, unoiled, began to roil.

In 1942 the Roosevelt administration, gearing up for war, abolished the CCC, and with it the Park Service's source of soft money for the Wild Life Division dried up. At the same time Interior Secretary Ickes persuaded Roosevelt to transfer the Biological Survey from Agriculture to Interior. The Wild Life Division was abolished and its staff sent to the Biological Survey, which then became the new Fish and Wildlife Service of the Department of Interior.[20]

A diaspora of Park Service scientists soon began. Rather than join the new bureaucracy, many quit the government altogether; but the remainder, including Sumner and Cahalane, were simply transferred to the new agency. Thompson stayed with the Park Service, but not as a scientist. He had been transferred to the Service's land-acquisition department. In the end, only three biologists remained in the Park Service.[21]

That the Wild Life Division was absorbed by the much greater Biological Survey was ironic, and for the animals, this merger spelled bad news. For forty years the Biological Survey had been devoted to exterminating animals, not preserving them. Wright's scientists, therefore, all recruited for the purpose of preservation, found themselves greatly outnumbered in the new agency. "We felt very alone," Cahalane told me. "There were only a few of us who wanted to protect animals, for the money [in the new agency] was in extermination, not protection."[22]

Meanwhile, the Park Service, having lost its best talent, entered a dark age. By the end of the war it had eight biologists, all classified as naturalists. As naturalists, however, their chief duties were educational; research had to be done in their spare time. Yet perversely, the Service tried to discourage their ad hoc attempts at research. In 1939 it formally changed the name of the division from Research and Education to Education, and in 1953 to Interpretation. This alteration fixed their role as tour guides, making it risky business for them to carry on research.[23]

Consequently, next to nothing was done in Park Service science for the seventeen years following World War II. No new publications such as the fauna series appeared. In 1945 a Park Service document inspired by Cahalane, "Research in the National Parks," identified seventy-seven biological research programs that needed funding, but little happened. In 1955 Cahalane quit in disgust over plans for Mission 66 that called for raising more than $1 billion for roads and visitors' facilities in our national parks, but not one cent for the study of wildlife. Not until 1958 did the Park Service, for the first time in its history, use its own appropriated funds for scientific research. The amount was $28,000, out of a Park Service budget that exceeded $40 *million*. "Meanwhile," as Sumner said, "in the parklands themselves, biological time bombs had gone on ticking through all the years of inattention."[24]

One such time bomb was the Yellowstone elk problem, and the great elk reduction there during winter 1961–1962 prompted the first review of Park Service science in thirty years. Secretary of the Interior Stewart Udall created the Advisory Board on Wildlife Management, whose report (the Leopold Report), as we have seen, urged that "a reasonable illusion of primitive America could be re-created, using the utmost in skill, judgment and ecological sensitivity."[25]

This goal they said required "ecologic skills unknown in the country today." But if restoring our national parks required such advanced ecological knowledge, how was the Park Service, which spent less than $30,000 a year on research and had no full-time scientists, to accomplish this task?

This was the question that Udall charged the National Academy of Sciences to answer. In response, the National Research Council of the National Academy created a committee under the direction of William J. Robbins of the National Science Foundation to evaluate Park Service efforts in science and make recommendations for "a research program designed to provide the data required for effective management, development, protection, and interpretation of the national parks; and to encourage the greater use of the national parks by scientists for basic research." The findings of this committee, known as the Robbins Report, was a devastating indictment of Park Service efforts.[26]

Robbins was less optimistic than Leopold that a reasonable illusion of primitive America could be re-created. While "viewing with sympathy the ideal of making a national park 'a vignette of primitive America,' " the writers of the Robbins Report observed, they pointed out the "difficulties in even approaching such an ideal. In some instances because of the paucity of historical records it would be impossible to determine what the condition of a particular park was when the white man first saw it. Changes, some irreversible, and current activities, in some instances impossible to control . . . suggest that the ideal, though admirable may not be fully attainable. . . ."[27]

By contrast, the Robbins committee emphasized that parks cannot be left to run themselves. Like Wright thirty years before, they urged that wildlife inevitably needs human help, and this task in turn requires our best scientific efforts. They stated, "No national park is large enough or adequately isolated to be, in fact, a self-regulatory ecological unit," thus "limitation of herds of elk, supervision of visitors in a park, control of water levels . . . controlled burning . . . are necessary functions of management if a park is to survive in anything like the condition which meets the purpose for which it was established."[28]

Thus, it concluded, "This Committee believes that management of national parks is unavoidable," where "management" was defined as "activity directed toward achieving or maintaining a given condition in plant and/or animal populations and/or habitats in accordance with the conservation plan for the area."[29]

"Carrying out these responsibilities requires," the Robbins Report continued, "knowledge about the parks and their problems and this can only come from research." Yet research, the committee declared,

> lacked continuity, coordination, and depth. It has been marked by expediency rather than by long-term considerations. It has in general lacked direction, has been fragmented between divisions and branches, has been applied piecemeal and has suffered because of a failure to recognize the distinctions between research and administrative decision-making, and has failed to insure the implementation of the results of research in operational management. Too few funds have been requested; too few appropriated. In fact, the Committee is not convinced that the policies of the National Park Service have been such that the potential contributions of research and a research staff to the solution of the problems of the national parks is recognized and appreciated. Reports and recommendations on this subject will remain futile unless and until the National Park Service itself becomes research-minded and is prepared to support research and to apply its findings. . . . There are simply too few research people and these few are inadequately supported. The Committee was shocked to learn that for the year 1962 the research staff [including the Chief Naturalist and field men in natural history] was limited to 10 people and that the Service budget for natural history

research was $28,000 — about the cost of one campground comfort station.[30]

"It is inconceivable to this committee," they concluded, "that property so unique and valuable as the national parks, used by such a large number of people and regarded internationally as one of the finest examples of our national spirit, should not be provided with sufficient competent research scientists in natural history as elementary insurance for the preservation and best use of parks."[31]

To remedy this disastrous situation, the committee recommended, among other things, that spending on research, then less than 1 percent of the budget and the lowest of any government agency, be raised at least to 10 percent; that "a permanent, independent, and identifiable research unit should be established," that research should be "mission oriented; that is, it should be concerned with the problems involved in the preservation of the natural features of a park"; that investigators be "free to pursue experiments which are in their judgment the most promising" and that the Park Service recognize "that the results of research cannot be predicted or prejudged" nor will they "always be pleasant"; that scientific work be of sufficient quality "that the results of research undertaken . . . should be publishable and should be published"; that local "research laboratories or centers should be established for a national park when justified by the nature of the park and their importance of the research"; that "consultation with the research unit in natural history of the National Park Service should precede all decisions on management operation"; that "interpretative staff should be well grounded in science"; that "universities, private research institutions, and qualified independent investigators should be encouraged to use the national parks in teaching and research," and that "a Scientific Advisory Committee for the National Park Service should be established."[32]

Udall praised the Robbins Report, and so too did the Director of the Park Service, who promised to implement the recommendations. "Those relating to organizational deficiencies will be corrected by the reorganization proposals now pending," the Director wrote to Udall on receiving the report. "The remaining difficulties," he added, "involving funding, will be corrected, insofar as possible, within the resources available to us."[33]

The Park Service, however, was preoccupied with other problems. The chickens sent out by Mather and Albright had come home to roost. By the mid-1960s the Park Service still had only two categories of personnel: rangers and naturalists. All top-management positions — from Superintendent to Director — came from the ranks of rangers. Yet according to law the ranger division — the 025 series as it is known — was still classified as nonprofessional by the Office of Personnel Management. There were no educational prerequisites beyond high school to be a ranger, and although most by that time had a college degree, top positions went to people with training in maintenance, law enforcement, or landscape architecture, not biology or history. Rangers were experts in "visitor safety and protection." They were park policemen, but they ran the service.[34]

By contrast, the naturalists were still classified by law as professionals, and entrance into that division required specialized higher education in biology, geology, history, archaeology, and anthropology. "The first and fundamental requirement for the performance of professional interpretive work," stipulated the 1962 Park Service classification guidelines, "is possession of a thorough knowledge of and professional competence in, the subject matter field to be interpreted." But despite their competence, interpreters had no hope for advancement through the service.[35]

This organization invited acrimony. One group ran the Park Service, but lacked the qualifications to do so; the other had the qualifications, but lacked the opportunity for promotion. Rangers continued to call naturalists posy pickers; naturalists called rangers gumshoes. Friction mounted.[36]

To cope with this problem the Park Service in 1966 established the Field Operations Study Team (FOST) to devise a solution. This study, promising a "shredding out" of the "non-professional tasks from professional positions," using in part a "psychological approach . . . to gain acceptance of the change with a minimum of resistance" from the rangers, did something nearly inexplicable. It abolished the naturalist classification, putting both interpreters and rangers into a newly defined ranger category. Henceforth all rangers (including naturalists) were to be called "Park Specialists." The FOS Team defined this new class as "professional work" that was "interdisciplinary in nature and scope."[37]

Yet this system was contrary to law. Despite what the FOS Team claimed, rangers, classifications specialist Raymond E. Moran of the Office of Personnel Management told me recently, "are not professionals. The law is very strict about what it calls 'professional.' That term is limited to those who have completed a specific course of academic training." Minimum standards for rangers, however, required no college at all. Although academic work in the liberal arts, police science, or park and recreation management could be counted toward qualification, work experience in the Service would do. Besides, anyone with true professional qualifications was automatically *disqualified* from being a ranger. "The following kinds of positions," stipulated the government's official classification guidelines for the Park Manager Series, "are excluded from this series: . . . positions which require . . . a thorough knowledge of, or professional competence in, a field of the biological or physical sciences . . . historical research . . . archaeology, geography, or other specific social science fields."[38]

The FOS Team, in short, had abolished the one true professional category in the Park Service, but obscured what they had done by illegally reclassifying rangers as professional. And to further obscure what it had done, the committee created a new "technician series": a category of lower-ranked workers that served as a blue-collar contrast to the new "professional" series. Although creating a standard that appeared lower than that of rangers made it seem that the latter category had been raised, in actuality — just like college professors inflating students' grades — Park Service standards had been lowered another notch.[39]

To no one's surprise, this "psychological approach" gained acceptance among rangers. Although the prerequisites for their work had remained unchanged or in some instances had actually been lowered, they had miraculously (if unofficially) been granted professional status. In this way the FOS Team achieved its objective. The conflict between gumshoes and posy pickers was resolved by obliterating the posy pickers. Henceforth naturalists would be just plain rangers, and thus, in theory, they had, for the first time, equal opportunity to climb the Park Service's career ladder.

But the price of giving naturalists the chance to succeed was lowering their qualifications. Contrary to the recommendations of the Robbins committee that "interpretive staff should be well grounded in science," they no longer needed professional competence in any field at all. Now a high school diploma in addition to appropriate work experience would suffice. Any course of study in the liberal arts was fine, but if anyone hoped to make superintendent some day, he or she had better have more than a smattering of courses in law enforcement. An individual with advanced educational training could still be an interpreter, but this training was no longer considered *relevant*. The past belief that it was important, according to the new Park Service textbook on interpretation, *Interpreting our Environment,* rested on "two debatable concepts. First, it is doubtful that people come to parks with the idea of being educated. Second, there is no correlation between possession of a degree in natural sciences and the ability to communicate."[40]

"Rather than total or overriding emphasis on the sciences," the textbook continued, "the formal education of the interpreter should be one stressing balance. If interpreters are anything, they are *communicators.*" As communicators, their personal traits counted more than knowledge and so personality became the most important criterion. The roster of qualifications began to read like a charm-school manual. Interpreters must have "sparkle," have "enthusiasm," "a sense of humor and perspective," "articulateness," "self confidence," "warmth," "poise," "credibility," and a "pleasant appearance and demeanor."[41]

For those then serving as naturalists, the FOST recommendations posed a dilemma. They had to choose (1) finding work as scientists, (2) joining the new cadre of ranger-interpreters, or (3) leaving the Service altogether. Because there was little money for research and because scientists' opportunities for advancement were nearly nil, most quit the Service or became rangers. Another diaspora of qualified Park Service personnel ensued. Many professional categories were decimated. Some, such as historians — whose work the Leopold Report urged should be "the first step in park management" — almost ceased to exist, for in every park they were replaced by technicians or interpreters with little or no scholarly training. Former Chief Historian Robert Utley explained to me recently that "FOST deprofessionalized historians at the field level. For the most part qualified historians simply ceased to exist in the national parks."[42]

With scholarly expertise abolished from interpretation, the purpose and methods of the branch were redefined. Interpretation was no longer called an educational activity. "Interpretation," stipulated the new Interpretation and Visitor Service Guideline "is not designed as an educational or entertainment service in its own right." Instead, it is "an integral function of overall park management," and a principal purpose was now "to promote public understanding and acceptance of the Service's policies and programs."[43]

Interpreters were no longer experts who instructed visitors on the natural history of the park, but "communicators" who carried the message of the Park Service to the public. Scholarship was not only irrelevant, but a handicap.[44]

Scholars were pictured as pedants and disparaged as specialists who knew much about little and had no ability to communicate. Old-time naturalists were described as quaint fuddy-duddies who, although well-intentioned, did not know the score. Freeman Tilden, a former newspaperman whose *Interpreting Our Heritage* became the Bible of the new vision of interpretation, wrote "To the specialist, the use of metaphor is calamitous, and the simile is almost an obscenity." He is a person, Tilden wrote, who is impatient "that the public does not show sufficient interest in his assemblage of information as such. He is likely to conclude the average person is somewhat stupid." The earlier generations of naturalists, Tilden condescended, "were unduly impressed by the word 'education.' "[45]

The division of naturalists, originally created as the field of research and education, later forbidden to do research, became an agency forbidden to engage in education as well. Instead, under the guise of a newly defined term — "interpretation" — they were to be "communicators," propagandists for the Park Service. To train these people, the Park Service built a center for interpretation in Harper's Ferry, West Virginia, where lesson plans and visitors' guides were written and printed, and where naturalists were sent for training.

Many naturalists rankled under this new regime. Having lost their last remaining bit of academic freedom, they objected to the cavalier way in which they were being relegated to serving management. And contrary to what the authors of the FOS Team had hoped, they did not lose the stigma attached to their calling. Few were appointed to a superintendency. Rangers still refused to consider them as true "greenbloods" (a Service expression for good rangers), and scientists privately ridiculed them as "barbershop biologists."

"Interpreters," one scientist told me recently, "exist only to apologize for Park Service mistakes." He continued,

Consider a recent incident that occurred in Hawaii: The Superintendent of City of Refuge National Park on the island of Hawaii was walking on the beach of the Park with his Chief Naturalist and a visiting Park Service scientist. The ground there was sacred to the natives, to whom it had been a sanctuary, much as churches were in the West. Its beach had been kept as one of the most pristine in the world. As the three men walked the shore they came to a two-hole outhouse. Beneath the outhouse and descending across the sand to the ocean was a yellow stream of urine. In the ocean where the two bodies of water met was a gigantic algae bloom.

"What the hell is going on here?" asked the scientist. The Superintendent, embarrassed, looked at his Chief Naturalist.

"That's all right," said the naturalist, "We'll interpret it for the public."

"Interpret, hell," replied the scientist. "Get it the fuck off the beach!"

That naturalists had been shorn of their qualifications in defiance of the Academy's recommendation that they be "well grounded in science" did not bode well for the future of research. Yet hopes remained high that a renaissance was in the offing. Indeed, there seemed reason to hope. In 1964 the Park Service created, as the Robbins commission had recommended, an Advisory Committee on Natural History Studies, consisting of Starker Leopold, Stanley Cain (then Assistant Secretary of the Interior), and Sigurd Olson; and it hired Dr. George Sprugel of the National Science Foundation to help set up a science program.[46]

During the next decade science research centers were created in Everglades National Park, Glacier, Hawaii Volcanoes, Saint Louis Bay, Mississippi, and elsewhere. Cooperative Park Study Units were established at several universities to encourage independent research and to promote professionalization among Service scientists. A new category of ranger called Resource Manager was created whose duties were specifically to write plans that would protect the parks. Expenditures for research reached $105,500 by 1965. *The Wolves of Isle Royale* was published in 1966, the first of the fauna series to appear since 1942.[47]

For those who hoped for renaissance, however, the bloom was soon off the rose. Sprugel found himself facing one roadblock after another. "The Robbins Report stirred a lot of resistance in the Park Service," Sprugel, now retired, told me recently. "They simply were not ready for change. The old-timers in the field never had any use for science and they were not about to give in." The problem, Sprugel told me, was money. He simply could not persuade the Park Service to put research in the budget. No separate science unit — like the earlier Wild Life Division — was created. Instead, although research monies increased, they came from soft and uncertain funds. Drawn from "overhead," they could disappear overnight.[48]

Nor could Sprugel get much help from the Advisory Committee. Leopold lived in California and Olson in Minnesota and neither had day-to-day contact with the Service. Although Cain was Assistant Secretary of Interior, his power for change was less than met the eye. "The Director controls parks in more than three hundred congressional districts," Dr. Robert Linn, then a research botanist in Sprugel's office, told me recently. "This often gives him political clout the Assistant Secretary does not have. For this reason it is difficult to accomplish any reform against the wishes of the Director." And directors, usually taken from the ranks of rangers, had little interest in science.

Sprugel, working with a handful of dedicated and frustrated scientists, felt very much alone. He finally quit in disgust. Linn succeeded him as Acting Chief Scientist and a year later Starker Leopold replaced Linn.

Despite Sprugel's frustrations, however, research personnel and funds continued to grow. A new classification, the Research-Grade Evaluation, was devised by which government scientists would be measured, not by their superiors in the Service (who were rangers) but by independent review boards of university professors. Park scientists, as Robbins had hoped, would enjoy academic freedom while on a regime of publish or perish. Their avenue to promotion would lie, not in conducting studies that would make the Park Service look good, but in publishing with scholarly journals where their work would be "refereed" by other scientists in the same field. This system would keep government research open to peer and public review, where its claims could be verified.[49]

By the mid-1970s, however, many of these early advances had turned into retreats. Research-grade evaluation was seldom used. "I was the only historian in the history of the Park Service to be on research-grade evaluation," Park Service Chief Historian Edwin Bearss told me recently. This classification never caught on, Bearss explained to me, "because rangers did not want a lot of GS-14s running around unsupervised."[50]

Such free-ranging scientists were seen as a threat. Because they were to be graded by university scholars, the Park Service would have no control over them. And it was perfectly possible that the review committee might promote a scientist to a rank higher than that of his own ranger-supervisor. In Everglades, for instance, the chief of the research laboratory had a higher rank than did the Superintendent of the Park.[51]

The research laboratories themselves, rather than being fore-runners of others, were the last of a kind. A proposal by Wyoming Senator Clifford Hansen in 1972 to establish one in Yellowstone was successfully resisted by the Service. Although at this time millions were being spent on Grant Village (and tens of millions more were in the pipeline), Yellowstone authorities rejected the project as too expensive. Besides, wrote Superintendent Anderson in 1972, "This proposal was made from outside the National Park Service, and . . . has not been a part of the Master Planning for Yellowstone National Park. Such construction would alter the natural area."[52]

Resource managers, far from the vanguard of a scientific revolution, became, following the recommendations of FOST, rangers, not scientists. These people, usually having little advanced scientific training, were poorly qualified for their work, and they knew their opportunities for promotion lay, not in scholarship, but in law enforcement. Their supervisor was the Chief Ranger, who, as the park's top policeman, had little appreciation of scientific knowledge in running the park. Yet rangers, whose pay and grade depended in part on the number of people they supervised, would not release their authority. Resource managers became captive pawns in a bureaucratic power struggle.[53]

In the end the FOST Report did as much damage to science as it had done to interpretation. "FOST," one official said to me, "was a palace revolution." It made official what had hitherto been unofficial: that rangers ran the Park Service. Police science was the training that counted most, and the concept of visitors' safety and protection became indelibly part of the condition many called "the ranger mentality." And when riots, for the first time in Park

Service history, broke out in Yosemite in 1970, emphasis on law enforcement was further sanctified. All rangers were required to take more than two hundred hours of law-enforcement training. For the first time they began to wear guns; their cars were outfitted with radar. They became full-fledged policemen from their gumshoes to their gumballs. Writes Bernard Shanks, an ex-ranger now a natural resources expert working for the state of Arizona, "The new rangers were more interested in firearms than flowers, more oriented toward hand-to-hand combat than mountain-climbing or hiking."[54]

The Robbins Report was soon forgotten. Interpreters and scientists were there only to serve management. For wildlife this reading meant trouble. By enthroning rangers as the guardians of our parks, the requisites of visitors' safety and protection became the guiding philosophy governing all management. Law-enforcement solutions would be sought to ecological problems. It was a situation that invited irony. "Imagine," one senior official said to me, "the government lining up all the security guards of the Smithsonian and saying, 'Listen up you guys, one of youse is going to be the Curator!' Yet that is just what has happened in the Park Service. Our national parks are outdoor museums, yet they are run by policemen, not by historians or scientists. Now you see why they are in trouble."

The policy of ecosystems management, as it was interpreted by the Park Service, became tailored to a ranger corps dedicated to visitors' safety and protection. For it was nothing more than a prescription for leaving nature alone — "the policy of noninterference with wild life," George Wright called it — a philosophy of management most thought had long before been consigned to the dustbin of scientific mistakes along with Ptolemy's theory of sun motion, the aether, phlogiston, and the wild evolutionary ideas of Lysenko. And so, although policemen were needed to keep people from hurting the ecosystem, scientists were unnecessary. Resource management became people management. In this way the ranger philosophy was consecrated with all the trappings of a biological theory. No one noticed that natural regulation was just another name for "visitor safety and protection."

Denied budget, authority, and leadership, scientists became a separate caste. One told me, "We can do anything we want to do so long as we do not interfere with management. If someone interferes, he is fired; if he can't be fired, he is transferred, and if he can't be transferred, the rangers build a wall around him."

The Park Service went to work building a wall around science. Researchers received little direction or supervision. The stricture of the Robbins commission — that all research be published — was preached more than practiced. Publication was simply not one of the criteria by which people were judged. As Bearss put it, "the Park Service is project oriented, not publication oriented." Those who did publish often found themselves in trouble with their superintendent when results did not confirm policy. No research was better than embarrassing the Service.[55]

Instead, most realized that their real role was — like interpreters — to be apologists for management. If the Park Service no longer wanted to kill and trap elk, biologists would find a theory demonstrating that reductions were unnecessary. If the Service wanted to close the garbage dumps and reduce bear problems, park scientists would find another "wild, free-ranging" population of grizzlies that never touched human food. If it wanted to build Grant Village and this project required destroying 5 percent of the cutthroat trout in Yellowstone Lake, park biologists would conclude that "loss of [this] spawning area would be acceptable."[56]

This role was an insufferable one for those accustomed to the clear air of academic freedom.

"I was the first and last independent government scientist to work in Yellowstone," Kittams told me. "I was hired in 1949 to do research independent of management. But when Lon Garrison came to Yellowstone [in 1957] he had firm ideas that all scientists should be under management. As I did not agree, I left. Bob Howe, who replaced me, was, therefore, a management biologist, not a research biologist."[57]

Similarly, other government scientists left their jobs, disillusioned. In Everglades, Dr. James Kushlan quit to join a university faculty after his ten-year study on the effects of water level on alligators was suppressed. Although it was conventional wisdom among park managers that the park was suffering from insufficient water, Kushlan found that the reverse was true: high water was flooding alligator and woodstork nesting sites, threatening these species. Proper management, he suggested, required manipulating the sluice gates in the levees in order to reenact fluctuations of water that existed in primeval times. But this, Kushlan found, was not a popular conclusion.[58]

"The Superintendent told me he did not even want to read my study," Kushlan told me. "If he read it he would have to design a management plan to satisfy it," Kushlan explained, "and manipulation of water levels was a lot more risky than doing nothing. The less information the Superintendent had, the easier it would be for him to sway with the political wind." Instead the Superintendent suppressed the study, and later the chief of the South Florida Research Center followed with an order prohibiting all scientists at the Center from publishing the results of their work.[59]

Finding these restrictions unacceptable, Kushlan, along with several professional colleagues, left the Service. "It is impossible to do any long-range study [two years or more] in the National Park Service," Kushlan said. "Yet alligators take fifteen years to reach maturity. There is no peer review and efforts are continuous to discourage publication. Consequently the Park Service has spent $1.5 million a year for seven years at the Everglades research facility and has nothing to show for it."[60]

In building a wall around scientists, rangers also built a wall around themselves. Insulated from the truth, they learned only what others wanted them to hear. Their management, Shanks said, became "amateurish and political." Indeed, according to a 1972 study of Park Service decision making by Daniel H. Henning of Eastern Montana College, "Wildlife rangers would not recommend anything concrete that the Superintendent might not support." For this reason, he concluded, "the role of the 'outside' wildlife specialist to recommend policy remained paramount."[61]

Indeed, independent scientists had — as we have seen — at crucial times played a decisive role in causing the Park Service to abandon harmful policies. Theodore Roosevelt, a talented amateur ecologist, and William Rush, an independent professional working for the state of Montana with private funds, were the first to see that the elk were overgrazing the range. Joseph Grinnell first blew the whistle on Park-Service practices of predator control. George Wright himself was quasi-independent, for he was (at first) required to do his research with his own funds. The Craigheads, and the team of biologists from the University of Montana assembled by them at the Cooperative Wildlife Research Unit at Missoula, spearheaded not only grizzly research but also the most detailed study of the northern elk range ever done in Yellowstone. It was their research that the Leopold Report, as we saw, praised as a model of wildlife study to be emulated.

These considerations in turn inspired the Robbins committee to urge the Service "to support and accommodate independent research effort" and to consider "contractual arrangements with private scholars" rather than attempting to solve "every type of problem requiring mission-oriented research" itself.

Yet paradoxically, independent research became one of the casualties of "reform." The language of the Leopold Report, urging that research "should be in the hands of skilled park personnel" was used as the rationale for government control. The byword, "mission-oriented research," recommended by both Robbins and Leopold, took on new meaning. Used by Robbins as a way of saying that research should be dedicated to the mission of preserving the parks, it was now a synonym for government control. In individual parks such as Yellowstone the power of the supervisory scientist over independent research was made nearly total. Absolutely no studies would be done over his or her veto.[62]

The cooperative wildlife study units established at various campuses around the country, rather than freeing scholarship in the Park Service, co-opted the university communities. Signing agreements limiting their research to support of "interpretation," they could have no official influence on management. "It is like kissing an orangutan," one Park Service scientist said to me. "If you can kiss an orangutan, it is already tame, and once tame, it will never be wild again." So too the independent researcher, once kissed by the Park Service, was never the same again. The rewards — in money, professional advancement, and opportunities for research — if one's work pleases the Service, were sufficient to enchant the most dedicated professional.

Because Yellowstone, in particular, was a rare and wonderful wildlife laboratory, those who hoped for permission to study there did not dare risk the enmity of the Department of the Interior by openly reaching conclusions that conflicted with the prevailing policy. Only those who, by virtue of their association with past policies, were already pariahs in the park, were free to speak their minds.[63]

In Yellowstone, rule by rangers and the decline of science were translated into practices that affected the future of wildlife. Keeping people away from wildlife and "fragile resources" became the solution to all problems. Poaching, though an insignificant factor in the decline of Yellowstone's wild species, was seen as a threat to wildlife because it was a problem easily amenable to law enforcement. Grant Village was justified as a way to keep people from the "resource"; garbage dumps were closed to keep grizzlies from people; the backcountry was closed when it became apparent that the grizzly population was declining; the boundary-line area was closed when the bighorn sheep were decimated by pinkeye. Yet that Grant Village might be as ecologically fragile as Fishing Bridge, that closing the dumps might cause more grizzlies and people to die, that closing the backcountry was an irrelevant response to an ecological crisis, that these were simply law-enforcement responses to biological problems, went unnoticed by a management composed of policemen.

Rather, natural regulation, dressed in the trappings of a biological theory, gave policy the patina of academic respectability while real science was carefully prohibited. Cast in the role of defenders of official policy, park scientists were in a position to ensure that no studies would be done that might raise difficult questions. As they made the animal counts and determined what studies were done, they could prevent or manipulate those studies which might prove themselves — or the Service — wrong. Cautions raised by independent scientists who argued that the problems were more complex and might require active management went unheeded. The Park Service simply did not have people in responsibility trained to appreciate the potential contributions of scholarship. Consequently scientists, whether in the Service or out, were seen as troublemakers if they criticized policy.

The reaction of Yellowstone managers to both such criticisms was characterized by the condition Shanks called the "fortress mentality." Park headquarters in Mammoth, he commented, was a company town "with a unique social life tied almost completely to the Park Service. Interaction with outsiders is minimized." Cut off from the outside world and trained to put the Service first, rangers, rather than accepting criticism as constructive, tended to see it as a personal affront. Any critic was viewed as an enemy and Park Service responses tended to be personal rather than substantive.[64]

"When I joined the Service," one Park Service scientist said to me, "my professor advised me to remember that my first loyalty was to the truth, then to the country, and last to the Park Service. I have found that in the Service, our loyalties are supposed to be the reverse. Loyalty to the Service comes before obedience to truth."

These priorities were reflected in the actions of the Yellowstone research office. Far from trying to test the truth of their work by offering it for review by independent scholars, Yellowstone biologists published little; what was printed was more likely to appear in a popular magazine than in a scholarly or "refereed" journal (where contributions would be evaluated by specialists).[65]

And although the park had a budget for research and the Service could point to scores of projects in progress, none touched on sensitive issues. Studies on planktonic rotifers, butterflies, thermal dragonfly populations, and mushrooms were allowed to proceed while work on black bear, moose, beaver, the water table, soil erosion, compaction, elk trampling — any project, in short, that might tell whether natural regulation was a successful policy or not — was disallowed.[66]

Numerous research proposals were routinely discouraged, either by denying funds for the work or by placing so many restrictions on research that the applicant, fearing his study would be emasculated, gave up.

"I and a student of mine once wanted to do a study on the moose," Jonas told me in 1984. "Cole told us they didn't need the study because they already knew what was happening to the moose. Then they put so many restrictions in the way that we gave up and decided to do our study in Canada instead."[67]

And last year, he continued, "I called Mary Meagher to propose a study on the beaver. Such a study had been done every thirty years in Yellowstone since Seton, and because the last — mine — was completed thirty years ago I thought it was time for another." Nevertheless, Meagher, whose office dispensed countless dollars for research, turned him down. "She told me we knew enough about the beaver," Jonas said.

The possibility of doing science on a large animal or a politically sensitive subject had become so remote that most university researchers gave up trying. "The university community is being specifically excluded," Professor Peek told me. "There is no encouragement there whatsoever. They will tell you differently, but if you go down there and try to do something, first of all there would be no money, and then there would be no serious support. There is example after example of this. You could go in for a short time on a subject — like behavior — that isn't controversial, but you can't do long-range projects on subjects that address their major problems. The thing has been perverted."[68]

"The situation in Yellowstone," another biologist (who wished not to be named) told me, "is one of intransigence. Some of the Park Service scientists there have the attitude the park belongs to them, and they don't want interference from the outside. It is an atmosphere of distrust."

Today the state of knowledge in the Park Service has sunk as low as it was in the pre-Robbins era. It still has no separate research division. Efforts to ensure professional standards of researchers — such as requiring peer review and publication — have ceased; crucial management decisions are made without benefit of research. Independent study, though receiving more dollars (through the Cooperative Wildlife Study Units) than before, no longer plays a role in advising management. Superintendencies continue to be nearly the sole preserve of rangers. Yet rangers remain the least qualified of government land and wildlife managers.[69]

Science still has a shaky position in the Park Service budget. There is still no line item for research. And although the government can point to nearly $100 million appropriated each year for resource management, much of this money, according to one scientist who helped to develop the resource-management system, is spent on "police cars and new roofs for fire stations." Less than $1 million is spent on training resource managers each year.[70]

Instead of focusing on biological problems that lie within each park and are therefore perhaps within the power of the Service to solve, authorities are increasingly seeking to place blame for deterioration of parks on global changes over which they have little or no control, such as acid rain. Indeed, acid rain, according to Director Dickenson, is now "the number-one threat to the national parks." And although this surely is a grave threat, it is — unlike the fate of bighorns and grizzlies in Yellowstone — not a problem that the Park Service can do much about. Yet in 1985 it spent nearly $4.5 million on studying acid rain and air quality — more than it spent on natural-resources management training and natural-science research put together.[71]

A few dedicated people have continued to labor against the prevailing tide of ignorance, yet reform always seems just out of reach. In the early 1970s an offer by the National Academy of Sciences to review Park Service science again was quickly rebuffed. A new resource-manager training program was established in 1982, and in 1983 it graduated thirty-two rangers. But most of these "students" had no teachers. Instead they were sent to parks that had no resource managers at all, where they were expected to train themselves.[72]

Indeed, in many ways conditions today are worse than ever. Interpreters face continual pressure to remain apologists for policy. In a March 1982 memorandum, Director Dickenson, urging all regional directors and superintendents to reassess interpretive programs "that are not basic to our mission," once again reaffirmed that a purpose of the program was "to help the public understand the reasons behind management policies and decisions."[73]

The research center at Saint Louis Bay is now closed, and all but one of the scientists at the Everglades facility have been replaced by nonprofessional technicians. Richard Briceland, the present Chief Scientist (now called the Associate Director for Natural Resources) was trained, not as a scientist, but as a mechanical engineer (having written his dissertation on airfoils), and only one member of his seven-person staff is a scientist. The person responsible for training resource managers is an attorney, Carol Bickley. Because the Park Service system is organized along feudal principles that give virtual autonomy to the regions, park superintendents and regional directors (all rangers) have nearly absolute control over science.

Not one person in the Rocky Mountain region, Superintendent Barbee told me recently, was on Research-Grade Evaluation; not one anthropologist or archaeologist in the entire Service is now engaged in research; there are only three anthropologists in the Service and they are all in management positions. Although the national park system has nearly three hundred administrative units, it has only around one hundred resource managers, and many of them have no formal education beyond college. By January 1984 there were 120 historians in the entire Service, of whom thirteen had doctorates; and this number has declined since. Within a year twelve of those with a Ph.D. have left the field. Only forty-five natural scientists in the entire Service have doctorates, fewer than on the faculty of any good, small liberal-arts college.[74]

Out of a billion-dollar budget, not one cent is spent on basic research. And far from approaching the 10 percent level recommended by the Robbins Committee, spending on research in the Park Service (in 1985) was $14.5 million — less than 1.5 percent. Of this amount $7.5 million was spent on research in the life sciences and $208,000 on the social sciences.[75]

As we'd expect in these conditions, Park Service management of resources has worsened. "Management," wrote Park Service Research Scientist J. Robert Stottlemyer, "can go on, often for decades, in the absence of any clearly defined policy, let alone an adequate ecological foundation."[76]

"The lack of research capacity," wrote Shanks, "grew into a major handicap in recent years; when the agency shifted policy, it was often only following an ecological fad or a politically expedient trend." Even Congress found planning appalling. One congressional report commented that, "The investigation staff believes any program resulting from a process as disrespected, as untrustworthy as the NPS planning process obviously lacks reliability for congressional use in the appropriations process." The writers of the report urged that "the NPS should take immediate steps to improve the quality of its management at all levels."[77]

Far from being improved, management standards are declining further. Rangers are appointed to positions — and given a GS rating — higher than the law allows. Director Dickenson wrote to Rick Davidge in August 1982, "I am somewhat perplexed by your note in which you indicate that 'NPS is using a standard that has not been approved by OPM as a part of the selection process for Superintendents and Regional Directors.'" Yet the Director confessed that this perplexing observation was accurate. The Park Management Superintendent Guidelines, he wrote to Davidge, "were not approved by OPM [internal document]; however, we have discussed with OPM staff the possibility of OPM validating our guidelines."[78]

Rather than validating Park Service guidelines as the Director had hoped, however, OPM found that they needed reform. In particular, they discovered that the division (introduced by FOST) between rangers and technicians was a distinction without a difference. Rangers were not, as the Park Service had insisted, a professional series, and thus the tasks they performed were really no different from those assigned to technicians. "The extent," OPM wrote to the Director, "of the 'managerial' knowledge, skill and related abilities required at the GS-9 level was/and is limited." Instead, "the work performed by the ranger and the technician both require field related knowledges *(sic)* and skills." They urged that the distinction be abolished, a recommendation that would reduce rangers in rank. Whereas the old series began at GS-5, under their proposed revised standards, the ranger series would begin at GS-1.[79]

The OPM sent their recommendations to the Park Service for comment, and the Director created an "025 Task Force" composed of rangers and resource managers to evaluate them. For members of the Task Force, this assignment seemed like Christmas. Having lived with Park Service incompetence in the field at close range, their reservations far transcended questions of classification. William Supernaugh, one of the Task Force members, told me, "During my career in the Park Service I have seen a decline in the scientific knowledge expected of rangers while the complexity of the resource problems has increased." The average ranger, he continued, "is good at handling visitors but would not recognize a pine-beetle infestation if he walked through it. We had a good track record of training biologists to be cops but a poor record of training cops to be biologists. We wanted to change that."[80]

Indeed, the Task Force's response went beyond responding to OPM's recommendations, commenting, "an increasing number of ranger personnel who lack educational backgrounds in fields which most directly relate to the management of park resources"; and "present [1969] 025 classification standards did not account for the increased complexity of park ranger responsibilities brought about through recent legislation and greater variety of NPS area."[81]

They urged, "Academic requirements should be modified by the elimination of police science, sociology and behavioral sciences as qualifying fields of study. Field oriented natural science, natural resource management, earth science, anthropology, archaeology, history and park and recreation management are retained as qualifying fields of study."[82]

Their recommendations were not only more than OPM wanted, they were also more than Director Dickenson had asked for. Seymour Kotchek, Park Service classifications specialist, told me, "The Task Force was mainly concerned with improving ranger qualifications; but the Director thought their recommendations too restricting. He wanted more freedom to make appointments than the Task Force recommendations would allow."[83]

Throughout the Service anxiety ran high over the entire issue of new standards. Many, reading the fine print, predicted that OPM's recommendations would lead to downgrading of positions, which would take money right out of their pockets. One ranger said to me, "There are people in the Service who would kill to prevent that from happening."

They needn't have worried. In 1985 the new Director of the Park Service, William Penn Mott, Jr., accepted OPM's new Classification and Qualification Standards, abolishing the technician series and absorbing these people into the ranger corps. Few now expect this revision will cause demotion of old-time rangers, yet, because technicians were unskilled, making them rangers as well was a tacit admission by the Park Service of the fact that the FOS Team had tried to hide: that rangers are not professionals. Indeed, rangers have in effect been certified as nonprofessional. But no one outside the agency seems to have noticed.[84]

A recent *Newsweek* article entitled "The Changing Rangers" reassures the reader that because it has become more specialized, the Park Service today is more professional than ever. "We used to just manage the land, and the people were incidental," they quote Grand Canyon National Park Chief Ranger Kenneth Miller as saying, "Now we have to manage the people first."[85]

The principal job of rangers today, the magazine tells us, is law enforcement: "protecting the parks from the people and the people from the parks." As for the other tasks, "a separate staff of 'naturalists' now handles the nature walks the rangers used to lead, and a team of 'resource management' scientists makes the important decisions on trees, rocks and creatures." Our national Parks, *Newsweek* was reassuring us, remain in competent hands.[86]

That *Newsweek* had been misinformed somehow escaped notice. Emphasis on law enforcement is not new but as old as the Park Service itself. The division of naturalists is more than fifty years old and less specialized than ever. And resource managers, rather than being scientists, are rangers with little or no postgraduate training. Yet these facts are less apparent to the public today than they were in Albright's time.

"The Park Service," Shanks said recently, "has a vast reservoir of public support it can mobilize at any time." Yet the persistence of this support is understandable. The image of the Park Service benefits from many things. Its mission — to preserve our national parks unimpaired for future generations — embodies, as Robbins stated, the finest aspirations of our national spirit. The parks themselves are wondrous places enjoyed by hundreds of millions of people and the Service is an expert host to its guests. Most visitors never meet a career ranger, instead seeing only "seasonals" — people like Professor Jonas — often knowledgeable college professors working summers but having no voice in management.[87]

Dr. Linn reminds us that national parks or recreation areas are in more than three hundred congressional districts, which gives the Service enormous clout with Congress. The parks themselves are spectacular settings for television media, attracting beneficial publicity; they are priceless laboratories for wildlife research and those who control them have great power to co-opt the fraternity of university scientists. The study of wildlife ecology is a new and complex field, involving subtle issues and much that remains unknown; problems do not show themselves quickly and when they do they are easy to obfuscate. And the connections between OPM standards and the death of bears, between archaeology and the ecosystem, between commitment to truth and preservation of a park, are complex and often hard to follow.

We can only guess what were among the casualties to come from that fateful collision on the lonely desert in New Mexico that clear February day in 1936. But perhaps truth was one. "If Wright had lived," Russ Grater told me, "he would have found a way to keep the Wild Life Division going." If he had lived, Connie Wirth suggested to me, "he would have gone all the way." If he had lived . . .

We can never know how things might have been had Wright not been cut down. But those with the opportunity and vision to make a difference are rare. Their appearance, no matter how brief, is a gift we must appreciate, for they give us reason to hope.

15

The Deep Hole Gap

Lying next to the road near LeHardy Rapids of the Yellowstone River, just six miles downstream from Yellowstone Lake, Mud Volcano is a caldron of boiling mud and steaming, odoriferous, poisonous hydrogen sulphide gas. Spewing mud sometimes hundreds of feet into the air and smelling like rotten eggs, it is not just another roadside attraction. "The greatest marvel we have yet met with," Nathaniel Langford wrote upon seeing it in 1870; and indeed, now next to a large parking lot, it attracts thousands of visitors each year.[1]

And yet few of these visitors noticed that during spring 1979, Mud Volcano was at the center of a remarkable occurrence: trees around the caldron suddenly turned brown and died. The ground, scientists discovered, was becoming hotter. It was cooking the trees. Even more curious, seismographs in the park during the previous summer and fall recorded hundreds of earthquakes too small to be felt within a short distance of the same spot. A seismic swarm, geologists called it. Chemical analysis of the gases coming from the volcano revealed traces of helium 3 — a rare isotope hundreds of thousands of years old — "a living fossil," one geologist described it.[2]

What was going on?

Yellowstone, geologists knew already, was one of the hottest spots on earth and the second most active earthquake zone in the country. More than 2,000 tremors vibrated through the park each year, and it had more geysers, hot springs, and fumaroles (steam vents) than the rest of the world put together. They were so many, in fact — ten thousand and more — that no one had been able to count them all. Through these vents enormous amounts of heat escaped, 1,800 milliwatts from every square meter, twenty times the continental average. Just two hundred feet beneath the surface the ground temperature was more than 300 degrees Fahrenheit.[3]

Thanks to this ubiquitous heat, trees parboiled by hot springs were not an uncommon sight in Yellowstone, and the brown conifers at Mud Volcano seemed to the eye quite ordinary. As long ago as 1871, explorers J. W. Barlow and D. F. Heap recorded that, in the Firehole geyser basin, "dead and withered trees bear evidence of the deadly effect of hot water which has flowed among them." And it was known that the activity at Mud Volcano ebbed and flowed through the years. Langford actually thought it had "not been long in existence; but that it burst forth the present summer. . . ."[4]

Scientists also had long known that the geothermal features of Yellowstone remained in constant flux. Geysers came and went. Some, such as the spectacular Excelsior geyser (which last erupted in 1888) may remain dormant for as long as a century, or may play out altogether. Others, such as Imperial (which first erupted in 1927, remained dormant between 1929 and 1966, and since has erupted regularly), were entirely unpredictable. Even Old Faithful geyser was less reliable than many supposed. It was probably less than three hundred years old, and the interval between eruptions was growing.[5]

And yet events at Mud Volcano turned out to be more than ordinary. They were extraordinary events that in turn elicited extraordinary reactions among men. Signaling that a natural cataclysm might be in the offing, these events — and the unusual forces that drove them — in turn piqued intense curiosity in the scientific community. Yellowstone — the same place that remained *terra incognita* to so many wildlife biologists — was to reveal its every shoal and reef to geologists.

This is the story of how curiosity created the park, and now may threaten it.

"What a field of speculation this presents for chemist and geologist," wrote trapper Osbourne Russell, after describing the geysers and hot pools he found during his trip across the plateau in 1838–1839. Indeed, ever since its discovery, Yellowstone had been an object of scientific curiosity, which shaped the character and development of the park.[6]

The thermal features of Yellowstone were always among its major attractions, and if they had not been there, the park almost certainly would never have been created. To trappers and explorers of the nineteenth century, wild animals were common; they were also food. Few thought it necessary or desirable to set up sanctuaries for them. The original Yellowstone Park enabling act, though prohibiting "wanton destruction of the fish and game," did not proscribe hunting altogether and did not provide for enforcement against poaching. It was not, as we saw, until the great slaughters of game on the Great Plains in the 1870s that pressure began to mount to make the park a wildlife sanctuary, leading to passage of the Lacey Act in 1894.

But the thermal features were another matter; they were special. "The geysers of Iceland sink to insignificance beside them," Lieutenant Doane told Congress on his return, "they are above the reach of comparison." He said they were the "new and perhaps, most remarkable feature in our scenery and physical history." These inanimate wonders were the first to attract people to Yellowstone and to inspire the idea of a national park.[7]

Indeed, the geysers, mud pots, and hot pools seemed so exotic that most conceived them as too fantastic to be believed. When John Colter, after visiting Yellowstone alone during late fall, 1807, told his friend William Clark of the "hot spring brimstone" he saw near the Gardner River, Clark was the only one to believe him. "His stories were not believed," wrote historian Chittenden; "their author became the subject of jest and ridicule; and the region of his adventures was long derisively known as 'Colter's Hell.' " Even though many trappers looking for beaver in Yellowstone in the 1820s and 1830s brought back accounts of "boiling fountains," or of "columns of water of various dimensions, projected high in the air, accompanied by loud explosions, and sulphurous vapors, which were highly disagreeable to the smell," these stories were still treated as the ravings of those who had lived in the wild too long. In fact, many accounts were apocryphal. A trapper named Joe Meek, attempting to escape pursuing Indians by crossing the plateau in 1829, claimed to have seen "larger craters, some of them four to six miles across," which "issued blue flames and molten brimstone."[8]

The trouble was that visitors to Yellowstone in those days were not the sort to inspire belief. Many were like Walter deLacy, who succeeded a suspected murderer known as "Hillerman, the Great American Pie-eater" as leader of a group of prospectors he called the "forty thieves," and who came back with accounts of "Hot Spring Valley" in 1863. To speak out about the wonders of Yellowstone was to risk being thought a liar. Charles W. Cook and David E. Folsom, who explored the Yellowstone in 1869, were, according to Langford, so "bewildered and astounded at the marvels they beheld, they were, on their return, unwilling to risk their reputations for veracity by a full recital of them. . . ."[9]

Yet paradoxically, considering the skepticism with which such accounts were received, the existence of the Yellowstone geysers finally began to gain credence, thanks to the efforts of the most celebrated liar of them all, Jim Bridger.

Explorer, guide, mountain man, and teller of tall tales, Bridger, "The Baron Munchhausen of the American West," historian Haines called him, described Yellowstone's petrified forest on Specimen Ridge to someone as "a petrified forest still standing with petrified birds in the branches singing petrified songs." He also told everyone he met about "the 'Great Springs' so hot that meat is readily cooked in them." But few believed. He tried to publish an article on Yellowstone, but it was rejected as too fantastic. He persuaded the editor of the *Kansas City Star* to write a piece about his account, but the article was suppressed as too incredible.[10]

He did, however, elicit curiosity among explorers. He inspired maps by Father Pierre deSmet in 1851 and Captain William F. Raynolds of the U.S. Army Corps of Topographical Engineers in 1859, which referred to many of the thermal features of the plateau. These maps in turn prompted more explorations of the Yellowstone by the army, the curious, and, more significantly, the Northern Pacific Railroad.[11]

Nathaniel Langford wrote,

in Virginia City, Mont., at that time, of the existence of hot spouting springs in the vicinity of the source of the Yellowstone and Madison Rivers, and said that he had seen a column of water as large as his body, spout as high as the flag pole in Virginia City, which was about sixty (60) feet high. The more I pondered upon this statement, the more I was impressed with the probability of its truth.[12]

Langford in turn received the backing of Jay Cooke, owner of the Northern Pacific, to mount an expedition to the upper Yellowstone to investigate the possibility of using these attractions to bring people west by train. On returning from his expedition of 1870, Langford was hired by Cooke to give twenty lectures on the wonders of Yellowstone, as part of a publicity campaign by the railroad. Among those who attended Langford's first lecture (at Lincoln Hall in Washington, D.C.), was the physician and amateur geologist Ferdinand V. Hayden, head of the U.S. Geological Survey of the Territories. Hayden, who also had strong friends in Congress and in the railroad lobby, was inspired by Langford's account to request funds from Congress for a geological survey of "the sources of the Missouri and Yellowstone rivers." At the same time he was quietly asked by Cooke to survey the possibility of a route for a railroad along the way.[13]

On a $40,000 appropriation from Congress, Hayden assembled the first official expedition to explore the Yellowstone. A party of twenty scientists, physicians, photographers, and artists, including noted photographer William H. Jackson and artist Thomas Moran (who was taken along "directly in the interest of the N.P.R.R. Company"), the expedition set off on June 1, 1871.[14]

By the end of August, the expedition returned to Montana and Hayden set off for Washington. Returning at the end of October, he found waiting for him a letter from A. B. Nettleton, an agent of the Northern Pacific, who suggested: "Let Congress pass a bill reserving the Great Geyser Basin as a public park forever. . . . If you approve this would such a recommendation be appropriate in your official report?" This was the first shot fired by the railroad in a lobbying blitzkrieg for the future of Yellowstone.[15]

On December 18, 1871, a bill was introduced in the U.S. Senate to make the area a national park. Hayden and Langford were enlisted by the railroad to help push the idea through Congress. Hayden organized an exhibit in the Rotunda of the Capitol of geological specimens along with Jackson's photographs and Moran's sketches.[16]

Just six years before, in 1865, Lewis Carroll's story, *Alice's Adventures in Wonderland*, had appeared, and the word "wonderland" was receiving widespread currency as a way to describe the land of geysers and hot springs. Copies of an article by Langford, "The Wonders of Yellowstone," which had first appeared in *Scribner's Monthly*, were sent to every member of Congress, along with Lieutenant Doane's report of the 1870 expedition to Yellowstone. On assurances from Hayden that concessionaire fees (principally from the railroad) would be sufficient to run the new park and that no public funds would be required, the bill passed Congress easily — the Senate approving it on January 30 and the House on February 27, 1872. President Grant signed it into law on March 1.[17]

"A thorough solution of the wonders of this valley can only be obtained by long and patient investigation," advised Captain J. W. Barlow while viewing the Firehole geyser basin in 1871. And no sooner had the park been created than this investigation was under way.[18]

In the same spring as President Grant signed the Yellowstone Park bill into law, the Secretary of the Interior appointed Langford as the park's first superintendent — without salary — and Hayden had coaxed $75,000 from Congress for another geological survey. Along on this second expedition was William H. Holmes, on leave from the Smithsonian Institution. Holmes's study, "Report on the Geology of Yellowstone National Park," appearing in 1883, was the first systematic account of Yellowstone geology. Holmes was followed by a team of the U.S. Geological Survey under the direction of Arnold Hague, which produced, in voluminous detail, a summary of the park's geology. Hague was followed by J. P. Iddings, who studied the park from 1888 to 1896.[19]

Yellowstone became a mecca for geologists seeking to unlock its mysteries. This intense interest continued after the Park Service was created in 1916. Three of the first four books published on Yellowstone by the Park Service's Division of Research and Education were on geology. The first woman ranger in the Park Service — Isabel Bassett Wasson, a graduate of Smith and Columbia appointed in 1920 — was a Yellowstone geologist. Researchers from the Carnegie Geophysical Laboratory, the Geologic Survey, and various universities, such as W. F. Foshag (1926); E. T. Allen and A. L. Day (1925–1935); N. L. Bowen (1935); H. A. Brouwer (1936); C. N. Fenner (1938–1944); and R. E. Wilcox (1944) focused on a variety of park wonders: the ancient glass quarry at Obsidian Cliff, the geysers, and the lava flows.[20]

What caused the geysers? they wondered. Were they products of ancient mountain building or the harbingers of a new holocaust? Were they waxing or waning?

Yellowstone, the early geologists concluded, was a relic of ancient mountain building. Like horseshoe crabs and Komodo dragons, it was a living fossil, a dying reminder of the great forces that created the Rocky Mountains. They knew that for millions of years parts of Yellowstone had lain under shallow seas; that during the time of the dinosaurs, in events known as the Laramide orogeny (which events we now know took place about one hundred million years ago), the floor of this sea began to crack and rise, forming the first of the Rocky Mountains. They knew that still later, during early Tertiary times (an era now established as about fifty million years ago), dramatic volcanic eruptions convulsed the region and that accumulations of lava from these eruptions made most of the mountains in and around the park — the Absaroka in the east, much of the Gallatin in the West, and the Washburn range in the park itself. They knew that glaciers thousands of feet thick had still later sliced across the plateau, scouring the terrain and remolding and obscuring signs of the earlier volcanism.[21]

The volcanic events, Holmes and Hague concluded, made the plateau as we see it today. The geysers and other thermal phenomena of the region were produced by the residue of heat from these ancient cataclysms. But they were gradually cooling. Hague wrote in 1893, "From a geological point of view there is abundant evidence that the thermal energy in the park is becoming extinct." A study of the geysers between 1925 and 1932 by Allen and Day of the Carnegie Geophysical Laboratory seemed to confirm Hague's observation. They concluded that although "the total fund of energy remained relatively constant, the permanency of a phenomenon like geysers seems less certain."[22]

Hague, Holmes, and Iddings, however, did mention something they could not explain. The volcanic soil was composed of two kinds of rock: One was a dark material, andesite, the typical congealed lava that flows from many volcanos; the other was a light-colored fine-grained rock high in silica, the basic constituent of glass, rhyolite. This too was a volcanic rock, but it was not often found on the surface of the earth. Yet in Yellowstone it was found in enormous quantities. Why was it there? And why were two forms of volcanic rock found in Yellowstone? Holmes, Hague, and Iddings never found the answer.[23]

Later research added to the mystery. It turned out that two forms of rhyolite lay in Yellowstone. One was clearly congealed lava — flow rock, some called it — and the other consisted of small grains fused into solid rock — a material known as welded tuff. How could these three very different kinds of rock, including two kinds of rhyolite, have been made by the same volcanic eruptions? And more curious, how were the welded tuffs formed? Were they the result of lava flows, or of some other, unknown mechanism?[24]

Events elsewhere began to supply the needed clues. On May 8, 1902, on the island of Martinique in the French West Indies, Mount Pelée, after fifty years of silence, erupted with tremendous violence, sending a hot cloud of ash down the mountain at more than 130 miles an hour, completely engulfing the town of Saint Pierre. It killed 30,000 people, all but two in the town (one a prisoner in a dungeon). Ten years later, in Alaska, Mount Katmai, thought to be an extinct volcano, also exploded suddenly, covering a wide area with red-hot lava that turned the rain to steam, producing the formation later known as the Valley of Ten Thousand Smokes.[25]

These were not ordinary eruptions. Unlike the matter ejected by most volcanos, the lava did not flow out of the ground as a slow-moving river of molten rock. Rather, it blew out with enormous speed as a hot cloud of ash — a nuée ardente, or hot cloud, the French geologists called it. Such incandescent or pyroclastic flows, as they were later called, moved down Pelée and Katmai like avalanches, so hot and so inflated with gas that they ran like water. When these came to rest they collapsed, forming tuff a hundred feet thick.[26]

During summer 1950, while the world's geologists were still puzzling about these events, Francis Boyd, a young Harvard graduate student, came west for the first time. His instructor, George C. Kennedy, had grown up on a ranch in Montana's Centennial Valley and was spending the summer looking for uranium in the Phosphoria Formation west of the park. He took Boyd along. While in the area, Boyd became interested in Yellowstone volcanism. Could the land of Yellowstone rhyolite have been formed by a hot cloud similar to that which had cascaded down Pelée and Katmai? The resulting tuffs were similar but not identical: in Martinique and Alaska they had not welded as they had in Yellowstone.

Still, Boyd found the idea attractive. Perhaps, he thought, when the incandescent flow came to rest it was hot enough to weld the tuff. But Kennedy was skeptical. Instead, he suggested that the Yellowstone tuff had been deposited by collapsed froth flows, hot magma inflated with gas that moved slowly from the crater like a rolling stream of soapsuds before gradually congealing. Others on the Harvard faculty and elsewhere held still another view. The Yellowstone tuffs, they insisted, were simply hot ash that had accumulated by settling over long ages.[27]

The skepticism of Kennedy and his colleagues was not hard to understand. Yellowstone had one of the largest formations of rhyolite on earth. Tuff covered 600 square miles of the park to a thickness of 1,000 feet. Flow rock covered another 1,000 square miles to a depth of 1,000 feet. If all this rhyolite had been placed there by one series of hot clouds such as those of Pelée, the explosions that propelled these clouds would have been far greater than any ever known. To smother a region with 400 cubic miles of ash in a matter of days or perhaps minutes would have been a cataclysm dwarfing any known in the 4.5 billion years of earth history. It would also have left an enormous crater, and no one knew of such a crater in Yellowstone.[28]

An eruption so stupendous, and a crater vast enough to vent that much ash, seemed altogether improbable. The famous eruption of Krakatau on August 27, 1883 was so loud that it was heard 3,000 miles away, and so powerful that it caused a tidal wave killing 30,000 people. Yet it ejected only twelve cubic miles of ash. Tambora, erupting on the Indonesian island of Sumbawa in 1815, killed 100,000 people and so filled the atmosphere with dust that it blacked out the sun for days and brought on the year without a summer, dropping global temperatures by up to five degrees and giving North America the coldest weather in two hundred years. And yet Tambora ejected from 24 to 48 cubic miles of debris. The recent eruption of Mount Saint Helens in the Cascades moved only half a cubic mile of material. A Yellowstone eruption would have been at least eight times as forceful as that at Tambora and thirty times that of Krakatau![29]

But Boyd remained curious. To solve the mystery, he decided

to write his Ph.D. dissertation on the "rhyolite plateau" of Yellowstone. He spent summers in 1951 and 1952 in the park, and his work became a landmark for geologists. Boyd told me recently that "many on the faculty at Harvard did not believe the Yellowstone tuff could have been put there by a pyroclastic flow. To convince them I had to prove it beyond the shadow of a doubt." And by careful and original thermodynamic analysis, Boyd did so.[30]

"All the welded tuff exposed in Yellowstone Park," he concluded, "has been emplaced by a single, rapid series of eruptions" — huge avalanches of hot gas. And the crater that had ejected these avalanches had been under their noses all along! Like Charles Addams's cartoon of scientists standing in a dinosaur footprint so large they did not notice it, the crater of this volcano was so vast, on a scale so surprising and so obscured by later glacial deposits, that two generations of geologists had failed to notice it. He wrote, "The Washburn Range has the form of a horseshoe. . . . The topography is suggestive of a caldera." Lying astraddle the center of the park, the crater or caldera encompassed an oval 46 miles long and 28 miles wide, stretching from Cache Creek on the northeast, under the north end of Yellowstone Lake, and to the Madison River in the southwest.[31]

The Yellowstone plateau, Boyd established, was made by forces quite different from those which made the rest of the West. He concluded that it was produced not by conventional volcanism, but by one huge explosion in relatively recent times. Rather than part of the Absaroka volcanic activity that came to an end fifty million years ago, as Hague and Iddings had supposed, the Yellowstone caldera had erupted, Boyd suggested, in the late Tertiary period around five million years ago. Four years later, when he published his findings, he gave this event a still later date. It belonged, he wrote, to the mid-Pliocene epoch — three million years ago.[32]

Boyd had solved a problem that had perplexed professional geologists for seventy-five years. But in a sense it was the amateur Hayden who first guessed the answer. Climbing to the summit of Mount Washburn in July 1871, Hayden wrote,

a bird's-eye view of the entire basin may be obtained, with the mountains surrounding it on every side without any apparent break in the rim. This basin has been called by some travelers the vast

crater of an ancient volcano. It is probable that during the Pliocene
period the entire country drained by the sources of the Yellowstone
and the Columbia was the scene of as great volcanic activity as that
of any portion of the globe.[33]

As it was to turn out, this volcano was far younger and more
active than either Hayden or Boyd had imagined.

Captain William A. Jones observed while visiting the Firehole
geyser basin in 1873, "Further elucidation must be the result of
careful observation and study." Yet Boyd's revelations, after seventy-five years of observation and study, did not elucidate all the
wonders of Yellowstone. Shortly after he had completed his dissertation and left to take a job with the Carnegie Geophysical
Laboratory in Washington, events exposed the extent of their ignorance.[34]

At Blarneystone Ranch west of the park, the water-softener
tycoon Emmett J. Culligan had just finished building the best bomb
shelter imaginable. Pouring hundreds of thousands of dollars into
steel and concrete, he had designed a redoubt that could withstand
the biggest nuke the Russians could throw at us. He had prepared
for the worst.[35]

But on the night of August 17, 1959, at exactly 11:37 P.M., the
ground under the survival shelter rose fifteen feet without warning,
splitting it in half down its 150-foot length. Simultaneously, not far
away, huge tidal waves surged down the entire length of Hebgen
Lake, sloshed over the dam, and — now walls of water twenty feet
high — continued to race down the Madison River.[36]

Just as the first wave reached Rock Creek, near the end of
Madison Canyon, the 7,600-foot mountain at the canyon entrance
split in half. Eighty million tons of loose rock came roaring down
on the several dozen campers dozing in their sleeping bags at the
Forest Service campground on the other side of the Madison River,
preceded by a hurricane-force wind that blew cars and trailers
away, tore trees from the ground, and pulled people from their
tents.[37]

The cascading water and lurching avalanche collided at the campground and continued up the opposite mountain, burying nineteen
people forever and carrying many away in a river of debris. When
the landslide had settled it formed a dam 240 feet high across the

Madison River, creating a new lake that trapped dozens of campers in the narrow canyon.[38]

At the same instant in Yellowstone, 298 geysers and hot springs erupted (160 of them for the first time), and continued to spout for days without interruption. After the initial shock, geyser activity had changed radically. Some, like Steady Geyser and Grand Geyser, quit altogether. But many new ones appeared. Still others, such as Daisy, Great Fountain, Castle, and Oblong, shortened their eruption intervals. The water temperature in all the thermal areas rose five and a half degrees. Water levels in wells as far away as New York and California dropped.[39]

Known in Montana as "the night the mountain fell," this was the Hebgen Lake earthquake, at 7.5 on the Richter scale the severest ever recorded in the Rocky Mountains and the second strongest in the history of North America. More destructive than two hundred atomic bombs, it brought new attention to the geology of Yellowstone.[40]

The Geological Survey rushed scientists to the scene to find the cause of the quake, and how it was related to the thermal and volcanic activity of the region. Earlier geologists had thought that the Yellowstone geysers were all independent of one another, and that they relied on shallow and local water sources. But the simultaneous eruption of nearly three hundred geysers and the effect of the quake on water level in wells throughout the country cast doubt on this theory. Clearly, the geysers of the park comprised one hydrothermal system, and this system was somehow related to the deepest movements of the continental plate.[41]

That Yellowstone was susceptible to quakes came as no surprise. The Hayden party had experienced many. Hayden observed in 1871, "I have no doubt that if this part of the country should ever be settled and careful observations made, it will be found that earthquake shocks are of very common occurrence." But during the intervening eighty-eight years careful observations had not been made. Seventy-six quakes strong enough to be felt by rangers were recorded, but no seismograph had been put in the park to detect quakes too small to be felt or to trace their origins. Now the Hebgen quake made many wonder. Just what was the extent of this activity? [42]

One of those arriving at the scene was Dr. Warren Hamilton of the Denver office of the Geological Survey. The cause of the quake,

Hamilton discovered, was not hard to guess. "Yellowstone National Park," he reported, "lies at the intersection of some of the most important tectonic elements of the Northwest." Rimmed by "the tectonically most active mountain masses in the northern Rocky Mountain region," it was the intersection of opposing movements within the plate, one heading northward, another northeastward, and a third at right angles to these, trending east to southeast. The floor of Yellowstone was splitting apart.[43]

Teams from the Survey began to map the region thoroughly. Dr. Gerald Richmond of the Denver office discovered that the plateau had been covered with ice in three distinct ice ages, which he called the pre-Bull Lake, Bull Lake, and Pinedale glaciations, respectively. By studying aerial photographs Hamilton was able to identify rhyolite flows that Boyd had missed. And by examining the relationships between glacial moraines and these flows he was able to give an approximate age to the latter.[44]

The ice ages of Yellowstone were relatively recent. Belonging to the Wisconsin glaciation, they ebbed and flowed during a 100,000-year period that came to an end with the retreat of the Pinedale ice 12,500 years ago. Yet viewing air photographs, Hamilton discovered rhyolite flows *overlying* some of this glaciation. Yellowstone volcanism, he concluded, was even more recent than much of the Pleistocene ice!

He told me recently, "I recognized that the youngest lava flows came out of young moraine — younger than Bull Lake and older than Pinedale. The youngest flows were post-Wisconsin — very, very young."

While examining air photographs Hamilton also noticed that the community of Island Park on Yellowstone's western boundary lay in a basin surrounded by a circular ridge of rhyolite. This ring was approximately eighteen miles in diameter. It was, Hamilton reported, another "huge caldera."[45]

Yellowstone geysers, Hamilton also discovered, were not dying as was once supposed. Nor were they remnants of water heated by shallow rock now cooling after some ancient explosion. Rather, they came from a deep volcanic source decidedly still active. He said, "The young rhyolites cannot themselves be the source of much of the heat. . . . Most of the heat must be of deeper origin, and the abundance, heat and composition of the steam seem to require a magmatic source." A huge reservoir of superheated molten rock, he suspected, lay under the ground in Yellowstone and it was this magma that kept the geysers going.[46]

"Unusual processes operate here," Hamilton mused. Indeed, the more scientists studied Yellowstone, the more complex and recent it seemed to be. Earthquake and geothermal activity was not decreasing as earlier geologists had thought. And each time the region was studied, a later date was assigned to the volcanic activity. Now they began to suspect that it was still active.[47]

These mysteries fascinated Arthur B. Campbell, United States Geological Survey director for the Northern Rocky Mountain region. Yellowstone, he recognized, was clearly special. But how could we learn more? A complete geologic study of the park would cost millions. Where would the money come from?[48]

It was, in a sense, to come from the moon. By the time Hamilton left the park in 1961, the space race had begun. President Kennedy had given NASA until 1970 to put a man on the moon, and plans were made for exploring Mars and other planets. Millions were being poured into new technology to explore the solar system from orbiting spacecraft. But how would we know whether this remote-sensing equipment would work? Now NASA needed a way to test it, and the best way was to find some place on earth that looked much like the moon or other planets — some place with a lot of volcanic rock, pockmarked with craters — and study this place both on the ground and from space. In that way scientists on the ground could verify how accurate were the data received from space.

To find a suitable spot for this study, NASA advertised. Campbell noticed the ad. Why not allow NASA to study Yellowstone, Campbell wondered? He met with his friend and former graduate-school classmate, John Good, Yellowstone's Chief Naturalist. Good liked the idea, and so did his superiors. Though NASA's goal was to get to the moon by 1970, theirs was to complete a high-profile, scientific project in time for Yellowstone's centennial in 1972. A high-technology mapping of the park would be a wonderful birthday present. The Geological Survey and the National Park Service jointly answered NASA's ad. Yellowstone would be the place to test NASA's new bells and whistles.[49]

By summer 1965 this project was in full swing. Five teams of geologists descended on the park to map it from the ground. The first group was assigned to study old rock in the northern half of the park; the second, old rock in the south; the third, early volcanism; the fourth, young volcanism; and the fifth, glaciation. They combed the backcountry, taking with them some of the most advanced technology ever used to study the earth: paleomagnetic dating and potassium-argon dating for measuring the age of volcanic rocks; obsidian-hydration dating for determining times for the glacial ice; trace-element isotopic geochemistry to learn the magmatic "parentage" of the rhyolite; and the whole panoply of hardware from NASA's toolchest: side-looking radar to seek lava flows, color aerial photography to identify alluvial fans, infrared imagery to detect thermal activity, airborne microwave, and doppler radio.[50]

A team headed by Robert Fournier of the Menlo Park office of the USGS sank test wells throughout the park to measure the heat flow. The park was, he discovered, one of the hottest places on earth. Energy generated by radioactive decay of various elements — radium, thorium, potassium 40 — produced, on a worldwide average, an increase in temperature of one degree with each hundred meters of depth. But Fournier found the ground below Yellowstone much hotter: "We hit 153 degrees Celsius (more than 300 degrees Fahrenheit) at two hundred feet," he told me recently. At the surface anywhere within the caldera the heat flow was 27 times above the world average. In the Firehole Basin it was more than 700 times the world average.[51]

The heat was too great to have come from rock cooling after an ancient volcanic eruption; it had to come from an active source. This theory was confirmed by the team assigned to study young volcanism. The team leaders, Robert L. Christiansen and H. Richard Blank, wrote, "The present rate of heat flow represented by this hydrothermal system cannot have operated for the last 600,000 years unless there is a large body of magma that continues to supply heat at the top. . . ."[52]

There was evidence, moreover, that this magmatic insurgence was cyclic. As tuff, cooling after an eruption, formed a "cooling unit," — a stratum subtly distinguished from others by the times and rates at which it cooled — Christiansen and Blank, by identifying separate cooling units, were able to catalogue distinct volcanic cycles. Three times in the last two million years, they established, the calderas in and near Yellowstone had exploded in the most cataclysmic events in the earth's history. The last eruption sent six hundred cubic miles of debris into the atmosphere and stratosphere, buried half the United States in ash, and blacked out the sun perhaps for years. It was an event, Christiansen and another colleague, Robert B. Smith, later observed, "on a scale known nowhere else in recorded geologic history."[53]

Yellowstone was, they concluded, the center of activity of a "resurgent caldron cycle." Beneath the park a plume of hot magma, piercing the earth's crust, nearly touched the surface. As this heat and magma pressed upward, the ground bulged, forming a dome. Gradually the pressure beneath the dome became so great that it burst. After eruption the caldera, empty of magma, collapsed, forming a gigantic depression or crater that in time became obscured by erosion.[54]

Along the rim of the caldera a ring of fractures appeared, venting the heat from the rising plume below. Rainwater and melted snow, entering these fractures, formed an enormous, interconnected hotwater plumbing system. The cold surface water, being denser than hot water, sank deep into the cracks of the crust. As it sank, it came closer to the plume and grew hotter, eventually reaching nearly 500 degrees Fahrenheit. It eventually became hotter and lighter than the water above it, and began to rise. These were the convection currents that drove the geysers and hot pools. Nearly all the geothermal activity lay along these ring fractures, or along the major earthquake faults such as that running from Hebgen Lake into the park.[55]

The caldron cycle formed, they theorized, a self-sustaining feedback system, in which the earth's crust, once thinned by melting, permitted more heat to rise from the earth's interior, melting the crust even more. If so, if this was a machine still running, then the caldera probably would one day erupt again. Indeed, they discovered, signs were already apparent that an eruption might be imminent. Three times it had erupted at approximately 600,000 year intervals, Christiansen and Blank found, the last 600,000 years ago. Twin domes were now within the caldera. Hydrothermal activity had increased, suggesting that the plume was growing as well. This increased activity, they cautioned, "conceivably could even be . . . the first stage of a new volcanic cycle." Indeed, they concluded, such a cycle "could even be under way at present."[56]

By 1970 NASA, having put a man on the moon and achieved its goals, pulled out. But this cutoff did not bring intensive geological study to a close. On the contrary, it had only begun. The Department of Defense also used the park as a testing ground for its intelligence satellites, and government and university scientists from around the country, intrigued by recent discoveries, flocked to the park.

Many large questions remained. The more physical scientists studied the park, the more they needed to know. What was the engine that drove this machine? How likely was a new eruption? What damage would it do?

These questions revolved around the "Yellowstone melting anomaly." The park was a soft spot in the earth's crust, but why? The answer lay in understanding the relationship between Yellowstone volcanism and the movements of the continental plate. Either plate motion caused the volcanism, or volcanism was one piston in the engine that drove the plate.[57]

The view that Yellowstone volcanism represented deep forces that drove the continental plate was known as the hot-spot theory. According to this view, a hot spot or "convection plume" lay under the earth's crust, where magma from the mantle melted away at its base. The continental plate traveled southwestward over this spot at about one and a quarter inches a year. And like a piece of paper moving slowly over a burning candle, this plume occasionally burned through the crust, causing the gigantic caldron eruptions. This was a popular explanation of Hawaiian volcanism, and increasingly it was applied to account for Yellowstone as well. But it left something important unexplained: why there were convection plumes at all. The thermal activity in Yellowstone, according to this view, remained a mystery.

That the Yellowstone soft spot was caused by movements in the continental plate became known as the propagating-rift theory. The continental plate, from the Rockies to California, was stretching, and as it stretched it split, according to this view, releasing pressures below and permitting heat to flow — to propagate — northeastward under the continental plate until it surfaced at Yellowstone. This was the explanation favored by Hamilton, Christiansen, and Blank. But if it was true, volcanism and earthquake activity were both effects of larger plate movements. And the key to understanding Yellowstone would be found in seismic studies.

Shortly after NASA left, a team of geologists from the University of Utah began to search for this key. Employing a variety of sophisticated techniques, they mapped epicenters of earthquake activity and other movements of the crust.[58]

Yellowstone, the Utah scientists discovered, was the second most active earthquake zone in the country. From five to more than a hundred tremors vibrated through the park each day, often in seismic swarms. Many were aligned along a fault running from the caldera northwest to Hebgen Lake — a pattern, the Utah team observed, which "represents ongoing tectonic north–south crustal extension." The continental plate was indeed stretching and splitting in Yellowstone.[59]

To understand further what was going on they detonated dynamite in various parts of the park and, using 225 seismographs established at various locations, measured the speed with which the waves from these blasts traveled through the ground. Because shocks will not travel through magma as fast as through solid rock, they could determine, by the depth and speeds at which these waves traveled, the thickness of the earth's crust.[60]

They established 900 monitoring stations in and around the caldera to measure gravitational acceleration. Because the gravitational pull was less through molten rock than through hard rock, the strength of gravity over the caldera would tell them how close magma was to the surface.[61]

They measured the magnetism of the park. Because rock heated above 960 degrees Fahrenheit loses its magnetism, they could determine at what depth rocks under the park reached this temperature by measuring the depth at which magnetism ceased — known as the Curie depth.[62]

These experiments confirmed what Christiansen and Blank had suspected: the crust was razor thin. Although the continental plate averaged at least twenty-five miles in thickness elsewhere, the dynamite (seismic-refraction) studies confirmed that in Yellowstone it was between two and six miles thick.[63]

Gravitational acceleration was also slower under the park, indicating that the crust rested on something at least partially molten. Further calculations by Jeffrey A. Evoy of the University of Utah revealed that this liquid was a magma chamber extending at least 155 miles into the earth. And though the Curie depth in the rest of North America averaged between ten and nineteen miles, under

Yellowstone, the Utah geologists found, it lay just three and one-half miles under the surface.[64]

The ground under the park was, they discovered, not only thin, but doming. Scientists had long known that the entire region was uplifting. This was the normal process of mountain building, which has, over the ages, lifted the Yellowstone plateau more than 1.2 miles above the surrounding area. But the Utah scientists found something far more surprising. Between 1975 and 1977, Dr. Robert B. Smith, leader of the Utah team, and his colleague, J. R. Pelton, measured benchmarks (bronze plaques) put into the ground by the USGS in 1923. The central part of the park, they discovered, had risen twenty-seven and a half inches (700 millimeters) during this fifty-three-year interval — on the average more than half an inch a year (14 millimeters) above the surrounding terrain. Smith and Pelton concluded, "The deformation in the region is fairly recent."[65]

Soon others began to suspect that the uplift was occurring even faster than Smith's measurements had indicated. A Park Service geologist, Dr. Wayne Hamilton, pointed out that the bulge's center, at Mud Volcano not far from the outlet of Yellowstone Lake, tilted the outlet upward, raising the water level. By carefully measuring changes in water level he was able to determine that since 1940 the water in the south end of the lake had risen nearly an inch (23 millimeters) a year. His results were later confirmed by Daniel Dzurisin of Cascade Volcano Laboratory in Vancouver, Washington.[66]

The work of Smith, Hamilton, and others confirmed the suspicions of the NASA-USGS groups. Tectonic forces were shaping the face of Yellowstone, and would surely shape its future as well. Movements of the continental plate were somehow producing volcanic activity that kept the hydrothermal system going. The crust was a thin and brittle layer over a large pool of molten lava. And now it was beginning to bulge.

The recent doming of the ground in the park proved that enormous pressures were building underneath. Scholarly consensus grew that another eruption might be coming. Smith and his colleague Pelton wrote in 1979, "There is a possibility that the uplift at Yellowstone represents a new magmatic insurgence heralding the start of a fourth volcanic cycle." Christiansen agreed; he wrote in

1984, "It is . . . possible that the volcanic system is heading toward another major ash-flow eruption."[67]

If this cycle came around again it would, Christiansen and Blank wrote, "be a major human disaster." Indeed, though its influence on life in the past was unknown, it had to have been catastrophic. It would almost certainly bury those living within a thousand miles of Yellowstone in gray, choking dust. And its effect on world weather, Owen Toon of NASA's Ames Research Laboratory told me recently, "would be on the level of the effects that are predicted to follow a large-scale nuclear war." The ash cloud from the explosion, according to Toon, one of the world's leading experts on weather, might reduce the light reaching the earth to 5 to 10 percent of normal for months, preventing photosynthesis in plants and bringing bone-chilling cold exactly like a "nuclear winter."[68]

In amount of energy the caldera might release, nuclear war would be pale by comparison. Mount Saint Helens, Dr. Christiansen explained to me, was equivalent to a bomb of ten to fifty megatons (millions of tons of TNT). Yet a major eruption of the caldera would release 100,000 *times* that — up to five million megatons — three times the explosive power of the combined nuclear arsenals of the United States and the Soviet Union.[69]

Yet, it seemed, the question was not whether this cataclysm would happen, but when.

On the morning of July 20, 1981, David A. Kerwin parked his car near Celestine Pool at Old Faithful. When he opened the door his dog jumped out, scurried across the basin and, without hesitation, dove straight into the 200-degree water. Kerwin, in pursuit, dove in after him.[70]

A friend, Ronald Ratliff, pulled Kerwin out. Suffering third-degree burns over 85 percent of his body, Kerwin was rushed to the University of Utah's Intermountain Burn Center. He died the next day.

In the confusion, everyone forgot the dog. Apparently he died quickly and his body sank to the bottom. There it began to cook, its fat seeping into the surrounding water, where it changed the chemistry of the pool dramatically.[71]

The boiling point of water is determined by surrounding pressure. At sea level, where the atmosphere is heaviest, it is 212 degrees Fahrenheit. At the higher Yellowstone elevations, where air is thin, it hovers around 199 degrees. But below the surface of the hot pools, where water is under pressure from the water above, the boiling point is far higher. In the depths it remains superhot but unboiling, kept from turning to steam in much the same way as in a pressure cooker. But the oil oozing from the dog upset this delicate balance. Lighter than water, it rose to the top. There it reduced weight on the surface just enough to lighten pressure on the water below, bringing the entire pool suddenly to a boil.

Without warning the pool erupted into a geyser, blowing the dog into the air and expelling him from the pool.

As Kerwin's tragic death and its macabre aftermath demonstrated, the geysers of Yellowstone were delicate mechanisms: lethal, but exceedingly fragile and unpredictable. The entire geothermal system seemed constructed around a tripwire, ready to explode, perhaps without warning. And yet the very danger and uniqueness of this land held its fascination, particularly for geologists. The dire possibility of a caldron eruption and the intriguing challenge it represented attracted tremendous attention from the scientific community. For very different reasons it also appealed to political and corporate interests. The raw energy of the caldera, spelling disaster for some, suggested opportunity to others. Our attempts to understand the earth, though they told us much about the ecology of the park, also raised the possibility of a threat quite different from that of the expectorations that come from mother nature.

"We had been unsuccessful in our search after the precious metals," wrote Edward deLacy, about his prospecting trip across Yellowstone in 1863, "but having lived in the country ever since, and seen the developments made within the last twelve years, I am still of the opinion that at some future time, when the country on the North and South Snake rivers becomes settled up, that both gold and silver mines will be found there."[72]

Fortunately for the American people, prospectors such as deLacy and his comrades came out of Yellowstone empty handed. Supposed to have nothing of value, the region could be protected from exploitation at no cost to the nation or industry. Its only value, other than for attracting tourists, was as a laboratory. No one supposed that one result of scientific research might be a new kind of prospecting. Yet the very "wonders" that first drew explorers to the plateau turned out a hundred years later to be attractions not only for scientific research, but for those seeking precious metals as well.

Shortly after NASA left, Yellowstone became a magnet, not for prospectors but for a new kind of alchemist, intent on finding out how mother nature made gold, silver, and other rocks of lasting value. And just as the sobriquet "wonderland," had been inspired by Lewis Carroll's story first entitled *Alice's Adventures Underground* (written two years after deLacy came to Yellowstone), so in the modern era it was to be the wonders of Yellowstone that inspired a new kind of adventure down a different kind of rabbit-hole.

In June 1974 the Carnegie Institute and the National Research Council of the National Academy of Sciences cosponsored a workshop at Ghost Ranch, New Mexico. The topic was "Continental Drilling for Scientific Purposes." The idea was not new. Since the early 1950s the government had pursued elaborate drilling explorations at sea. Beginning with "operation Mohole" — an attempt to drill through the ocean floor near Hawaii — such searches had been made possible in part by the quantum leap in undersea technology made during World War II. In 1968 the Deep Sea Drilling Project was begun, sending the *Glomar Challenger* to probe deep below the world's seas.[73]

Drilling on land lagged behind these efforts. Continental plates were younger than those under the oceans, and far more complex. Less was known about them and about how to drill through them. But now the exciting new theory of plate tectonics had matured, permitting interdisciplinary studies of the continents and making exploration of their crusts more attractive. And drilling at sea had perfected the technology for such endeavors, making them more feasible. The workshop participants urged a large-scale program for piercing the continental crust.[74]

While those at the Ghost Ranch labored on an agenda for deep drilling, the nation was gripped by a fuel crisis. This shortage cemented the conviction, held by many, that we were running out of the nonrenewable substances that provided for defense and sustained our standard of living. All the easy places in which precious metals and strategic minerals could be found had been exploited. The price of gold and silver was on the rise. And though some saw this scarcity as a sign that we had finally reached the age of limits, others were inspired to search for new frontiers. Growing economic and political incentives demanded more sophisticated technologies for exploration that would delve into the deepest, most obscure nooks and crannies of the earth. As these enticements grew, so too did governmental and corporate interest in deep drilling.[75]

Soon the Departments of Defense, Energy, and Interior were each pursuing their own deep-drilling projects, spending together half a billion dollars and more a year: drilling boreholes; developing high-temperature cements and packers; testing "elastomers," open-hole packer systems, and high-pressure roller cone or stratapox bits to prevent blowouts; experimenting with drilling fluids, fracture detectors, borehole-to-borehole communications, thermal conductivity, directional drilling techniques, mud-pulse telemetry systems, and corrosion studies; building downhole motors, high-temperature fluid testing, and magma simulation facilities; conducting feasibility studies on molten-rock drilling technology, rock mechanics; and much, much more.

In a race to reach the continental basement, these agencies, sometimes in competition with each other, and generally unnoticed, cheerfully poured increasing amounts of public monies down "dedicated research holes" and "holes of opportunity."[76]

As the second fuel crisis threatened in 1978 and the price of gold soared, the U.S. Geodynamics Committee of the National Research Council, fearing possible proliferation and redundancy of uncoordinated efforts, sponsored a second workshop on deep drilling. Held at Los Alamos, the committee urged that the various agencies engaged in underground exploration together establish a body to coordinate our deep-drilling efforts, to be called the Board on Continental Drilling for Scientific Purposes (later the Continental Scientific Drilling Committee — CSDC).[77]

Since Los Alamos, the CSDC has overseen an expanding underground research effort. Exploration of the earth, it argued, would help us understand and predict earthquakes and volcanos, develop geothermal energy, map the earth's crust, and find sites for disposal of radioactive waste. But most important, deep drilling would aid in discovering valuable ore deposits. The committee wrote, "Deep drilling will improve understanding of the formation of important national resources such as gold, silver, copper, lead, and zinc deposits. Such understanding will lead to better insight into how to explore for such deposits. . . ."[78]

The committee began to look for "targets." The best place to drill to uncover these secrets was in active hydrothermal systems. "Many important metals," they observed, "including gold, silver, copper, cobalt, molybdenum, tin, and tungsten, are concentrated in mineral deposits formed by hydrothermal activity driven by heat from subvolcanic intrusions." Unfortunately there were only three such systems in America: Valles Caldera in New Mexico, Long Valley and Mono craters in California, and Yellowstone. They concluded, "The Yellowstone caldera is the outstanding geothermal target in the United States for deep research drilling."[79]

But there was a problem: Yellowstone was a national park. It offered, the committee said, "no opportunity for supportive industry drilling and the area is environmentally sensitive." It might also be difficult to gain National Park Service approval.[80]

Indeed, hydrothermal systems were exceedingly fragile. Permitting just a small leak could upset the balance enough to destroy an entire geyser field. Putting a hole into plumbing under enormous pressure risked a blowout in which large amounts of water could escape. And in every place in which man had tampered with such systems, he had destroyed them. All over the globe lay geyser fields ruined by curiosity and acquisitiveness. The Wairakei and Tauhara geyser basins in New Zealand, among the largest in the world, were destroyed by geothermal development. So too were El Tatio, Chile, Lardarello, Italy, and Steamboat Springs, Nevada. Exploratory drilling alone forced others, such as the Beowawe Geysers of Nevada — once the second-largest thermal area in North America — to extinction when water was allowed to escape from experimental drill holes.[81]

Park Service fears that a similar tragedy might befall the Yellowstone geysers led the U.S. Forest Service, in a 1980 Environmental Impact Statement, to withhold permission for commercial exploration of the Island Park, Idaho, area as a possible site for geothermal development. Although this possible geothermal area lay outside the park, neither Forest Service planners nor Park Service geologists could be sure that its exploration and development might not accidentally tap the park hot-spring or geyser system and destroy it.[82]

The Island Park EIS warned, "Geysers depend on a dynamic, yet fragile system which can be easily disrupted. . . . Any alteration of heat or water that flow through these natural systems can cause the 'plumbing' to dry out, disintegrate, and no longer produce geyser action."[83]

Nor was destruction of the geysers the only potential damage. The thermal basins, the EIS cautioned, were also important to wildlife. "Bison, elk, trumpeter swan, Canada geese, and many other waterfowl congregate in the thermal areas or on the rivers during the winter months."[84]

Yet even exploratory drilling could upset these and other animals in countless ways. A blowout or other accidental discharge could lead to "poisoning of terrestrial invertebrates, soil flora and fauna, vegetation and wildlife." It could cause "alteration of surface water quality which affects wildlife through loss of food, habitat, interference with feeding and behavior." It could encourage the growth of "nuisance organisms," change groundwater quality affecting "the diversity, productivity, and quantity of vegetation" dependent on the water table. It could lead to thermal pollution that would modify the atmosphere and disturb wildlife.[85]

The increased number of people brought to the well sites, the EIS continued, would multiply human–wildlife conflicts while forcing some species to avoid these areas. These animals would lose habitat for "feeding, security, nesting, wintering, migration." The clearing and excavation necessary to prepare a drill site would crush small species; cause others to avoid the area; reduce cover and food for resident wildlife; and disrupt or eliminate breeding, nesting, brooding, resting, and rearing activities. It would increase stress, interfere with migration of big game, and alter predator–prey relationships. The increased road traffic would lead to more collisions with wildlife. The loud noises associated with drilling would interfere with predator–prey relationships, reproduction, resting or hibernation, feeding, and migration.[86]

Subsequent erosion at the cleared drill site would reduce "breeding and nesting sites, cover, and/or other important wildlife habitat." Gas escaping from the wells would result in "modification of atmosphere and dependent wildlife," and "unpleasant odors may impair certain wildlife functions: hunting by smell, individual recognition."[87]

Whatever the effects following geothermal exploration in Island Park, more dramatic consequences were likely to follow the more ambitious program that some hoped was possible for Yellowstone. Although the Island Park geothermal area was relatively near a settled community far from the park's geysers, the CSDC hoped to drill near Yellowstone Lake, in the middle of the Yellowstone wilderness, in prime wildlife habitat, in the heart of hydrothermal activity. Then there was the matter of the caldera itself. No one fully understood it. Some wondered whether drilling might accidentally trigger an eruption. Might the government be digging a Temple of Doom?

Estimated by the Los Alamos participants to cost $12.4 million, this project would require at least a five- to twenty-acre clearing, a building to house the drilling rig, a large diesel-electric generator, a place to stack many miles of pipe, and a completely sealed storage facility in which to put the excavated — and probably toxic — sludge. It would be in effect an industrial park, floodlit around the clock, broadcasting the thumping noise of drill rig and generator through the backcountry, and fed by huge trucks carrying pipes, which would have to share the narrow park roads with recreational vehicles, trailers, family cars, bicyclists, and hikers.[88]

And no matter how careful they were, danger of a blowout and of a spill of toxic waste could not be ruled out. Indeed, the Los Alamos participants mentioned this possibility: "Geothermal exploration and exploitation have already destroyed some of the other major geyser areas in the world. The disturbing effects can be far-reaching. . . ."[89]

Technology for drilling into molten rock, they conceded, had not yet been developed. "Above 350 degrees Celsius [660 degrees Fahrenheit] concepts and methods for drilling and completing wells do not currently exist." Yet drilling in Yellowstone called for reaching temperatures of 500 degrees Celsius (930 degrees Fahrenheit) — a task, they admitted, which would require "revolutionary" advances. "The hazards and risks of drilling into such a high-temperature environment have not been thoroughly identified. One might anticipate, for example, dangers from 'blow-outs' when overpressured steam zones are unexpectedly encountered, as well as dangers from explosive interaction between downhole fluids and high-temperature magmatic intrusions."[90]

Despite these reservations, they did not discard the plan. Yellowstone was too important a target to be easily abandoned. "We shall never be able to obtain information about this remarkably large and important hydrothermal system except through a major program of scientific drilling," they persisted. "A national scientific drilling program should therefore emphasize what a unique natural laboratory Yellowstone can provide."[91]

To help gain public acceptance for the idea, the committee urged that "any such drilling must also be totally dissociated from any conceivable implication of economic exploitation." Instead, they suggested that the project be presented as an extension of the Park Service's own interpretive program. "The beauty and uniqueness of Yellowstone Park result from the igneous activity that underlies the Park. The detailed knowledge of the subsurface environment would complement existing park information and make Yellowstone a more meaningful national park." Some on the committee suggested that, after the hole was dug, the housing containing the drill rig be made a park museum.[92]

In 1984 the CSDC established a subcommittee to explore ways of drilling in the park. Robert Fournier chaired the group, and John Varley, Research Administrator for Yellowstone Park, represented the Park Service. Their report is not yet published but its participants assured me that Yellowstone was still under active consideration.[93]

"The committee will not recommend drilling in Yellowstone," Fournier told me recently, "until it can be assured that all potential risk is removed. Any drilling in the park will be done with utmost sensitivity to the environment." Varley concurred. "I laid down conditions for any such drilling project," he told me. "We identified a site near Lake that is already disturbed. We insisted that all drilling be done in winter when the bears are in hibernation and few visitors are there, so that the noise will be a minimal disturbance and trucks will not add to traffic congestion." At this writing, however, they had not ruled out drilling in the park, and no one had attempted to answer the many questions posed by the Island Park EIS.

Fournier remains enthusiastic. Drilling into the caldera would not accidentally trigger an eruption or quake, he reassured me, and new technologies could be developed to prevent a blowout. "It can be done safely," he said, "and it will enormously aid our understanding of the ways mineral deposits are formed."

This enthusiasm of their colleagues on the CSDC threatens to split the usually close-knit fraternity of geologists. While the CSDC keeps Yellowstone on its wish list, others silently worry.

Wayne Hamilton told me, "Permitting such a large, potentially disruptive project in the park would be a bad precedent. The dangers are real, and once there, the project might be hard to terminate. We could end up with a permanent drill operation in the park."[94]

"Probably they could take sufficient care to ensure that no major disaster occurred," a member of the CSDC (who asked not to be named) told me. "But being humans, anyone can make mistakes. I can't help feeling Yellowstone is not the place for this kind of project." Dr. Warren Hamilton thinks it is just a waste of money. "There is nothing we can learn from it," he told me, "that we cannot learn by studying surficial geology." And Robert Smith continues to doubt whether it can be done safely: "We do not know enough to do it yet," he told me.[95]

Despite such reservations, deep drilling shows signs of picking up momentum. "The national program receives strong support from the Hill," Robert S. Andrews, the senior staff officer of the CSDC, told me. "There has been a lot of enthusiasm, both in Congress and the White House," Fournier concurred. Perhaps too much enthusiasm, he hinted. "We sometimes have to slow some people down," he said. And indeed, the bandwagon seems to be rolling.

Some worry that the Russians may be ahead of the United States in this Stygian technology. As long ago as 1970, on the Kola Peninsula 150 miles north of the Arctic Circle near Murmansk, the Russians began to drill a hole. By 1984 the drilling complex was a building twenty-seven stories high and the hole was more than seven miles deep, far deeper than any hole we had. Could a deep-hole gap be developing? Some apparently thought so.[96]

"Mr. President, the United States has a lot of 'catch-up' work to do in this area," said South Dakota's Senator Larry Pressler in introducing a congressional resolution in favor of deep drilling in September 1984. "Russia and many of our European counterparts are far ahead of us in this important scientific technology."[97]

Thanks to Senator Pressler's efforts, President Reagan signed, on October 12, 1984, a joint Senate-House Resolution supporting the Continental Scientific Drilling program, to be implemented through the Department of Energy, the National Science Foundation, and the Geological Survey, and to encourage participation by the private sector.[98]

Senator Pressler continues to work for appropriations for the program, as well as for making the South Dakota School of Mines and Technology the center for these studies. President Reagan's Office of Science and Technology also strongly supports the project. Its spokesman, Bruce Abell, told me recently, "Obviously it has high priority."[99]

Indeed, it does. In its most recent report, the CSDC urged "that this area continue to be considered for possible research drilling." Apparently the park, already used for the exploration of space, may become the launch pad for a voyage to the center of the earth.

The geology of Yellowstone is part of its ecology. The same forces that move continents, build mountains, sustain the magnetic field, and change weather also affect the cycle of life and death in the park.

The soil determines the kinds of plants that grow on the plateau. Rhyolite, being little more than fine-grained granite, is acidic and lacks magnesium and potassium, two constituents vital for most life. Its soils provide poor wildlife habitat, where little more than lodgepole pine and huckleberry can survive. Andesite — covering many mountains surrounding the park — is by contrast more alkaline and contains many of the nutritive elements that support the succulent plant species on which bears, ungulates, and other animals depend. The limestone that lines the bottoms of many valleys surrounding the park is more fertile still; its highly alkaline environment offers rich fertilizer for all things from grass to trout.

The hot springs succor life as well. Animals use them for warmth and as places to find grass when the snow is deep. They also harbor a fantastic variety of microflora and insects. In the Yellowstone and Firehole rivers this heat inspires midwinter hatches of caddis and mayfly, insects that make the trout grow fat.[100]

For millennia people too have found that the thermal areas made the plateau — otherwise cold and forbidding — habitable in winter. They cooked in the hot pools and came to collect obsidian left by volcanic eruptions to turn into Clovis points and grinding tools. Then the geysers became the focus of the white man's fascination, exciting the curiosity of explorers, the dreams of visionaries, the avarice of railroads, and the generosity of Congress.

Yet this underground activity not only gives life; it also takes life away. The caldera did kill and will do so again. And if death is a necessary prelude to life, then it is a source of renewal. Every six hundred thousand years or so the slate is rubbed clean. For millennia deer and sheep, wolves and bears came from great distances to recolonize this plateau after their cousins had been swallowed in a holocaust.

Indeed, scientists are becoming convinced that meteorite bombardment, global weather changes, and volcanic eruptions have an important part in evolution. Massive extinctions caused by these catastrophes create vacant niches that are an incentive for evolution to fill them. Speciation fills voids left by the disappearance of earlier creatures. And so today, we might assume, continues a pattern of "punctuated equilibrium." Long periods of relative stability are followed by major ecocatastrophes and massive extinctions, making room for the evolution of whole new kinds of life as the end of one species becomes another's beginning.[101]

Yet although geology and biology are interdependent, in Yellowstone one has flourished while the other languished. In the same spring when Congress gave Hayden $75,000 to study rocks in the park, it refused to give Langford a cent for protection of wildlife. Holmes, Hague, Iddings, and an entire generation of geologists had completed studies in the park before the first cent of taxpayers' money went to biological research. The bison was saved in part by private grants. The park's first wildlife specialist, Milton Skinner — like George Wright — supported research with his own independent wealth. By the time Walt Kittams was hired as the park's first full-time biologist in 1948, geologists from the Survey had enjoyed seventy-five years of subsidized research.[102]

While Yellowstone authorities were denying the Craigheads permission to use helicopters to track grizzlies, they were inviting NASA and the Department of Defense to traverse the park with every flying device known to man. While they objected to exclosures as conflicting with wilderness values, they permitted the Geological Survey to drive its drilling rigs throughout the park to dig boreholes. While they denied biologists the use of radio-collars, they permitted geologists to detonate enormous quantities of dynamite throughout the park (one in the middle of Scaup Lake, possibly causing extinction of a salamander known to be native to the waters). While they prevented Professor Hornocker's proposed census of mountain lions because it would require tracking dogs, and while they discouraged Jonas from studying the beaver and moose, they were about to embark on a major deep-drilling project.[103]

Only two things, it seems, threaten the park: insufficient curiosity and too much of it.

Although the Park Service itself conducted little geological research in Yellowstone, the door has always remained open to the USGS and university scientists. Geology, one Park Service scientist explained to me, "enjoys two advantages over the life sciences: it has no management implications and it has great potential commercial value." Biologists, unlike geologists, might embarrass managers, just as Adolph Murie did when he found that killing coyotes was a mistake.

And so while government learns what it can about the inanimate, the study of life is left mostly unexamined. Unlike geology, it can never promise to turn base metal into gold.

The Environmentalists

Religions are born and may die,
but superstition is immortal.
— Will and Ariel Durant

16

The New Pantheists

THEY were dressed in their neatly creased designer jeans, Orvis flannel shirts, and L. L. Bean Maine rubber moccasins — a kind of unisex uniform for the trust-fund cowboys of the Rocky Mountain West. Seated in folding chairs or sprawled on the ground, surrounding the featured speaker, to whom they listened with rapt attention, were every color of the environmental rainbow: representatives from the Idaho Conservation League, Montana Environmental Information Center, Wilderness Society, Wildlife Society, local and national politicians, philanthropists, activists, and foundation officers.

I was sitting on the lawn with my wife, Diana. We had been invited to this gathering in recognition of our small contributions to the cause of environmental quality. For seven years we had run a summer educational program in natural history for young people on our ranch at the Smith River in Montana; and I had been the founding director of the Northern Rockies Foundation, an early effort to raise money for environmental, educational, and cultural activities in the region.

We were sharing the grass with the movers and shakers, the cutting edge, the cream of the American environmental movement. As I sat in the bright sun of this Jackson Hole dude ranch I could see Drummond Pike, President of the California Tides foundation and active supporter of countless public interest groups; Huey Johnson, founder of the Trust for Public Lands, former Western Regional Director for the Nature Conservancy, and former Director of Resources for Governor Jerry Brown of California; Bob

Bonine, former program officer for the Northwest Area Foundation of Saint Paul, Minnesota; Albert Wells, philanthropist and Project Consultant to the San Francisco Foundation; Maryanne Mott, summertime Montana resident, heir to the General Motors fortune, and trustee of the Mott Foundation (assets: $420,000,000); Bill Cunningham, Director of the Montana Wilderness Society; and Scott Reed, Idaho attorney and former member of the board of the National Audubon Society.

The occasion was a celebration of the tenth birthday of the Northern Rockies Action Group, the primary seed of the environmental movement in the northern Rockies. Styling itself a "Public Interest Support Group," for a decade NRAG (pronounced "En-rag") created or supported nearly every citizens' action group in the northern Rockies, including the Montana Environmental Information Center, Northern Rockies Foundation, Idaho Conservation League, Wyoming In-Stream Flow Coalition, Northern Plains Project, Montana Conservation Congress, Greater Yellowstone Coalition, Colorado Open Space Council, Montana Wilderness Association, Powder River Basin Resource Council, Northern Plains Resource Council, Montana Wilderness Association, and state and local chapters of the Sierra Club and many other national organizations. Since 1973 it had been in the midst of every battle, fighting strip mines, dams, power plants, coal-slurry pipelines, timber sales, hardrock mining in wilderness, ski resorts, geothermal development, and oil and gas exploration in national forests. Its staff had been Senator Metcalf's personal representatives monitoring the dispute between the Craigheads and the Park Service.[1]

With offices in Helena, Montana, not thirty miles from the continental divide, NRAG was where the two coasts met. Founded by William L. Bryan, Jr., a refugee from Maine; supported by the Rockefeller Brothers Fund of New York, the Conservation Foundation of Washington, and the Heartline Fund, Shalen Foundation and others from California; run by a board that included, at one time or another, Jon Roush (Ohio and Washington, D.C.), former Executive Vice President of the Nature Conservancy; George Klingelhofer (Pittsburgh and Sun Valley), industrialist and philanthropist; Tom Mudd (California), son of mining entrepreneur Harvey Mudd; Tom Silk (California), attorney; Leonard Sargent (Connecticut), philanthropist and now Montana rancher; and Leeda Marting (New York), Executive Director of the John Hay Whitney

Foundation, it brought bicoastal affluence to America's heartland.

Yet — although few of its board members were natives of the region — NRAG maintained close ties with the local state governments. It supported politicians such as Montana's Governor Ted Schwinden and Senator Max Baucus, and, not coincidentally, its graduates — people like David Hunter, former NRAG staffer and now Montana Director of Employment Security — sometimes found themselves appointed to important positions in state governments in the region.

The NRAG was, in short, the heart of intermountain environmentalism. It was the power establishment of the public-interest movement, a place where eastern and western influence met. In campaigns to save the environment in the Yellowstone region, we would not be exaggerating if we said that either NRAG was involved, or the campaign did not happen.

What, then, did we talk about at this tenth anniversary, as we sat on the ground ninety miles south of Yellowstone? Strip mining? Timber harvests? Wilderness preservation, the plight of the grizzly? No.

We talked about noetic sciences. Our discussion leader, Dr. Willis W. Harmon, Director of the Center for Study of Social Policy at the Stanford Research Institute and cofounder of the Institute of Noetic Sciences, led the participants in a quest for "alternative futures." He predicted demise of the "industrial-era paradigm," urging that we build a "transindustrial paradigm" by "connecting our social, spiritual, and ecological visions."[2]

Harmon and other participants railed against "positivistic science" and extolled the unified vision of a sacred reality, leading to "a natural fusion of science and religion."[3]

We must, Harmon insisted, "reinvent the world," in our search for the "esoteric core." We must "recognize our essential connectedness." He talked about the world's spiritual and shamanistic traditions, about how those peoples recognized, as we do not, that the powers of the mind are nearly limitless. He talked about his "Human Consciousness Research Program," which provided scientific proof of near-death experiences of the afterlife, of "remote viewing" (seeing what is on the other side of the mountain), of Tibetan meditation, and of the "values explosion."[4]

What did all this have to do with Yellowstone?

The presence of this California futurist that afternoon, the discussion of "paradigms" reflected the reality that, despite the serene setting, a crisis was brewing.

After two decades of unbroken political victories the movement was in danger of stumbling over its own success. Congress had enacted nearly its entire agenda into law. But where would environmentalists go from here? They still knew what they opposed, but were no longer sure what they were for.

Although they were a spiritual movement aimed at reforming the attitudes of Americans toward nature, they could not agree on a creed. Although they were an ecological movement dedicated to scientific understanding of natural areas, they were not certain they had a science. And although they were a political movement working through lobbying, litigation, and grassroots community organizing, they had no agenda.

And these uncertainties would have an enormous effect on Yellowstone.

Perhaps the first signs of spiritual crisis appeared on the day after Christmas, 1966. On that day Lynn White, Jr., a professor of history at the University of California, delivered an address in Washington, D.C. to the American Association for the Advancement of Science (AAAS) entitled "The Historical Roots of Our Ecologic Crisis." White's message was simple: Judeo-Christianity was destroying the earth. He argued that Christianity, in making man in the image of God, "is the most anthropocentric religion the world has ever seen." And by putting human beings at the center of things, this religion destroyed the sense of the sacred that had been part of all pagan attitudes toward nature. He observed, "By destroying pagan animism, Christianity made it possible to exploit nature in a mood of indifference to the feelings of natural objects."[5]

This exploitive attitude, White suggested, derived directly from the command in Genesis to subdue the earth, and in turn was the inspiration behind Western science, whose aim was to control and consume natural objects. In this way Western culture was dedicated to exploiting nature; Christianity was its rationale and science its tool. And because this attitude was so fundamental, so much a part of our past, there was little hope.[6]

It was our inability to cope with destructive forces embedded in our culture, White suggested, that was the heart of the ecologic crisis. That is why the environmentalists seemed so negative. They knew how to stop destructive changes, but not how to initiate constructive ones. He urged, "The 'wilderness area' mentality advocates deep-freezing an ecology, whether San Gimignano or the High Sierra, as it was before the first Kleenex was dropped. But neither atavism nor prettification will cope with the ecologic crisis of our time."[7]

The reaction to White's address was electric. Printed and reprinted countless times, it had enormous influence on environmentalists: For them, Christianity would never be the same again, *for he had given them someone to blame.* He had said out loud what preservationists had whispered for more than a hundred years, a secret suspicion they feared to make public: Christianity was the culprit. Like Martin Luther nailing his ninety-five theses to the door of All Saints Church in Wittenberg, White had brought the issue out into the open. The environmental movement now had an epistle for spiritual reform.[8]

While scurrying to blame Christianity for the fate of the earth, however, few noticed that White's thesis had badly missed its mark. Most failed to see that in calling Judeo-Christianity the most anthropocentric religion the earth has ever known, White overlooked Zoroastrianism and Confucianism, among others — faiths so social that they almost failed the test of religiosity. They did not notice the menagerie of all-too-human classical Greek and Roman Gods such as Zeus-Apollo, for whom nature was at best a backdrop for their peccadillos. They ignored the human-centered ancestor worship of the Shinto, product of a Japanese culture whose agriculture had plowed the earth for millennia without depleting it.[9]

And by blaming the Bible for man's rape of nature they failed to account for Christianity's own quarrel with science, epitomized in the Catholic Church subjecting Galileo to the Inquisition; they forgot the Western tradition of natural law, an attempt to found our legal system on nature, and the Christian philosophies of Locke, Rousseau, and Hobbes, who sought legitimacy for temporal authority in "the state of nature." They apparently forgave White for failing to explain adequately why the Eastern Orthodox Christian peoples lagged behind the West in exploiting nature and for not considering alternative explanations for the West's destructiveness: that possibly it was the heritage of a very utilitarian Roman culture, or secular Greek philosophy, or an accident of geography, demography, or even weather.[10]

It also escaped their notice that much of the horrific damage done the earth had been committed by people of other faiths. Long before the first word of Genesis was written, Pleistocene people had perhaps forced to extinction hundreds of species, and Taoist and Confucian Chinese had deforested much of their country; before the time of Christ, Phoenicians had stripped the coast of Dalmatia of its trees, allowing its soil to pour into the Mediterranean; other Near Eastern cultures had turned the Fertile Crescent to desert and destroyed the Cedars of Lebanon; even today cubic miles of Himalayan soil are dumped into the Bay of Bengal each year — thanks to the rapid deforestation accomplished by the Buddhist people of Nepal — and desert is right now spreading with the wind in Hindu India and Moslem Pakistan.[11]

The flaws in White's thesis, when noticed, had little effect on its power and popularity, for his timing was perfect: people were ready to believe him. The enthusiasm with which this thesis was received demonstrated only that a deep-seated antipathy to Judeo-Christianity already existed. White had proved that the movement rested on radical religiosity, and, as the quick and uncritical acceptance that greeted his remarks demonstrated, this religious strain was not new. In fact, it was as old and American as apple pie.

"A thousand Yellowstone wonders are calling. 'Look up and down and 'round about you!' And a multitude of still, small voices may be heard directing you to look through all this transient, shifting show of things called 'substantial' into the truly substantial spiritual world whose forms flesh and wood, rock and water, air and sunshine, only veil and conceal, and to learn that here is heaven and the dwelling-place of the angels."[12]

So wrote John Muir on visiting Yellowstone in 1885, penning another psalm for the movement that had already become a new, distinctively American religion. For Muir — founder of the Sierra Club and father of modern environmentalism — wilderness was more than a pleasant place to be. It was infused with the sanctity of God.

A refugee from traditional Christianity and modern society, Muir was raised in Scotland in an educational system, as he put it, "founded on leather." Forced as a boy to memorize passages from the Bible and whipped if he failed a lesson, he understandably grew up to have a low opinion of both Christianity and education. It was the outdoors, not formal learning, that mattered.[13]

Nature became his religion. Haunting the high country around Yosemite, piecing together a living in the backcountry so that he might stay in his beloved mountains, Muir never missed an opportunity to proselytize on their behalf. "I like preaching these mountains like an apostle," he exclaimed. And indeed he did, using the emotive power of biblical cadences to make his point.[14]

While preaching these Yosemite mountains in 1871, Muir met Ralph Waldo Emerson, acknowledging, with churchy prose, his affinity with this celebrated Yankee philosopher. "I invite you to join me in a month's worship with Nature in the high temples of the great Sierra Crown beyond our holy Yosemite . . . in the name of all the spirit creatures of these rocks and of this whole spiritual atmosphere Do not leave us now," Muir wrote, urging Emerson to stay.[15]

Indeed, Muir, along with his older friend Emerson and their contemporary, Henry David Thoreau, were the first apostles of a loose collection of ideas that formed a new religion of nature. Although they never became a formal church, never broke openly with Christianity, and did not share exactly the same views, these men and their followers did accept a vaguely defined spiritual unease about established creeds and a definite sense of reverence toward a sacred nature. Theirs was an eclectic faith of the wilderness. Emerson was its Moses, leading the faithful out of the grasp of established religion; Thoreau its Isaiah, the prophet; and Muir its David, guardian of the Promised Land.

Perhaps this awakening began in Massachusetts in February 1831. In that month Ellen Louisa, Emerson's eighteen-year-old bride, died of tuberculosis. Distraught over his loss, Emerson — then a Unitarian minister — had a crisis of faith. Unconsoled by the cold reason of the Yankee church, he began to search for a new faith. He abandoned his ministry and took a trip to Europe. What was he looking for? "He recognized from the start what he did not want," Emerson scholar Carl Bode wrote. "He rejected Judaism as austere, Buddhism as esoteric, Christianity as fossilized. Gradually he clarified what he did want. The essentials were simple: a kindly God, a kindly universe, and a few universal laws."[16]

Collecting the ideas of mystics and romantics — Goethe, Swedenborg, Coleridge, Wordsworth, Blake, Novalis — Emerson returned to formulate a faith based on intuition, mysticism, and a profoundly romantic attitude toward nature. He called his new philosophy transcendentalism. Emerson suggested that a Godlike spirit was in or behind everything. The material world was a bad copy of the spiritual. "Nature," he wrote, "is the symbol of the spirit . . . the world is emblematic." To know spirit, therefore, is to know the truth. But this knowledge can be achieved only through intuition and mystical experience. The intuitive contemplation of nature, therefore, is the highest human activity.[17]

Transcendentalism soon became a fashionable philosophy among intellectuals in New England. It was also made to order for Emerson's young friend and neighbor, Thoreau. A pencil-maker by trade, Thoreau, by his own admission, spent more than four hours a day tramping the fields and farms around his home in Concord, Massachusetts. Nature, Thoreau was convinced, was infused with the sacred. "Is not nature, rightly read," he asked rhetorically, "that of which she is commonly taken to be the symbol merely?"[18]

For Thoreau, nature was nearly synonymous with God. Modern urban industrial society, therefore, in cutting people off from nature, was denying God. The spiritual was in everything. Historian Stephen Fox wrote that Thoreau, in an essay written in 1858 for the *Atlantic*, "declared his faith in the immortality of a pine tree and its prospects for ascending to heaven. The *Atlantic* ran the piece without the offending sentence, thereby eliciting a furious letter from the author." Indeed, for Thoreau the way to salvation was through reunion with the primitive. In his celebrated classic, *Walden*, he extolled the virtues of returning to nature, and rejecting all things modern. He wrote in 1862 a sentence that was to become the rallying cry of the modern environmental movement: "In wildness is the preservation of the world."[19]

From the time of Emerson, Thoreau, and Muir this distinctive environmental religious perspective continued to grow. Slowly it was distilled into beliefs characterized by antipathy to progress, science, and the Judeo-Christian religious tradition. Instead, borrowing from sources including aboriginal cultures and Eastern religions, it preached compassion for all things.

"The quality of Muir's vision has undeniably colored my own moods of response and clarified the statements of my camera," remarked Ansel Adams, renowned photographer, conservationist, and early president of the Sierra Club. And indeed Adams was one of a long line of influential and distinguished intellectuals who refined the environmental perspective, keeping it alive and giving it respectability; a line that included Joseph Wood Krutch, *New York Times* drama critic and outdoor writer; Robinson Jeffers, the poet whose line, "organic unity — love that, not man apart," became the motto of Friends of the Earth; William O. Douglas, Supreme Court Justice and environmental activist; Charles Lindbergh, aviator and conservationist; Sigurd Olson, north-woods conservationist, philosopher, and writer; Aldo Leopold, founder of the Wilderness Society; Rachel Carson, whose *Silent Spring* ushered in the environmental awakening of the 1960s; and David Brower, president of the Sierra Club from 1956 until 1971 and founder of Friends of the Earth.[20]

Krutch, experiencing a midlife conversion to conservationism, decried the Hebraic desacralizing of nature and openly rejected Christianity after seeing that harbinger of winter's end, the spring peeper: "On the first day of spring, something older than any Christian God has risen. The earth is alive again." Justice Douglas became enamored of Buddhism after staying at a monastery in Tibet in 1951: "I realized that Eastern thought had somewhat more compassion for all living things. . . ." Charles Lindbergh, having made his reputation with a machine, turned in later life to the natural world and the wisdom of Eastern religions and aboriginal peoples. Perhaps the first prominent American to discover Taoism, Lindbergh became fascinated by preindustrial cultures. He wrote, on visiting this native Philippine tribe, "The major difference between an Agta hunter and myself lies in the invisible mass of knowledge my culture has crammed into my head." Olson, reminiscent of the transcendentalists, saw God in everything: "When He is symbolical in everything, then divinity is the highest goal of existence."[21]

Aldo Leopold, searching for a metaphysic that would explain the connectedness of things in an ecosystem, became a convert to the Russian mystic Peter Ouspensky. This philosophy, which Ouspensky called organicism, urged, along with transcendentalism, that all things in the universe are infused with spirit. Rachel Carson felt herself a disciple of Albert Schweitzer, the German physician who saw himself drawn to the Buddhist visions of reverence for all forms of life. David Brower also favored Buddhism. "We are in a kind of religion, an ethic with regard to terrain, and this religion is closest to the Buddhist, I suppose," he told writer John McPhee in 1971. Jeffers saw all things as infused with spirit. He preached through his poetry a view he called inhumanism: "Yourself, if you had not encountered and loved / Our unkindly all but inhuman God, / Who is very beautiful and too secure to want worshippers, / And includes indeed the sheep with the wolves, / You too might have been looking about for a church."[22]

For these people, Judeo-Christianity had less meaning. It was too anthropocentric; the idea of a personal God seemed out of place, for it permitted all the evils of science, technology, and the illusions of progress. Rather than seeing human beings as at the center of the universe, these people saw humanity at best as part of an interconnected whole and at worst as the temple destroyers who desacralized nature. Nor did the Judeo-Christian view that only man partook of the sacred satisfy them. Instead they saw the sacred in even the smallest things of life.[23]

Their ideas became a wilderness religion, but a religion with many names: Buddhism, Taoism, inhumanism, organicism, mysticism, transcendentalism, animism. Yet however seminal and inchoate, however eclectic a collection of offbeat non-Western and nonmodern theories, these variegated ideas did coalesce into one theme. In denying the existence of a personal god, most agreed that their metaphysic was (as Ansel Adams described his faith), "a vast, impersonal pantheism": the view that the universe is one interconnected whole and that every atom in creation is part of the sacred being of God. Krutch openly adopted "a kind of pantheism." According to Fox, Muir was a pantheist as well: "In his pantheism, he felt closer to pagan European and native American religious traditions." Pantheism, wrote ecologist (and former Catholic monk) William Everson, became "the secret religion of America."[24]

Borrowing this new pantheism from various non-Western, nonindustrial peoples, these environmentalists not only rejected modern science but also became suspicious of traditional canons of reason. It was, they suggested, only by mystical communion with nature that we truly understand her. We must rely on our primitive abilities of intuition. "Possibly, in our intuitive perceptions," wrote

Aldo Leopold in 1923, "which may be truer than our science and less impeded by words than our philosophies, we realize the indivisibility of the earth — its soil, mountains, rivers, forests, climate, plants and animals, and respect it collectively not only as a useful servant but as a living being." To one desirous of knowing the earth, even books and numbers were suspect. "I have a low opinion of books," Muir wrote in his journal. John McPhee wrote that "Brower assured me that figures in themselves are merely indices. What matters is that they feel right. Brower feels things. He is suspicious of education and frankly distrustful of experts. He has no regard for training per se."[25]

In searching for new ways of knowing nature, many were drawn to the occult. Emerson, following Swedenborg, believed that communion with the dead was possible. Muir claimed to have had telepathic experiences. Muir's friend, George Dorr, a leading eastern environmentalist, claimed to have communicated with his dead brother through automatic writing, consulted mediums, and believed in extrasensory perception. Charles Lindbergh, according to Fox, "delved into clairvoyance, believed in telepathy, sought out a medium in London, and flew to India to study yogis." Sigurd Olson claimed to have found a lost boy by clairvoyance, a power he thought common among primitive people. And William Brewster, a prominent preservationist with the Massachusetts chapter of the National Audubon Society, even claimed to have visited his dead Irish Setter in Hell! The dog was there by the River Styx, Brewster wrote, "with wagging tail and smiling loving eyes. . . ."[26]

Although they remained suspicious of science and reason and infatuated with the spiritual and transcendent, however, environmentalists grew ever more ambivalent about the ways in which nature was to be known. To understand wilderness, to manage it, to have an ecology, they knew, depended upon the possibility that nature was knowable; yet to give it proper respect required recognizing its intrinsic spiritual value. But how, if nature was infused with spirit, could it remain accessible to our understanding? What of the role of science in wildlife management? What must we make of the blossoming study of ecology? If we could know nature only by ways other than the quantitative and mathematical, was science to become something mystical and intuitive? *How was nature to be known?*

Meanwhile, environmentalists had a more pressing question: What was to be man's place in nature?

On March 3, 1891, Congress passed a bill creating fifteen national "forest reserves," totaling more than thirteen million acres. The beginning of our national forest system, this virgin land had been set aside — but for what? The law did not say. The National Academy of Sciences established a commission to decide on the purpose of these new lands. Conservationists began to choose sides. Some — among them Muir — wanted simply to protect these lands from development. Others wanted to put them to use, practicing "scientific" or "wise" management.[27]

The leading proponent of the latter approach was Gifford Pinchot, whom President Theodore Roosevelt appointed to the position of Chief Forester. A graduate of Yale, Pinchot did graduate study in Europe, where he came to learn the old and distinguished science of forestry known as silvaculture. This discipline itself had grown out of an environmental crisis several hundred years earlier.[28]

In the 1660s the British navy discovered that England was running out of wood with which to build ships. In response, the Royal Society commissioned a study of the king's forest reserves by one of its founding members, John Evelyn. Evelyn's report, published in 1662, entitled *Silva: A Discourse of Forest Trees and the Propagation of Timber in His Majesty's Dominions*, inspired a new science. Thereafter forests were to be farms, dedicated to growing trees to ensure sustained yield. By the nineteenth century all major forests of Europe were managed by this new science of silvaculture, and their efficiency much impressed the student Pinchot.[29]

Silvaculture was the philosophical ancestor of the policies of "sustained yield" and "multiple use" that guide National Forest Service management today. It was a commercial and utilitarian approach to the management of forests. But it held little attraction for most environmentalists. Treating these virgin lands as tree farms was hardly attractive to those who sought solitude and mystery in wilderness. And yet, at first, they sought to compromise. "The forests must be, not only preserved, but used," wrote Muir in 1895, "and . . . like perennial fountains . . . be made to yield a sure harvest of timber, while at the same time all their far-reaching [aesthetic and spiritual] uses may be maintained unimpaired."[30]

Gradually, however, Muir and his colleagues came to believe

that preservationism and commercialism would not mix. The national forests, they concluded, under the control of Pinchot's policies, would remain threatened by commercial exploitation. Gradually they lost faith in compromise. Slowly giving up the hope that our national forests could be saved as wilderness, Muir turned to promoting national parks. If national forests could not be havens protected from road and axe, then at least, it seemed, national parks were safe.[31]

But this hope too was soon dashed. The site of the disillusionment was a place called Hetch Hetchy. A beautiful valley of the Tuolumne River, a twin of the exquisite Yosemite, Hetch Hetchy lay secure, it seemed, in the Yosemite National Park wilderness preserve. But even national parks, it turned out, were not entirely safe.

The city of San Francisco was running out of water, and its mayor urged that a dam be built at Hetch Hetchy, supplying the city with both water and electricity. After the 1906 earthquake, the city, making use of national sympathy, pushed its case with increasing success. The lines were quickly drawn. A classic fight between preservationists and progressives ensued, with no room for a middle ground: Either a dam would be built in a national park, destroying priceless natural beauty, or it would not. A national political controversy for a decade and longer, Hetch Hetchy was also a personal battle between Muir and Pinchot. In the end Pinchot won. "As to my attitude regarding the proposed use of Hetch Hetchy by the city of San Francisco," Pinchot wrote, ". . . I am fully persuaded that . . . the injury . . . by substituting a lake for the present swampy floor of the valley . . . is altogether unimportant compared with the benefits to be derived from its use as a reservoir."[32]

In 1913 President Woodrow Wilson gave final approval to build the dam, and Muir was devastated. He wrote: "These temple destroyers, devotees of ravaging commercialism, seem to have a perfect contempt for Nature, and instead of lifting their eyes to the God of the mountains, lift them to the Almighty Dollar. Dam Hetch Hetchy! As well dam for water-tanks the people's cathedrals and churches, for no holier temple has ever been consecrated by the heart of man."[33]

Although Muir lost that battle, he won the war. The loss galvanized support for his protectionism. After the American public had seen the priceless wild jewel, Hetch Hetchy, disappear beneath the rising surface of the reservoir, few who deeply cared for wilderness remained convinced by the utilitarian ideas of Pinchot.

Hetch Hetchy became the watershed, for it taught the dangers of compromise. From then on the issue was simple: it was between preservation and destruction. Just as it was impossible to be half a Christian, so too was it impossible to be half a preservationist. Henceforth environmentalists would focus on those things — such as expanding and protecting the national park system — which would further the cause of preservation.[34]

And yet, thanks to the apparent simplicity of the Hetch Hetchy issue, other problems, more important to wildlife, were submerged. Preservation, it was assumed, was synonymous with protection. Few asked, as George Wright did, whether protection from development alone would ensure survival for wildlife in areas no longer in original condition. Few wondered whether, in preserves already disturbed by elimination of predators and suppression of fires, such a passive policy would save habitat. Once the issue had been cleaved so finely, environmental debate for the rest of the century would concentrate on the question of protection versus development, not on the means of preservation. Once Hetch Hetchy had forced environmentalists to choose, they were forcing us all to choose as well. Those espousing management of wilderness would be dubbed descendants of Pinchot, out to exploit nature. The idea that one could be a preservationist and yet advocate active measures to save habitat seemed a contradiction.

In this way the growing spiritualism of the environmental movement, encouraged by events such as Hetch Hetchy, nurtured a wilderness ethic of protectionism. If all nature was sacred, environmentalists said in effect, then nothing should be disturbed. The prescription for preservation was benign neglect. Nature must be protected from the depredations of man.

But in saying so, defenders of the environment moved away from pantheism. In a pantheistic system all is one and no one thing can stand out over others; there can be no division between the natural and the unnatural. Such a universe can have no enemy, because an enemy would be something outside nature, which according to pantheism is impossible. Whatever happens, then, is natural and right. To a true pantheist, therefore, man can do no wrong. Whatever he does, he is part of nature, part of the sacred, and therefore part of the actions of God. If human beings pollute themselves into oblivion, if they ignite the globe with thermonuclear fire, if they kill off every living thing, to a pantheist these are only the workings of God and were meant to be.[35]

In the environmentalists' scheme of things, however, man can stand apart from nature, and when he does he often commits the worst sin of all, the sin the ancient Greeks called *hubris* — usurping the role of God. By playing God he stands apart and thereby does evil. Yet they also seemed to suggest that human beings *should* stand apart from nature as well. For the ethic of protectionism was nothing more than *protecting nature from man.*

Human beings, according to this growing idea, could do nothing right. They were damned if they did stand apart, and damned if they didn't. On the one hand we were all part of an indivisible, sacred reality. On the other hand, we must be kept away from nature. Rather than pantheism, therefore, the new environmental perspective was something else; in fact, it was hopelessly confused. For it still had not decided: *What was man's proper role in nature?*

These were the invisible and conflicting currents that filled the air when Lynn White made his address to the AAAS. An environmental spiritualism was evolving, forming the basis for a distinctly American brand of preservationism. Because everything was sacred, believers in this ethic said, nothing should be touched. But at the same time, they left crucial questions unresolved.

White's warning summed up this growing spiritual unease. For more than a century the faith of environmental activists had been coalescing around a collection of ideas implying rejection: rejection of Judeo-Christianity, traditional science, technology, and civilization. In the void created by this iconoclasm they had assembled a jumble of ideas picked up from all over the world: from Buddhism, Tao, Russian mystics, pre-Socratics, romantic poets, Indians, and pessimistic German philosophers. And although each believer followed his own star, a family resemblance connected their ideas. Many quietly called themselves pantheists, espousing the view that all things were connected and sacred. They urged noninterference with spiritual nature. But they were still not sure how nature was to be known, and what was to be the place of human beings in the scheme of things.

These uncertainties, however, went unnoticed, because they had never been brought up for open debate. Realizing that in rejecting Christianity they were far ahead of the rest of the country, few announced their apostasy. "Conservationists," noted Fox, "in doubting the national myths of technology and progress, already had enough problems on their hands. To question the implications of *Christianity*, in an overtly Christian society, only complicated their efforts." Thus an open break with traditional religious beliefs was avoided. Instead, leaders of the new awakening conducted their own private searches for substitutes to the mainline faiths. In fear of offending our Judeo-Christian culture, they often confined these heresies to private correspondence or cloaked them in the patina of anthropological interest in primitive religions.[36]

By openly attacking Judeo-Christianity, White brought these heresies into the open and made them grist for public debate. But in doing so he drew attention to their problems. The ambiguities of their new pantheism would no longer wait for resolution.

In calling attention to "the ecologic crisis of our time," therefore, White had set environmentalists a threefold challenge: to find a religion, replacing Judeo-Christianity, which would resolve the question of our place in nature; to find a science, replacing the one that had produced our destructive technology, which would show us how nature could be known; and to construct a social agenda, replacing the one based on unlimited growth, which would change our culture before it had destroyed the earth. And by the time William Bryan came west to Montana in 1972 the quest for all three had begun.

17

The Subverted Science

THE item on the agenda was the budget, but the topic discussed was poison.

It was Squirrel Island's annual town meeting. The small community church was packed with leaseholders — elderly ladies and gentlemen who had the right to vote in this summer Maine community. The number of ladies present was surprising. Wearing dresses seemingly made of endless rolls of cloth, and floppy, flowery hats, their presence gave the meeting the atmosphere of a ladies' club luncheon depicted by *New Yorker* cartoonist Helen Hockinson: "OLAFs," the men called them (for "Old Ladies Are Funny") — an acronym that none of the ladies found funny at all. In fact, they rightly pointed out, the men were pretty funny themselves, with their knobby knees and bowed legs looking like bleached drumsticks below their plaid Bermuda shorts. Such was the genteel war between the sexes in 1961.

The question before the gathering was appropriation of funds for the annual war on mosquitos. For more than a decade the island had, without debate, employed a plane to douse the little bugs with DDT. Now, for the first time, there was debate.

One lady was speaking: "Rachel says that spraying will kill birds as well as mosquitos." Another chimed, "Rachel says we will be poisoning ourselves as well." Then another: "Rachel says it could pollute our water supply. . . ."

"Rachel says . . ."?

I was sitting in a pew at the back of the church. Visiting my parents, who summered on Squirrel, I had just returned from studies at Oxford University in England. The logic in the ladies' arguments seemed overwhelming to me, and indeed, convinced the other menfolk as well. The motion to spray was defeated. Squirrel Island went on record as perhaps the first community in the world to reject DDT.

Who, I wondered, was Rachel? A neighbor who lived on the mainland not far away, I was told. But how did she know so much about the evils of insecticide?

By the next year the whole world knew who Rachel was, and why she knew so much. Rachel Carson's *Silent Spring* was published, documenting just how we were poisoning ourselves with DDT. The book's publication, more than any other event, announced the modern environmental revolution. It became, Fox said, "one of the seminal volumes in conservation history: the *Uncle Tom's Cabin* of modern environmentalism."[1]

Silent Spring was perhaps the ordinary person's first encounter with ecology: "the web of life — or death" she called it. For although the idea that all living things affect one another was an ancient one, the scientific study of this maze of connections was less than a century old. It was not until 1866 that the German biologist Ernst Haeckel conceived the study of *ecology* — forming a contraction of the Greek words *logos*, meaning study of, and *oikos*, meaning household, or living relations — to describe "the whole science of the relations of the organism to the environment including, in the broad sense, all the 'conditions of existence.' "[2]

Such a late birth was not surprising. Ecology was a daunting idea. The study of an almost infinitely complex web of relationships connecting nearly all living things and their environment required synthesizing nearly every known science: chemistry, physics, and every kind of biology. And because human beings were certainly members of ecological communities, surely the sciences in which they were studied — anthropology, archaeology, history — should somehow be incorporated into any complete study of life. How, in this age of increasing specialization, was such a discipline possible? A radically new *method* for studying life was needed, permitting us to combine our knowledge of the parts to compose a coherent picture of the whole.

In 1935, Oxford botanist A. G. Tansley made a suggestion that he thought might accomplish this synthesis. He introduced the *ecosystem*. "Though the organisms may claim our primary interest," he wrote, "when we are trying to think fundamentally we cannot separate them from their special environment, with which they form one physical system.

It is the systems so formed which, from the point of view of the ecologist, are the basic units of nature on the face of the earth. Our natural human prejudices force us to consider the organisms . . . as the most important parts of these systems, but certainly the inorganic "factors" are also parts — there could be no systems without them, and there is constant interchange of the most various kinds within each system, not only between the organisms but between the organic and the inorganic. These *ecosystems*, as we may call them, are of the most various kinds and sizes.[3]

The idea was pregnant with possibilities, for in treating nature as a system it supplied to biologists mathematical tools for studying systems that had already been developed by physicists. Indeed, the idea of ecosystems had actually been borrowed from physics. It was a model that explained energy flow as a feedback mechanism, picturing nature as operating much like a thermostat. Just as a thermostat keeps a house at a constant temperature by turning the furnace on when the temperature drops and cutting it off when the temperature rises, so Tansley suggested that nature too was a system that maintained specific internal conditions despite fluctuations in external environments. An ecosystem tended toward equilibrium: it was *self-regulating*.

For ecologists, the ecosystem concept promised to accomplish the synthesis necessary for studying nature mathematically. An ecosystem could be pictured in the form of a model like an electrical circuit, by which energy, derived from the sun, was transferred by means of chemical processes through soil and grass up the food chain, and then, by decomposition, through the cycle again. Discovering this concept was an exciting breakthrough, for it allowed ecologists to examine living things with the powerful tools of physics and chemistry. "The ecosystem model," environmental historian Joseph M. Petulla noted, "added a mechanical dimension to ecology, giving access to the laws of thermodynamics, mathematics, and suggesting biogeochemical cycles. . . . Tansley's approach offered scientific respectability to ecologists."[4]

This seminal idea quickly caught on. It made possible dramatic advancements in community ecology — the study of any species that share an area. And it had countless uses: for studying physiological relations between things and their environment; competition; adaptation of species; niche overlap; self-regulating recycling systems of energy flow; distribution, hierarchy, and spatial relationships between animals; evolutionary histories; genetics; and dispersal. Better than any widget, this new universal tool seemed suited to answer the gamut of questions about living things.

It could also be used to examine our relations with nature. Indeed, Tansley insisted, human beings were part of any ecosystem, and excluding them from the study of ecology "is not scientifically sound."

Tansley wrote, "Human activity finds its proper place in ecology. We must have a system of ecological concepts which will allow of the inclusion of all forms of vegetational expression and activity. We cannot confine ourselves to the so-called 'natural' entities and ignore the processes and expressions of vegetation now so abundantly provided us by the activities of man."[5]

Most later ecologists concurred with Tansley. Ecology incorporated the activities of humans. "The components [of ecosystems]," wrote renowned ecologist Howard T. Odum, "often include humans and human-manufactured machines, units, or organizations such as industry, cities, economic exchanges, social behavior, and transportation, communication, information processing, politics and many others."[6]

Aldo Leopold carried this fruitful idea one step further. In his *Sand County Almanac*, which appeared in 1948, he advocated "the land ethic." The balance of nature, following the ecosystem model, is based on "the biotic pyramid":

Plants absorb energy from the sun. This energy flows through a circuit called the biota, which may be represented by a pyramid consisting of layers. The bottom layer is the soil. A plant layer rests on the soil, an insect layer on the plants, a bird and rodent layer on the insects, and so on up through various animal groups to the apex layer, which consists of the larger carnivores. . . . Each successive layer depends on those below it for food and often for other services, and each in turn furnishes food and services to those above. . . . Land, then, is not merely soil; it is a fountain of energy flowing through a circuit of soils, plants, and animals.[7]

Leopold saw, however, that we have "no ethic dealing with man's relation to land and to the animals and plants which grow upon it. The extension of ethics to . . . human environment is . . . an evolutionary possibility and an ecological necessity."[8]

Leopold suggested that the land ethic, in short, was a recognition that land was not merely soil, but a community of which mankind is merely a member. "The land ethic simply enlarges the boundaries of the community to include soils, waters, plants, and animals, or collectively: the land." It changes our role "from conqueror of the land-community to plain member and citizen of it. It implies respect for his fellow-members, and also respect for the community as such."[9]

In developing the land ethic, Leopold drew the boundaries of the ecosystem so that it would include mankind. For the future of this ethic rested on our having a complete science of life: Unless we understood our relations with nature, we could not be good biotic citizens. Such a complete science did not yet exist, but thanks to the bountiful ecosystem concept, it was now a distinct possibility.

Following publication of Carson's book, the idea of ecosystem became the byword of preservationists, and *The Sand County Almanac*, enjoying an unprecedented revival, was the Bible of the movement. Establishing a land ethic became a principal goal of the movement. Few realized, however, that its possibility rested on further advances in ecology, and that these advances were not immediately forthcoming.

True ecosystems, scientists soon discovered, were hard to find. In fact, as even Tansley had acknowledged, they did not exist! Tansley wrote,

> The whole method of science . . . is to isolate systems mentally for the purposes of study, so that the series of *isolates* we make become the actual objects of our study, whether the isolate be a solar system, a planet, a climatic region, a plant or animal community, an individual organism, an organic molecule or an atom. Actually the systems we isolate mentally are not only included as parts of larger ones, but they also overlap, interlock and interact with one another.[10]

The ecosystem, in short, was a tool by which scientists artificially separated their subject of study from everything else. It was an idea, a fiction, for in reality no "system" was isolated. The collection of things that counted as an ecosystem was a matter of convenience to the scientist: it was defined only to serve the purposes of analysis at hand. "What is a system," noted Odum, ". . . is an arbitrary characteristic of one's point of view." Although ecosystems might incorporate some parts of the earth, they had only fictitious boundaries. And once biologists began to study ecosystems, these boundaries were increasingly difficult to define.[11]

The first applications of Tansley's idea to nature were studies of ponds as ecosystems. Having neither inlet nor outlet, they were thought to be segments of the earth closest to his ideal of a closed system. For they seemed, at first inspection, to be closed circuits in which interaction among members of their biotic communities could be studied in isolation from the rest of the world.

The classic work of this kind was done in 1942 on the Cedar Lake Bog in Minnesota by R. L. Lindeman, entitled "The Trophic-Dynamic Aspect of Ecology." For Lindeman the bog was exactly like the thermostatic model, a closed-loop self-regulating system maintaining itself in equilibrium. Gradually, however, later scientists began to see that the system was incomplete: Even areas as apparently isolated as ponds were not closed systems. They were connected with the rest of the world in countless ways. Rather than isolated bodies of water, they were found to be merely part of the water table exposed above ground. The various forms of life that ponds contained were affected by underground currents, spring runoff, migrating waterfowl, trace elements dropped by rain, airborne spores, and the sun.[12]

In search of a complete, independent ecosystem, therefore, scientists began to look elsewhere. They studied other distinct geographic entities — such as islands or river drainages — but wherever they looked they could find no totally isolated systems. On a planet where some birds migrated tens of thousands of miles each year, where countless chemical substances were carried by the wind, where patterns of habitat were in constant flux, everything, it seemed, was connected to, and dependent on, everything else.

Community ecologists found that places such as Yellowstone seemed to be less isolated than they had once thought. "Yellowstone is hooked into the whole world," community ecologist Professor James H. Brown of the University of Arizona told me recently. "The more we learn the more we realize that large faunal species such as wolves and bears needed large areas — half a continent or more — to evolve. It is ridiculous to take any plot of ground and say it is a natural system."[13]

As science moved into the space age, the small, self-contained ecosystem such as the concept developed by Lindeman at Cedar Bog Lake receded even further from sight. A few, still searching for the elusive isolated ecosystem, resorted to constructing artificial ones, placing communities of microorganisms in one-liter, hermetically sealed "microspheres" for study. Many others cast their nets over wider areas, hoping that larger systems would prove to be more self-contained than the smaller ones. For these the ecosystem concept took on new meaning, referring not to a small spot on the earth like Yellowstone, but to the earth and its atmosphere as a whole. And more and more advances in nuclear physics and space exploration set the agenda for the study of ecology.

The atomic-bomb tests in the 1950s gave the world the first real idea that the earth was one ecosystem. When the Russians exploded a bomb in Siberia and the jet stream dropped its radioactive particles on Detroit, the whole world got the message: World weather was one system. Measurement stations were set up to trace the flow of airborne debris. Discovery of the radioactive isotope strontium-90 in cows' milk led to a national scare: where else might it show up? In response, the Atomic Energy Commission gave millions in grants to university plant ecologists to trace the route of this deadly radiation.[14]

By the 1960s, studies done by NASA in their search for life on other planets and their use of satellites to examine the earth's biosphere from space, further cemented the mainline scientists' opinion that everything on earth was interconnected. The earth's atmosphere, they discovered, was a complex and marvelous machine that kept all the chemicals vital to life in dynamic equilibrium. The ocean, former NASA scientist James Lovelock observed, was a "conveyor-belt system" for returning nutrients that had been washed to sea back to the land where they could sustain life. Trace elements essential for all life — selenium, dimethyl sulphide, halogens such as methyl iodide — once washed from the continents down rivers to the sea, were, by complex and fascinating processes, recycled into the atmosphere, where they returned to land again with rain. And during this recycling they were purified: the compounds threatening to life were diluted or transformed just as though they have been run through a giant filter.[15]

Many came to believe that only the earth itself, taken as a whole, could be considered self-contained. Only the earth was an ecosystem!

Yet even this judgment was open to question. Life on earth was directly affected by sunspots, meteoric bombardment, and interstellar phenomena such as cosmic radiation. The many mass extinctions that have occurred in earth's history — such as the final Cretaceous event, when much of life on earth disappeared sixty million years ago — were products, many thought, of just such extraterrestrial forces.[16]

Could it be that only the universe taken as a whole was a complete ecosystem? Coincidentally, pursuing this scientific idea to its logical conclusion ran head-on into the new theology. For pantheism, as we have seen, was just such an idea. The universe, pantheists were saying, was indivisible. Everything was interconnected. And now, increasingly, scientists agreed.

Further doubts about ecosystems nagged serious scientists. Some began to wonder whether the assumption — such a fundamental part of ecology — that there was a "balance of nature" had any basis in fact. And others began to suspect that much of community ecology was not good science.

"Ecology," noted biologist John A. Wiens at a recent symposium, "has a long history of presuming that natural systems are orderly and equilibrial." Indeed, ever since Tansley wrote, ecologists have assumed that ecosystems were self-regulating. This assumption was the basis for the National Academy of Sciences' claim that "compensatory mechanisms" would ensure that the grizzly population would rebound from the excessive mortality caused by closing of dumps and depredations by aggressive grizzly management. It was the idea that "self-regulatory mechanisms" would ensure that elk or bison would not overpopulate the range.[17]

Self-regulation, in fact, was the *whole idea* of natural regulation or ecosystems management in Yellowstone: If left alone, nature could take care of itself. But why, many scientists began to wonder, should we believe in this theory? Why should we believe that some invisible hand always guarantees that natural systems, if left undisturbed, will, despite continual fluctuation, remain roughly stable? Why should we believe that nature always knows best?

This assumption, it turned out, was just that — an assumption.

It came, Wiens wrote, "in part from a world view derived from Greek metaphysics, which proposes that nature must, ultimately, express an orderly reality, and in part from our theory, which is largely founded upon equilibrium or near-equilibrium mathematics." In fact, empirical studies by Wiens and others found that nature did not always do so. Wiens found that "the bird communities we have studied in grassland and shrubsteppe habitats are non-equilibrial."[18]

The vaunted self-regulating ecosystem, in short, the supreme assumption that guided management of wildlife in Yellowstone, more and more scientists began to suspect, was just a fictitious axiom, conceived to satisfy philosophical and mathematical conceptions of symmetry.

Likewise, doubt was growing that community ecology was good science. Nor was this worry new. Serious scientists had been apprehensive that ecosystems science dangerously oversimplified issues ever since Tansley first conceived the idea. In 1944, in a symposium sponsored by the British Ecological Society on whether or not competition for food was the major factor determining the structure of animal populations, some participants wondered how we could know the answer. One participant charged that "the mathematical and experimental approaches had been dangerously oversimplified and omitted consideration of many factors," including "predator attack" — the predator's role in controlling numbers of animals. Just as we discovered in Yellowstone, there was, he suggested, "little direct evidence" that limited food supply affected such matters as interspecific competition, for in the past "other factors usually kept populations below the point at which serious pressure was developed." Community ecologists were conceiving theories the evidence for which had disappeared generations ago.[19]

Throughout the 1950s, 1960s, and 1970s, community ecology was increasingly debated by wildlife biologists. The problem they faced was: How do we know when a hypothesis in community ecology is true? How do we know, for instance, whether populations in ecosystems are limited by "density-dependent" mechanisms or not? How do we test these theories? The study of ecological communities was so new and complex, the relations between living things and the land they occupied so changing, the differences between ecological communities so great, that verification turned out to be far more difficult than they had thought. The idea of a "crucial experiment," in which a theory is subjected to a test that either conclusively proves or disproves it — seemed remote.

Similar doubts plagued wildlife population ecologists. Biologist Wyatt W. Anderson said, at a colloquium at Oregon State University in 1980,

> The fusion of concepts in population biology has been based largely on mathematical models which explore relationships between such quantities as levels of genetic and environmental variation. . . . The problem here is a lack of experimentation to accompany the theory, to test predictions from it, and to suggest corrections. Critical experiments on rather fundamental aspects of population biology remain to be carried out. Theory has outpaced experiment. . . ."[20]

The community of biologists continued to fret, holding symposia at twelve-year intervals to share their misgivings: at Cold Spring Harbor, New York in 1957; at Brookhaven, New York in 1969; and most recently at Wakulla Springs, Florida in 1981.[21]

As the study of ecology progressed, these fears grew. The sponsors of the Wakulla Springs symposium concluded that

> community ecology does not have a strong tradition of using the diversity of evidence that has been so powerful in other disciplines. Sciences that have progressed rapidly (physics, chemistry, molecular biology) have made great use of sorts of evidence that ecology has not, of vigorous hypothesis testing and experimentation. . . ."[22]

They decided that "much of the evidence in community ecology to date . . . is flimsy."[23]

That no true ecosystem existed in nature, that natural regulation was an a priori idea conceived to satisfy philosophers and mathematicians, that evidence for ecological theories was both flimsy and difficult to obtain, were not the only problems facing this new field. For if modern ecology was to show us how we could, following Leopold's advice, become biotic citizens, it needed to become a science that would demonstrate the proper role for humanity in nature. We needed a truly interdisciplinary perspective combining not only chemistry, physics, geology, and biology, but the social sciences as well — especially, perhaps, history, anthropology, and archaeology.

But no such true interdisciplinary science appeared. Blending the social and natural sciences seemed a Sisyphean task. Neither the appropriate scientific method nor academic organization for developing such a unified theory existed.

Ecosystems science quickly became a subfield of wildlife biology, using a scientific method that applied to nature but not to people. The social sciences, however — the sciences that examined human behavior — studied only human behavior; they had no method for studying this natural world. *There was no science for studying relations between humanity and nature*, between natural processes and human institutions and behavior.

But with no way to examine the relations between man and nature, scholars did not have the tools necessary to help us carry out Leopold's fine advice. Petulla expressed the quandary: "If social scientists have ignored the role of the natural environment, ecologists have tended toward reductionism, and both natural and social scientists have worked in isolation from one another. For these reasons methodological problems loom large and will not be solved overnight."[24]

Lacking a conceptual foundation for a unified theory of life, university academic departments resisted development of a broad, truly interdisciplinary study of ecology. After *Silent Spring* appeared, to be sure, many colleges around the country introduced programs in ecology that they touted as "interdisciplinary"; but few went beyond an attempt to combine the various physical sciences with biology. And after a few years most of these programs collapsed for lack of faculty support.[25]

For scholars, the word "interdisciplinary" was nearly synonymous with "dilettante." Established academic departments determined what was respectable research and what was not, and interdisciplinary programs, not falling within the power of established departments, were seen, at best, as not serious science, and, at worst, as a threat. Professors quickly learned that they would advance little in their careers by jumping on the ecological bandwagon. Instead, at most universities, these programs either disappeared or were eventually absorbed into biology departments. Community ecology became just another part of biology, studying such things as competition in spider communities, structure of fish communities on coral reefs, size differences among sympatric, bird-eating hawks.[26]

For these reasons, a true science of ecosystems never developed.

Nature was far more complex than Tansley's model suggested. No union of the natural and social sciences was accomplished. And the knowledge necessary to heed Leopold's advice — understanding our place in nature — was not yet achieved.

Few in the environmental movement were aware of these methodological and institutional impediments to a true science of ecology. For the ecosystem idea was powerfully attractive. It suggested to the lay public the fundamental insight of environmental awareness: that all things were interconnected. It gave substance to the growing sense — articulated by Thoreau, Muir, and other theologians of the environmental movement — that nature was a web of mysterious, hidden forces that the contemplation of wilderness stirred in them. Consequently, the idea quickly became misunderstood. It became a name for those mysterious forces which bind natural objects together, and identified with that which evokes awareness of these forces: the apotheosis of unsullied nature, wilderness itself.

Caught in a war over progress, ecology found itself in a no-man's-land between territories that C. P. Snow called the two cultures, science and letters: between scientists who remained captivated by the quantitative allure of mathematics on one hand, and environmentalists who sought in nature an escape from reason. The scientists and environmentalists were moving further apart, and as they did, understanding each other less well.[27]

Although not noticed by environmentalists, the physical science they loathed had already co-opted their lovely study of nature. The ecosystem, Pedulla stated, "utilizes concepts taken from contemporary physics such as energy flow in systems." Its very power was in reducing wildlife problems to numbers. The study of ecology became — next to molecular biology — perhaps the most mathematical of the biological sciences. Today, biologist E. C. Pielou recently wrote, "the fact that ecology is essentially a mathematical subject is becoming ever more widely accepted."[28]

This mathematical emphasis even threatened the legitimacy of the new field, deemphasizing the search for evidence, the need for experiment. It became perhaps too abstract, a discipline attracting deskbound number crunchers more than those who liked to tramp about the woods in wool shirts counting deer scat. "Much of community ecology," concluded the organizers of the Wakulla Springs

Symposium, "has often been content with generalized mathematical theory and passively (rather than experimentally) collected observations."[29]

For their part, loving the *idea* of ecosystem, most environmentalists even so remained woefully ignorant about it. They had no desire to believe that this science, bringing us insights about "the balance of nature" and "self-regulating ecosystems," was the product of mathematicians' bias. Instead they began to turn science into religion.

"The science we call ecology," wrote ecological historian Theodore Rosak in 1972, "is the nearest approach that objective consciousness makes to the sacramental vision of nature which underlies the symbol of Oneness." It is, he says, "the subversive science. . . . Its sensibility — wholistic, receptive, trustful, largely non-tampering, deeply grounded in aesthetic intuition — is a radical deviation from traditional science. Ecology does not systematize by mathematical generalizations or materialist reductionism, but by the almost sensuous intuiting of natural harmonies on the largest scale. Its patterns are not those of numbers, but of unity in process."[30]

Thanks to the new environmental awakening, the gulf between ecological theorists and natural sciences was widening. A view of ecology that rested on the "sensuous intuiting of natural harmonies" would never accept the satellite telemetry the Craighead brothers used for mapping grizzly habitat, or any of the other tools scientists currently use that might help them understand an ecosystem: sophisticated electronic radio-tracking equipment, the huge pharmacopoeia of drugs for catching and manipulating animals, helicopters, guns, shackles, traps, snares, computers, FORTRAN and PASCAL programs, infrared photography for mapping vegetation zones, side-looking radar for searching out lava flows, tree-ring fire-interval studies, chemical analysis of soil, air, and water, grass-plot grids, horizon lines, pollen counting, radiocarbon-14 dating, uranium-isotope decay dating, fission-track dating, magnetic-polarity dating, potassium-argon dating, paleomagnetic dating, thermodynamic dating, amino-acid dating, nuclear magnetic resonance scanners, differential equations, exponential functions, binomial parameters, Lloyd's Indices of Mean Crowding and Patchiness, quadrats, sparse- and dense-grid cells, distance sampling, randomness testing, statistical deviations, sensitive seismic sensors, logarithmic tables, and nearly every other mathematical technique wildlife population experts, plant ecologists, soil scientists, limnologists, fire ecologists, anthropologists, archaeologists, and geologists could get their hands on.

Instead, just as David Brower used numbers when they "feel right," many of those who came to ecology felt a justified discomfort with formulas. Nearly all the pioneers of preservation — Emerson, Thoreau, Muir, Krutch, Adams, Olson, Douglas, Brower, Jeffers — were humanists. Other than Leopold and Carson, they were people at home with letters, not numbers. For three centuries and more, religion and philosophy had been losing a war with science. Since Galileo first dropped balls off the Tower of Pisa in 1589, humanists had been conceding the high ground to science, allowing themselves to be bombarded by the new Janissaries of mathematics, analysis, and experiment.

But now humanists were proving that they had been on the side of truth all along. Having exposed the awful consequences that science had unleashed on us — poison, pollution, excessive consumption, awful weapons of war and devastation — they had shown the world that science was leading us down the primrose path to extinction! All along we would have been better off if we had not been seduced by the promises of progress, had not followed the experts in physical science.

Now, armed with this insight, humanists were making their stand. Ecosystems, unquantifiable, must be protected from science. If "nature knows best," if the complexity of ecology surpasseth our understanding, then human beings must be kept away. Having proved themselves right, they were retreating to the mountains, to the sanctuary of their wilderness cathedrals, and daring scientists to enter. "The nature-linked collective consciousness of magic and mysticism, of totemic thinking and tribally ritualistic participation of most primitives," Yellowstone Master Plan architect Sigurd Olson wrote, showed us "that mystery and the unknown were truly the lure of wilderness travel." Wilderness was the last refuge of religion and humanities against dread positivism and technology. And so they had no use for statistics and mathematics. Whereas the scientist's ecosystem was an abstract idea without location or size, the environmentalist's was nature itself. It represented the fundamental cosmological insight that all was interconnected. Its very complexity demanded our humility. It remained ineffable, unknowable. Rather than a mathematician's fiction, it was part of the earth itself.[31]

The science of ecology in which Leopold, Carson, and others had such high hopes was falling short of its goal, even as the environmental movement was reaching its stride. A true science of life foundered on the rocks between the Scylla of mathematics and the Charybdis of metaphysics. While scientists moved ever more narrowly into the abstract realms of mathematical manipulation, environmentalists traveled more widely in search of a spiritual cosmology of nature. As scientists learned more and more about less and less, environmentalists were learning less and less about more and more. The study of places like Yellowstone fell into this widening gulf. Although few were aware, Leopold's land ethic — now part of the creed of contemporary environmentalism — rested on no foundation at all.

18

The Hubris Commandos

THOSE who climbed the stairs above the art movie theater at 9 Placer Street in Helena, Montana, ten years ago and first visited the offices of the Northern Rockies Action Group (NRAG) — three sparsely furnished, unpainted rooms — had to be impressed with what they saw. Combining folksy charm with a sense of impending revolution, the place reflected the best of the old and new. No person, it seemed, was over thirty and no piece of furniture under it. A young secretary, wearing an ankle-length granny dress and carrying her baby embedded in a Snuggly, worked her typewriter with terrifying speed. Longhaired men and women clomped about in their Swiss or Italian hiking boots — waffle-stompers, they were called — urgently conferring about Allenspur Dam, Colstrip power, stopping the awful strip-mine rape of the Great Plains. The atmosphere was as old as America and as new as California.

If not the world today, then the future, it seemed a certainty, was these young people's oyster. The environmental awakening just ten years previously had come crashing down on an unsuspecting America, and now they, recently out of college, had tramped west to pick up the cudgel. They were the vanguard of the new enlightenment, intent on washing away the sins of their fathers. They were going to rescue the northern Rockies — the last virgin part of America — before corporate greed had a chance to ruin it. And what better feeling than knowing one was on the side of angels! That they were paid less than $5,000 a year mattered little. This work was what grass roots was all about!

Although they did not know it, things were not to work out as they hoped. The early 1970s were a turning point in environmental America. A crisis was brewing. The sense of purpose that guided these people was based mainly on illusion. The newly energized superstructure of the public-interest movement lacked a sense of direction, a political agenda. Without a compass, it was in danger of losing its way.

In 1976, hiking in the Bridger-Teton Wilderness south of Yellowstone Park along a trail the locals called Interstate 90, and forty miles from the nearest road, I encountered a small city of tents hidden among the trees. Dozens of horses grazed in a nearby field. A complete and well-stocked bar stood under a large marquee. At a convenient distance his and her chemical toilets stood side by side. It was, I learned, the annual outing of a well-known national environmental organization, intended for the pleasure of its board of directors and more generous contributors.

This scene dramatized tensions that had long been building within the environmental movement. While challenging the establishment, it depended on the munificence of the rich. While striving to protect wilderness from the public, it depended on public support for its success.

Environmentalists had long recognized the critical importance of numbers. Unless Americans could get to natural areas, they could not be expected to recognize their beauty and value. They would not realize what was being threatened. And without a ground swell of public support, environmental organizations, no matter how enlightened, would not have the power to thwart corporate or bureaucratic plans. They would have no political base. Yet with sufficient public accessibility, natural areas could be saved. "If the public could be induced to visit these scenic treasurehouses," noted Allen Chamberlain of the Appalachian Mountain Club in 1909, "they would soon come to appreciate their value and stand firmly in their defense."[1]

The developing tragedy of Hetch Hetchy only reinforced the need for a sound base of public support. Many, including Chamberlain, believed it could have been saved if the valley had not been so inaccessible. "Take Yosemite Park as an example," he wrote in 1910. "Everyone is herded into the great valley, and little is done to encourage people to go into the magnificent country farther back in the mountains." Consequently, few ever saw Hetch

Hetchy. He urged that a road be built to the valley, and that "hotels and boarding camps" be established there.[2]

If wildernesses were cathedrals, in short, they needed congregations — a hard political fact that continued to preoccupy the movement. For their purpose was conversion. As Joseph Sax, arguing in 1980 for the national parks' "ideal," said,

> The preservationist is not an elitist who wants to exclude others. He is a moralist who wants to convert them . . . his claim is that he knows something about what other people ought to want. . . . The weight of the preservationist view, therefore, turns not only on its persuasiveness for the individual as such, but also on its ability to garner the support — or at least the tolerance — of citizens in a democratic society to bring the preservationist vision into operation as official policy. . . . For that reason I have described them as secular prophets, preaching a message of secular salvation.[3]

It was this preoccupation that encouraged Muir to accept the necessity for tourism. It was this same recognition that led to the alliance between preservationists and railroads, creating the national park system, and to the alliance of conservationists, Park Service, and automobile groups that inspired Mission 66 and the construction of Grant Village. It was, in short, for environmentalists, a kind of pact with the Devil, a catch-22: Saving the wilderness required promoting it, *using it.* But allowing millions to use places legally defined as areas "untrammeled by man" was not only contradictory but potentially devastating. To save the wilderness, it seemed, they would risk destroying it.

This contradiction seemed especially acute during the early era of the automobile. How else could millions of people be brought to the wilderness except by car? Horse-packing in the high country was expensive. It remained a definite preserve of the rich. But roads destroyed wilderness.

The only other way to reach the outback was by walking, yet backpacking in the early days was an exercise in pain. When sleeping bags weighed nine pounds, tents were made of tar-impregnated canvas, the only food that could be kept came in cans, and backpacks were pack boards, hard and unforgiving as slabs of cement, this form of travel was macho at best and backbreaking at worst. Requiring the dedication of an Eagle Scout, it was daunting to all but a few.

Lacking amenities, backpackers had to live off the land. To lighten their loads, they carried no cookstove but set wood fires; they walked in with little food but killed game and fish as they went; they built shelters of boughs. Such travel, in short, demanded *woodcraft*. The hiker of the 1940s and 1950s lived like early explorers such as John Colter, who set off on his six-month exploration of Yellowstone with little more than a couple of blankets, gunpowder, and supplies of lead and salt.

This was the ethic of self-sufficiency taught to generations of Americans by the Boy Scouts: how to light a fire without a match, how to make things of wood, how to skin a squirrel and cook it at the end of a stick, how to make a pouch from its hide, how to make a shelter, how to use leaves for toilet paper, how to bake a potato in the ground. It was an ethic perfectly suited to the era of true wilderness, and it was *good ecology*. Woodsmen consumed natural, biodegradable, renewable resources — trees, fur, trout. Living within the ecosystem, they simply became part of the local flow of energy.

But it was not and could not be the way to the wilderness for the millions the environmentalists hoped would come into the country. No matter how ecologically sound, woodcraft would destroy the woods if too many practiced it. Just as populous tribes in the sub-Sahara of Africa had turned savanna into desert by overuse, so too would a hundred thousand Scouts clean out a forest like a plague of locusts. Cutting boughs for bedding, lighting campfires, killing game, or catching trout — activities that done by a few are benign — would, when practiced by millions, become a threat. How, then, could wilderness be made accessible without being destroyed?

The environmental movement found a silent partner in the recreation industry. Combining space-age technology and the economy of scale affordable with a mass market, these companies, most of them relying on mail orders, opened up the backcountry in ways that the car never could. L. L. Bean, for years a small Maine sporting-goods store that sold funny-looking rubber-bottomed boots, became one of the largest mail-order houses in the world, grossing $240 million annually. Following in its tracks came Orvis, Eddie Bauer, Eastern Mountain Sports, Early Winters, Gokeys, Recreational Equipment, Kelty, Gerry, Northface, Sierra Designs, and countless other stores selling outdoor comfort. By 1970 these companies were doing $7 billion worth of business a year.[4]

Gone were tweedy stores such as the old Abercrombie and Fitch that provided a $50,000 shotgun or a hand-glued Leonard Tonkin cane fly rod to the Far Hills executive who wanted to spend a week chasing grouse or salmon in Scotland. In their place were purveyors of plastic to the rest of us, making use of the latest technology to provide every conceivable device for taking to the mountains: Dacron fiber-filled sleeping bags; rainwear made of Gore-tex (developed from Dupont Teflon frying-pan coating) or Bion II (from artificial-heart research); lightweight tents made of extruded aluminum, polypropylene, Mylar, polyester, or rip-stop nylon, and coming in every conceivable shape — geodesic domes, A-frames, pyramids, tunnels, cylinders, hypars; fishing rods of graphite fiber (developed for jet fighters); chemically processed, freeze-dried food; fiberglass cross-country skis; backpacks with internal or external stays made of carbon fiber (from NASA research); reflective Texolite (from space capsules); gold-plated Sierra drinking cups and night lamps, and so on endlessly.[5]

As the recreation industry grew, so too did the national parks and the wilderness system. Between 1964 and 1972, thirteen natural areas (including five national parks), four National Monuments, one National Scientific Reserve, twenty-nine historical areas, and twenty recreation areas were created, giving the national park system a total of 284 reserves. From a proposed 9.1 million acres in 1964, the wilderness system continued to climb, reaching, by the 1980s, 19.3 million acres, which included 187 areas in forty-one states, and another 10 million acres remained candidates for inclusion.[6]

And as wilderness grew, so too did the environmental movement. Membership in the National Audubon Society climbed from 45,000 in 1966 to 193,000 in 1972, and would exceed a half million by 1983. Sierra Club membership zoomed from 35,000 in 1966 to 137,000 in 1972, on its way to 300,000 by 1983. Indeed, as the years passed, this growth accelerated. "It took us eighty-nine years to reach 200,000 members," reported a spokesman for the Sierra Club in 1982, "and another thirteen months to reach 300,000." The Wilderness Society, National Wildlife Federation, and many other organizations experienced similar growth.[7]

As their membership grew, environmental organizations continued to promote use of wilderness. Organizations such as Audubon and the Sierra Club sponsored expeditions and outdoor educational programs to ensure that succeeding generations of Americans would understand the intrinsic value of wilderness. "Let Audubon show you the world as you've never seen it," chimed an advertisement for National Audubon Society Tours in 1981, listing fifteen itineraries to every part of this country from New England to Alaska. Sierra Club Outings boasted trips by bicycle, burro, ski, boat, horse, and bus, and scores of wilderness expeditions. More modestly, Defenders of Wildlife Expeditions offered trips to places such as Kenya, the Galápagos, and Alaska.[8]

Survival programs such as the National Outdoor Leadership School in Lander, Wyoming, and Outward Bound, inspired by a survival program run by the Royal Air Force, sprang up around the country. Offering quasi-military courses in "confidence building," they became finishing schools for the children of the upper middle classes — some the very same young people who missed the ultimate survival course in Vietnam. Fly-fishing schools, such as those run by Orvis and Fenwick, cloned tens of thousands of rubber-suited sportsmen who would spend discretionary income for the rest of their lives in search of the Right Fly.

Natural areas grew as well, but not perhaps in the way that Muir had hoped. A relative few sought out the solitude of the backcountry. Although national park attendance increased from 97 million in 1962 to 172 million in 1970, this expansion was due more to the added number of parks than to the growth of backcountry use within individual parks. Yellowstone visitation, which first reached 1.9 million in 1962, would still be at that figure in 1974 and would actually drop to 1.8 million by 1979.[9]

Likewise, though exact figures on wilderness visitation were not recorded, the use of these areas, though growing somewhat, was less than spectacular. Several studies revealed that giving an area wilderness designation resulted in only a slightly higher rate of use. Even in the relatively accessible Sierra of California, according to a recent Sierra Club study, "the majority of the wildland area . . . receives little visitation and is in good health."[10]

Nor did these areas attract serious backpackers. Instead, nine out of ten who entered wilderness did not spend the night, but turned around before they had gone six miles from their cars. "Four to six thousand people begin hikes from the trailhead at East Rose-

bud, but probably not more than 400 will pass by Fossil Lake"
(twelve miles up the trail), George Shaller, Supervisor of the Red-
lodge, Montana entrance to the 580,000-acre Beartooth-Absaroka
Wilderness told me recently. That few wanted to walk far was also
found in Yellowstone, where distances were great and where many
trails, built by the cavalry for horses a hundred years ago, were
difficult for backpackers. Park attendance remained stationary for
a twenty-five-year period at around 2 million, and backcountry use
(hiking and horseback riding) remained small and would grow
slowly, from 36,000 visitor-use nights in 1973 to 49,000 in 1982.[11]

Rather than serious backcountry use, these areas bore the brunt
of the explosion in recreational technology as Americans discovered they were wonderful places in which to play with space-age
toys. They became national *playgrounds:* places in which to hunt,
fish, climb mountains, ride a bicycle, cross-country ski, canoe, raft,
and hang glide.

For, paradoxically, this was an explosion in *consumptive* uses of
wilderness. Between 1965 and 1970, the number of hunters in the
country increased by 3 million, but birdwatchers actually decreased
by the same amount. In 1965, fewer than 600 people rafted on the
Colorado River through the Grand Canyon; by the mid-1970s nearly
15,000 did so. Indeed, between 1955 and 1975 the number of hunters and fishermen more than doubled (from 31 million to 74 million); their spending grew sevenfold (from less than $3 billion to
more than $21 billion), and the days they devoted to their sport
increased fourfold (from around a half billion to nearly two billion
recreation-days). Similarly in Yellowstone, while visitation remained stationary and backcountry use relatively small, trout
fishing — always popular — grew enormously: from 230,100 angler-
days in 1973 (the first year in which such records were kept) to
363,403 in 1981.[12]

As collectives of these recreationists, some environmental groups
became captive to the recreation industry, more interested in access
for play than in preserving habitat. Trout-fishing organizations,
supported by the recreation industry, would lobby and sue for
stream access while paying less attention to the consequences that
this access might have for trout. Wildlife organizations with large
constituencies of hunters would push for protection of game animals but remain indifferent to the fate of other species.[13]

Yet recreational activities, no matter how small in relative numbers, continued to threaten wilderness. Campsites — places of concentrated use — became "sacrifice areas," where grass was trampled; trees hacked, initialed, and stripped of limbs; streams cluttered with trout entrails; trails rutted by waffle-stompers; and where the surrounding land became dotted with "cat holes" of human waste. How was the land to be spared?[14]

Thanks to the new technology, there was, it seemed, an answer. It was called minimum-impact camping. In place of woodcraft, it was a new and nonconsumptive backcountry ethic. Promoted by survival schools, the National Forest Service, and conservationist organizations, its message was: "leave the wilderness as you found it."[15]

The enlightened camper would not build a fire, consuming the local fuel and leaving a "fire ring"; instead he or she would cook on a tiny gas stove. He would not bury his trash, but take it with him. He would be careful where he left his wastes and would burn his toilet paper after he used it. He would, in short, take everything into the wilderness he needed and take everything out he didn't. Entirely self-sufficient, he would travel in an ecological cocoon, leaving no footprint.

Fortunately this ethic quickly caught on and did much to minimize the imprint of millions on wilderness. It did not, however, make the problem go away. Much wilderness was simply too fragile for the numbers of visitors it received, and these people hurt it in the most surprising ways. Animals, researchers discovered, preferred plants that had been saturated with human urine, so that vegetation on which people had relieved themselves ran the risk of being overgrazed. The proliferating cat holes continued to be a problem. Backcountry diseases such as giardiasis, first brought west by the mountain men, became epidemics in the West. The West Fork of Rock Creek, an apparently pristine mountain stream coming directly out of the Absaroka-Beartooth Wilderness north of Yellowstone, became contaminated by the microorganism, causing an epidemic in the town of Red Lodge, which obtained its drinking water from the stream.[16]

Moreover, nothing was especially "ecological" about minimum-impact camping. Using highly processed metal from Sweden, plastic from France, or food from San Francisco, and burning fossil fuels from Saudi Arabia was, environmentally speaking, robbing Peter to pay Paul. The camper borrowed energy from elsewhere to consume in the mountains; he burned nonrenewable resources taken from another part of the globe in order to spare the renewable resources of his own backyard.

This was, in fact, precisely the criticism many in the third world made of American civilization in general: that it maintained a high standard of living at the expense of others. And ironically, it was akin to the artificiality that everyone seemed to deplore about garbage dumps for grizzlies. Garbage, as we have seen, was expelled because it was an exotic source of energy. Eating at the dumps, bears were consuming energy that had not been produced by the ecosystem. But the animals suffered from a double standard: That which was corrupt behavior for bears was enlightened behavior for human beings. Minimum-impact camping, though also importing energy, was hailed as a way of saving the wilderness.

Such ironies and imperfections in the new wilderness ethic, however, passed unnoticed. All that mattered was that the age of plastic had arrived just in time. Minimum-impact camping saved the wilderness and the conscience of environmentalism. Now the movement could, it seemed, eat its cake and have it: Millions could be invited to enjoy primitive America and new conservation practices would minimize the effect these numbers had on the land.

More than fifty years had passed since that cold, drizzly day in Maine in 1913, when Joseph Knowles, a part-time illustrator, shed his clothes in front of reporters and plunged naked into the wilderness in search of the ultimate primitive experience, returning two months later to tell the world: "My God is the wilderness. . . . My church is the church of the forest." Now, space-age technology gave new momentum to the flight from progress.[17]

The infatuation with preindustrial cultures that stirred the hearts of the new pantheists from Muir to Brower spread as the environmental movement grew. Environmentalism became, wrote sociologist Bill Devall (a self-described "deep ecologist") "the most conservative social movement in the 20th century." Some even called themselves "neolithic conservatives." They looked back to stone-age cultures, said Devall, "to the mind of the primitive as a standard of human sanity and social stability." Now, as recreation technology brought more Americans close to the land, this flirtation was assimilated into mainline American culture. The national excesses of conspicuous consumption and industrial waste further incited a search for a simpler life.[18]

The invention of polyester gave birth to primitive chic. Those of us who could went in search of a simpler life, seeking self-sufficiency close to the land. We moved to remote areas, lived in log cabins built by axe, practiced organic gardening, eschewed tractors for horses, spun thread for clothes. The *Whole Earth Catalog* rather than the Sears, Roebuck catalog was the source of our inspiration. We heated our houses with Ashley woodstoves, swept our floors with the nonelectric "Hokey" sweepers, lighted our rooms with Aladdin kerosene lanterns. We read the *Mother Earth News,* not the *Old Farmer's Almanac,* as we sought advice on building root cellars, turning washing machines into potters' wheels, building a safe skunk trap, selling nightcrawlers, growing mushrooms, or recycling old tires.

If we could afford to do so, we installed solar collectors for heating and windmills for electricity. But this was the rub: The simple life was expensive — too expensive for most Americans. And the ways of making a living in the outback were limited. Horse-drawn plows could not turn enough soil to turn a profit as well. The simple life was a luxury few could afford. It became a lifestyle reserved for writers, legatees of trust funds, marijuana farmers, and other noncommuters who needed no visible means of support.

It was also of questionable ecological value. For attractive as self-sufficiency was, it was not the best use of the earth's finite resources. It was in fact the analogue of the Boy Scout philosophy of woodcraft: Burning wood, for instance — a renewable resource — was a good thing when only a few did it, but a menace to public health and the future of forests when it became a national craze. Although not a renewable resource, a conventional fuel such as natural gas was far less of a polluter. Towns such as Missoula, Montana, where woodstoves were common, became plagued with persistent woodstove air pollution. Likewise, log houses consumed far more trees than the frame design did, and were less well insulated.[19]

Despite its problems, however, primitive chic was here to stay. For its attraction lay, not in its ecological integrity, but in its rejection of all that was modern and synthetic. It was diffused throughout our culture as "natural" became the favorite word of

Americans. We ate natural cereals and looked for the tell-tale "all-natural" claim on everything from food to fibers. If we could afford to do so we turned our backs on polyester and sought to surround ourselves with cotton, wool, leather, goose down, and fleece, while womenfolk anointed themselves with the essence of avocado, peach, mink oil, jojoba, rosemary, rose, aloe vera, safflower, sesame, sunflower, peanut, olive, wheat, and every other source in the cosmetician's garden.

Nostalgia, not what it used to be, was growing stronger. Spurred by the emphasis civil rights had placed on the past of native and Afro-Americans, millions of whites as well went in search of their roots. And if, as historian Frederick Jackson Turner suggested, it was the frontier experience that melded us as a nation, then our nation's roots, our sense of national identity, lost somewhere in Southeast Asia, could be regained by reviving our past. If the frontier was dead, the illusion of it could be recreated.

People such as the buckskinners found themselves riding the crest of this new wave of nostalgia. These loosely organized groups sought to relive the lives of their ancestors as vividly as possible. While some modeled themselves upon Indians, others upon the fur trappers of the West during the 1820s, and still others upon soldiers during the French and Indian wars of the early eighteenth century, all insisted on absolute authenticity: men dressed in deerskins, women in handspun ankle-length dresses. "This country was settled by people who didn't own anything but a muzzle-loading rifle and axe," wrote John Baird, publisher of the *Buckskin Report*, official publication of the National Association of Primitive Riflemen. And following this philosophy, they made everything for living by hand, from clothes to tallow candles, often from animals shot with handmade, muzzle-loading replicas of the black-powder guns used in this country between 1600 and 1865.[20]

Meeting at "rendezvous" throughout the country, they relived history by living in tepees, sleeping on buffalo robes, cooking in cast-iron pots and feasting on rattlesnake meat and other early American delicacies. And attendance at these events grew hugely. Four thousand people put up 1,100 tepees at the 1984 western rendezvous, an event that had attracted fewer than two hundred only a few years ago.[21]

Although not as eccentric as buckskinners, visitors to the backcountry shared this primitive mystique. Just as Starker Leopold sought to recreate the illusion of primitive America in our national parks, others attempted to relive it in wilderness. True wilderness — dangerous, fecund, surprising, which even Thoreau described as "savage and dreary" — had gone with the mountain men, but that did not matter.

Areas that entertained millions of visitors a year could hardly be called places "untrammeled by man," but if we could be on a mountaintop alone for just one evening, we could briefly imagine that magnificent solitude endured by those who once lived in an almost-empty continent.

We might cook our freeze-dried, chemically processed Rich-Moor beef Stroganoff on a Svea stove with highly refined fossil fuels and sleep in a Dacron Fiberfil sleeping bag, but we could pretend we were following the footsteps of "Liver-eating" Johnson as he roamed the West in search of beaver. The lake by which we camped may in primeval times have been barren of fish; the brook or rainbow trout now rising may have been descendants of fingerlings brought west from Maine or east from California and dropped in plastic bags by helicopter into the lake, but in their artificial fecundity they permitted us a moment of atavistic delight, as we feasted on the bounty of nature. The surrounding countryside, now nearly bereft of original fauna, may, in John Colter's day, have teemed with mountain sheep and have been regularly patrolled by wolves and grizzlies, but the prismatic beauty of the alpine landscape was satisfying enough, the original fauna not missed, and the declining probability of a sudden encounter with an angry carnivore a private relief.

No, truth mattered little to most backpackers. They accepted the altered landscape as wilderness because they wanted to believe it was so and because the law had so defined it. They were like buckskinners, playing a game. They were playing at being primitive, imagining themselves following the footsteps of Jim Bridger or John Colter. The wilderness, therefore, though an illusion, had a real effect. If people wanted to escape the twentieth century, this artifice would be their time machine.

The breakthrough in recreation technology and the birth of primitive chic occurred just as the boom babies were entering college and campuses began to implode with war protest, civil rights, feminism, and environmentalism. The young people at this time were ready recruits for a preservationist ideology. Politically active, environmentally aware, with the money and equipment necessary to get close to nature, they were brought up on an ethic of commitment and authenticity of experience.

Having spent their undergraduate days fighting authority, they found their identity in opposition: The war was the fault of military-industrial-academic collusion; suppression of women and minorities was the natural outcome of a culture dominated by white males. And the environmental crisis was the fault of nearly everyone over thirty. They were against modern urban society, capitalism, communism, technology, science, academe, Christianity, population growth, and traditional politics.

Many of these college students whose skills were honed in the crucible of campus antiwar politics flocked to the Rocky Mountains. But borrowing a leaf from the book of Ralph Nader, they were interested in more than saving the environment. They also intended to change politics in ways that would make future eco-catastrophes impossible. They would cross lances with those who would play God. They would put an end to our overweaning pride: They would be *hubris commandos*. The past social-political-economic order had done damage to nature because it represented special economic interests. Until the true public interest had a voice, no environmental reform was possible. There had to be organizations that would represent the public interest.

This was the prevailing atmosphere when William Bryan came to the Rockies in 1972. Having recently earned his Ph.D. in Resource Planning and Conservation at the University of Michigan and using funds from the Point Foundation (derived from sales of the *Whole Earth Catalog*), he set to work personally organizing environmental foundations in Montana, Wyoming, and Idaho. In the following year, receiving grants from the Rockefeller Brothers' Fund and others, he established his own organization, calling it the Northern Rockies Action Group.[22]

The NRAG became the prototypical public-interest organization. Its purpose was to be an organizer, helping the people of the northern Rockies mobilize in favor of the public interest. Keeping a low profile, it offered advice on "organizational start-up, structure, planning, evaluation, issue-strategy development, community organizing, media strategy, access and production, membership recruiting, fund raising, financial management, office management, conference organizing, volunteer management and staff management."[23]

Bryan's project was a dramatic success. Many in the northern Rockies at that time felt under siege, and NRAG arrived just in time to prepare their defense. As the nation entered the first fuel crisis, and as natural resources became depleted elsewhere, the mountain states were the next target for national economic development. Sitting on top of the world's largest coal reserves, rich in timber, uranium, phosphate, gold, silver, and copper, these states seemed ripe for the picking.[24]

Because the federal government controlled more than half their lands, a simple change in national policy could devastate the region. Gargantuan coal strip-mining operations were already under way and others were in the offing. Plans were drawn to make southeastern Montana an electricity-generating center for much of the country, with a complex of huge coal-fired electric generating plants spewing tons of sulphur into the pristine mountain air. Phosphate mines were to be built in fragile regions of Idaho. Activity was afoot to criss-cross America's last-remaining virgin mountains with power lines and oil and gas pipelines. The region's rarest and most valuable commodity — water — was to be pumped in coal-slurry pipelines to the Mississippi. The nation's last free-flowing river — the Yellowstone — was to be dammed near Livingston, destroying rich farmland, animal migration routes, and the best scenery and trout fishing in America. Towns such as Butte and Missoula, already choked by pollution from copper mining and refining or paper manufacture, faced uncertain futures.

Thanks to NRAG, public-interest groups sprang up throughout the region like mushrooms after rain: the Northern Plains Resource Council and Powder River Basin Resource Council, Idaho Conservation League, and Environmental Information Center. Local chapters of national conservation organizations benefited in countless ways from NRAG support. But always behind the scenes, this support attracted little notice. For NRAG was based on a fundamental truth: Issues are ephemeral, but organizations need not be. Today the topic might be strip mining but tomorrow it might be nuclear war. Bryan's interest was in building a political network in the northern Rockies capable of responding to changes in the public's interest.

But he had a problem: What was the public interest, and what assurance was there that the network NRAG created — organizations that had appointed themselves as guardians of the public interest — would truly represent it? The movement, both radical and reactionary, faced a crisis in identity.

Anarchistic, committed to decentralization, participatory democracy, direct action, and toppling the economic-political power structure, such groups were in part anti-authoritarian and populist. Run by young people who had served their apprenticeship as political activists in the antiwar effort, they were more at home with issues than with research, with litigation rather than education. Steeped in the cause of civil rights, they were more comfortable with the tactics of confrontation than with long-range ecological studies, with a sense of righteous indignation than intellectual curiosity. And having spent their undergraduate years fighting a university faculty they believed hopelessly compromised by the military-industrial complex, they had little use for authority or scholarship. Theirs was the politics of high moral dudgeon. They seemed to have an electric instinct to polarize the world.

Yet they were also conservative. These ideological descendants of the new pantheists, many living lives of the primitive chic, may not all have been "neolithic conservatives" but they were at war with progress. And despite calling themselves a public-interest movement, they were a handful of WASPS, most of whom came to the region from elsewhere. The boards of directors of these organizations, like those of country clubs, were "self-perpetuating": they chose their own members. And their overlapping structure — the "network" — ensured that a few people, sitting on several boards, could exert great influence on the movement's direction. Establishment foundations, such as Ford, Rockefeller Brothers, Mellon, and Hewlett, were also a source of support. The sons and daughters of America's wealthiest families, scions of commercial empires, were active members supplying both money and time.

Indeed, in this conservatism the northern Rockies network mirrored the state of contemporary environmentalism everywhere. For decades on the outside of the power structure looking in, they had become more and more part of the establishment.

National environmental organizations began to look like clones of the corporations they had been created to resist. Explosive growth had turned groups such as Audubon and Sierra into bureaucracies, wherein vested lobbyists replaced rough-edged activists. They practiced politics as usual: lobbying Congress to appropriate more funds for the Fish and Wildlife Service, filing lawsuits, organizing letter-writing campaigns, raising money from the brie and chablis set. They opened a revolving door with both government and establishment. Prominent environmentalists such as Leopold, Allen, Olson, and Ansel Adams advised the Park Service, and conservation organizations put on their boards, or hired as staff, former bureaucrats such as Cecil Andrus, who had been Secretary of the Interior (Audubon); former Assistant Secretaries Nathaniel Reed (Audubon, Nature Conservancy, and Boone and Crockett), Curtis Bohlen (World Wildlife Fund), and Robert Herbst (Trout Unlimited); former chairmen of the President's Council on Environmental Quality Russell Train (World Wildlife Fund) and Russell Peterson (Audubon); former Senator Gaylord Nelson (Wilderness Society), former Associate Director of the Fish and Wildlife Service Joseph P. Linduska (Audubon), and former Yellowstone Park historian Paul Schullery (Federation of Fly Fishermen).

Ambivalence remained the curse of environmentalism. Along with equivocation on the human role in nature and the role of science, now came a third, the role of activists. They were either a radical movement with a reactionary agenda, or a reactionary movement with a radical agenda. What should they do, and what should their tactics be? Should they join the establishment, changing it from within, or should they take to the streets in an effort to topple it from without? These were difficult questions, for although partly revolutionary, they had profoundly conservative goals; although at the grass roots, their aim was to change prevailing cultural norms; although made up of young people radicalized in college, they were funded by rock-ribbed conservative foundations; although attracting wealthy supporters, their aim was to thwart corporate greed; and although enjoying explosive growth during the age of plastic, they were infatuated with the primitive.

By the mid-1970s environmentalists everywhere were preoccupied with resolving this ambivalence. Sharing only negative beliefs, they could not say what they were for. Their theology, science, and politics were reaching an impasse. They rejected traditional faiths, but had none to replace them. They rejected traditional science, but had not yet developed a new one. They rejected the old political ways, but lacked a new plan.

"Ecology," wrote Theodore Rosak in 1972, "stands at a critical crossroads. Is it too, to become another anthropocentric technique of efficient manipulation, a matter of enlightened self-interest and expert, long-range resource budgeting? Or will it meet the nature mystics on their own terms and so recognize that we are to embrace nature as if indeed it were a beloved person in whom, as in ourselves, something sacred dwells?"[25]

Environmental writer Murray Bookchin wrote on Earth Day, 1980,

> It is necessary, I believe, for everyone in the ecology movement to make a crucial decision: will the eighties retain the visionary concept of an ecological future based on a libertarian (anarchist) commitment to decentralization, alternative technology and a libertarian practice based on affinity groups, direct democracy, and direct action? Or will the decade be marked by a dismal retreat into ideological obscurantism and a "mainstream politics" that acquired "power" and "effectiveness" by following the very "streams" it should be seeking to divert? . . . The choice must be made now before the ecology movement becomes institutionalized into a mere appendage of the very system whose structure and methods it professes to oppose. It must be made consciously and decisively — or the century itself, and not only the decade, will be lost to us forever.[26]

The movement needed a new philosophy, a new map to take it into the 1980s. It needed a complete ideology, a system of ideas that would offer something to believe in, not just something to oppose. In fact the search for this new philosophy had begun. The environmental movement needed, most agreed, a new "paradigm."

The expression "paradigm change" was coined by philosopher Thomas S. Kuhn in 1962 as a way of explaining how science progressed. A paradigm, Kuhn suggested, was a consensus among scientists about the problems in their field and the methods for resolving them. And so a new paradigm was an original way of seeing an old problem, a revised consensus among scientists on the proper perspective for viewing the world. It was, suggested Kuhn, characteristic of the way in which science advanced. A new and competing theory was introduced to account for phenomena that the old theory had difficulty explaining. Eventually, if the new theory was sufficiently persuasive, the scientific community embraced it and rejected the old, thus endorsing a new paradigm. But because a paradigm represented a consensus among scholars, such changes occurred only in fields in which nearly universal agreement was possible. These were, suggested Kuhn, the mature sciences, in which many questions of method and purpose had already been resolved.[27]

Environmentalists, however, could not wait for their field to mature. "Do we need," wrote philosopher Alan Drengson in *Environmental Ethics* in 1980, "a shift in paradigm in order to understand the continuing crisis of character and culture which has occurred in the modern era?" Most answered, Yes.[28]

"The topic of shifting paradigms in contemporary societies," Devall commented in 1981, "has received increasing attention from philosophers and social scientists during the last few years." That was an understatement. The search for a paradigm had become almost obsessive among environmental thinkers. Devall himself called for "a radical critical analysis of the dominant social paradigm." Ecophilosopher George Sessions saw environmental debate as "separated by competing paradigms." Willis Harmon thought the old paradigm had already begun to give way: he celebrated the "breakdown of the industrial-era paradigm and its replacement by another." Writer Fritjof Capra claimed that a paradigm shift of cultural values was beginning: "The paradigm that is now shifting has dominated our culture for several hundred years. . . ." Writer Marilyn Ferguson claimed that a new one had already arrived: "The paradigm of the Aquarian Conspiracy sees humankind embedded in nature. . . ."[29]

Of course what these people were looking for was not what Kuhn meant by paradigm, for, according to Kuhn's account, their search was impossible. A paradigm was a consensus among scholars in a highly developed science. It was not a theory, attitude, cultural value, or economic system. It could not emerge from the tossing flux and dissension of ecological woolgathering.

Rather, these ecophilosophers were seeking to discover for themselves a view of the world, a new cosmology. But where was it to come from? Like many new ideas in America today, it was to come from California.

19

The California Cosmologists

B Y 1972 the search for a New Philosophy of Nature had begun. Different kinds of forty-niners, bitten by a different kind of gold bug, sought to assay the richest veins of metaphysical, scientific, and political ideas they could find. And although they were not sure what to look for, they knew it had to be new. "Why not," asked Sessions, "*really think the unthinkable,* which means challenging the whole paradigm?" (italics his).[1]

"We are long past the time," Rosak wrote in the year during which Bryan came to Montana, "for pretending that the death of God is not a political fact." Yet he remarked a religious fervor in the land, a searching of a radically new kind: "There is a strange, new radicalism abroad which refuses to respect the conventions of secular thought and value. . . . The religious renewal we see happening about us . . . I accept as a profoundly serious sign of the times, a necessary phase of our cultural evolution. . . ." He called for the "creative disintegration" of modern industrial society.[2]

"Modern physics has opened up two very different paths for scientists to pursue," wrote Fritjof Capra, physicist at the Lawrence Radiation Laboratory in Berkeley in 1976, "They may lead us — to put it in extreme terms — to Buddha or to the bomb."[3]

"Human consciousness," wrote the poet M. C. Richards in 1973, "is crossing a threshold as mighty as the one from the Middle Ages to the Renaissance."[4]

"People," she continued, "are hungering and thirsting after experience that feels true to them on the inside, after so much hard work mapping the outer spaces of the physical world. They are gaining courage to ask for what they need: living interconnections, a sense of individual worth, shared opportunities. . . ."

The new philosophy would be too revolutionary for traditional Christianity, humanism, or atheism. All Western ideas were suspect. By the late 1960s, Devall said, "more and more intellectuals were questioning the premises of humanism." Rosak agreed: "Humanism is the finest flower of urban-industrial society; but the odor of alienation yet clings to it and to all culture and public policy that springs from it. . . . Humanism, for all its ethical protest, will not and cannot shift the quality of protest in our society. . . . Indeed, it stands full square upon the stone that must be overturned."[5]

"Pragmatism, Marxism, scientific humanism, French positivism, German mechanism: the whole swarm of smug antireligious dogmas emerging in the late eighteenth and nineteenth centuries," wrote philosopher Pete Gunter, "and by now deeply entrenched in scientific, political, economic, and educational institutions *really do not, as they claim, make man a part of nature*. If anything, they make nature an extension of and mere raw material for man."[6]

If Christianity was to survive, wrote Jeremy Rivkin in his *Entropy*, there must be a "second Christian Reformation." If Christians do not embrace "a New Covenant vision of stewardship, it is possible that the reemerging religious fervor could be taken over and ruthlessly exploited by rightwing and corporate interests."[7]

But where would such an original cosmology come from? Not from the button-down institutions of Massachusetts and Connecticut. For this was the burgeoning of a movement that challenged, not only the Eastern establishment, but the entire Western tradition. Its prophets would be scruffy and alienated refugees from Middle and Eastern America. As Paul Ehrlich wrote in 1968, "Much more basic changes are needed, perhaps of the type exemplified by the much despised 'hippie' movement — a movement that adopts most of its religious ideas from the non-Christian East. It is a movement wrapped up in Zen Buddhism, physical love, and a disdain for material wealth. . . ." And it was to come from California.[8]

"Sooner or later, we are going to have to understand California — and not simply from the motive of predicting the future of

the rest of the country. . . . Something is struggling to be born
here," wrote Jacob Needleman, professor of philosophy at San
Francisco State University, in 1973.[9]

Indeed, something was. The search for a new paradigm became
the cottage industry of California. Theodore Rosak (Long Beach),
Bill Devall (Humboldt State), George Sessions (Sierra), Willis
Harmon (Stanford), Jacob Needleman and Alan Watts (San Francisco), Gary Snyder (Nevada City), Marilyn Ferguson (Los Angeles), John Rodman and Paul Shepard (Claremont), Roderick Nash
(Santa Barbara), Lynn White, Fritjof Capra, Carolyn Merchant,
J. Peter Vajk, William Everson (all Berkeley), Robert Brophy (Long
Beach), and Raymond Dasmann (Santa Cruz) were just a few of
those who joined in the search for a radical ecology, and in so
doing continued the tradition of Muir (Yosemite), Jeffers (Carmel), Ansel Adams (San Francisco), George Wright (Oakland),
and Brower and Starker Leopold (both Berkeley). For the environmental awakening and this burgeoning state were made for each
other.

"The flashing and golden pageant of California," Walt Whitman
wrote, "the new society at last . . . clearing the ground for broad
humanity, the true America." A classless society, rich in a diversity
of life-styles, egalitarian, fluxing, seminal, nurturing a radical tradition from Hetch Hetchy to Berkeley to Diablo Canyon, a people
without a past, numbed and dazed by change that turned avocado
groves to ticky-tacky overnight, a place blessed with weather that
proved the good of nature's God, had the mix to make a new
cosmology. "Traditional patterns of relating, based on locality, are
askew," wrote Zen Buddhist scholar Alan Watts. "What people
in the east can't see is that new patterns are being developed."[10]

Rejecting a personal God, Western science, anthropomorphism,
traditional canons of reason, technology, and Calvinistic morality,
they sought a new vision that sacralized nature, incorporated the
passivity of Eastern religions, espoused mysticism or the virtue of
drugs, preached tolerance of personal life-styles, provided a new
science that relied on intuition, proved the interconnectedness of
things, explained the validity of psychic phenomena, established a
new relationship between humanity and nature, and founded a
new ideology of radical politics that would lead us back to primitive
truths.

By the early 1970s the search had begun in earnest. Diablo

Canyon, a nuclear reactor built along an earthquake fault, became the center of a controversy that polarized the state, making nearly every kind of person into an ecologist, and heating the debate on tactics and direction. It was, wrote Devall, a "spiritual watershed," galvanizing the proponents of radical politics and linking a variety of social movements "in the affirmation of natural diversity and resistance against the kind of energy policy, centralized political authority and nuclear terror that Diablo Canyon symbolizes."[11]

Poets, philosophers, economists, and physicists joined the ecologists in a search for a new beginning. Soon everyone took to the chase: Native Americans, drug cultists, astrologers, feminists, gay and lesbian rights activists, mystics, mainline clergy, sexual and animal liberationists, vegetarians, religious cults of all persuasions saw the ecology movement as a metaphysical basis for their own life-style.

Like a swarm of bees turned loose in a greenhouse, they buzzed around a flowerbed of exotic religions and an eclectic cornucopia of offbeat ideas — Tao, Hinduism, Zen Buddhism, Hua-Yen Buddhism, Mahayana Buddhism, Gnosticism, Manicheanism, Vedanta, Sufism, Cabalism, Spinozistic Pantheism, Whiteheadian metaphysics, Heideggerian phenomenology, Jungian archetypal symbolism, Yoga, biofeedback, transcendental meditation, psychedelic drugs, self-awareness exercises, psychotherapy, pre-Socratic philosophy, the "Inhumanism" of Robinson Jeffers, Gandhian pacifism, animism, panpsychism, alchemy, ritual magic — and through eclectic pollenizing inseminated an endless variety of hybrid ideas — Buddhist economics, fossil love, planetary zoning, transformation technologies, holistic medicine, the future primitive, bioregionalism, deep ecology, shallow ecology, reinhabitation, biocentrism, ecosophy, ecological primitivism, feminist physics, chicken liberation, Earth National Park, stone-age economics, neolithic conservatism, species chauvinism, Yin Yang and the androgynous universe, Gaia, global futures, Spaceship Earth, coevolution, re-choosing, the rights of rocks, ecological resistance . . .

And yet through this swirl of chaotic, primeval theorizing, patterns began to form, and themes resonated. A California Cosmology materialized, coalescing around three overlapping ideas. The search for a new religion led to the insight that *Everything is sacred*. The search for a new science led to the principle: *Everything is interconnected*. The search for a new politics of commitment centered on the belief that *Self-transcendence is possible through authentic experience.*

All searches for a new religion shared belief in the principle that *everything is sacred.* The Western tradition, these new cosmologists agreed, had "desacralized nature," and the task ahead was to "re-sacralize" it. That nature is sacred was the insight of primitive cultures forgotten by an urban society dominated by Western science.

"Things divine," writes Rosak, "crumble and vanish before the first breath of an easy scientific skepticism. They become epistemological embarrassments, the stuff of archaic superstition, the fraudulent masks of ignorance or unjust privilege, nothing more. Therefore — in the name of Reason — away with them one and all! And on with the great business of research and development."[12]

The path to true understanding came not by progress or reason but by embracing the primitive and mystical. "If man is to remain on earth," wrote ecopoet Gary Snyder, "he must transform the five-millennia-long urbanizing civilization tradition into a new ecologically sensitive harmony-oriented wild-minded scientific / spiritual culture."[13]

"The fascination with Native Americans, Arctic Eskimos and Australian aborigines by philosophers and poets," wrote Devall, "indicates much more than just romantic sentimentalism. The search for the 'future primitive' is the radically conservative statement of deep ecology." The New Philosophy of Nature, he continued, "reaffirms the integrity of Nature and seeks understanding of the 'tao in ten thousand taos.' Those living in modern societies can begin understanding, as opposed to just explaining, the mystery of Nature through ritual as much as through the science of ecology."[14]

Because all nature was sacred, everything had the right to be left alone. Attacking attitudes that they called anthropocentrism and "species chauvinism," many urged liberation of animals or even inanimate things.

Devall urged, "Other species have a right to follow their own evolutionary destinies without human intervention." It is, said philosopher Tom Regan, a "moral imperative" to follow the "Preservation Principle . . . a principle of non-destruction, non-interference, and, generally, non-meddling." Snyder advocated "fossil love." Arguing against the view that "there is a pecking order in this moral barnyard," ecologist John Rodman suggested that forests and rivers be given the same ethical and legal status now enjoyed by corporations.[15]

Philosopher Edward Johnson preached chicken liberation: "The claim that chickens are docile, tractable, stupid, and dependent may or may not sell chickens short but . . . it is worth remembering that similar charges used to be made about the character of slaves." Writer Christopher Stone, in *Should Trees Have Standing?* published in 1975, advocated the rights of trees, a position that the Sierra Club Legal Defense Fund carried all the way to the Supreme Court. A similar position was taken by Justice William O. Douglas in the Mineral King case of 1972. The Santa Barbara historian Roderick Nash asked, "What, after all, does a rock want?" Then, reflecting a sensitivity of which Carmel poet Robinson Jeffers would have approved, he concluded, "One is to suppose that rocks, just like people, do have rights in and of themselves."[16]

Climbing out of the pretty catacombs of California, these ideas became the creed of an environmental activism. Actually, they were not entirely new, but rather similar to notions held by Muir, Leopold, Carson, Krutch, and others of the earlier generations. But what kind of notions were they? Were they the "vast, impersonal pantheism" that Adams supposed? Many commentators did cite particular pantheistic religions as the source of the new faith, such as Buddhism, Hinduism, Tantra Yoga, and the Tao; others credited more obscure pantheists such as the seventeenth-century Dutch-Jewish philosopher Benedict Spinoza, the Greek pre-Socratic philosopher Parmenides, Saint Francis of Assisi, or twentieth-century philosophers such as Martin Heidegger (Germany), Alfred North Whitehead (England), Teilhard de Chardin (France), and so on.[17]

But this new faith was so eclectic, borrowing ideas from so many places, that no classical definition would fit precisely. It was not exactly Buddhist, Hindu, or Tao. The old ambivalence was still there. It had not resolved *where to place man in nature*. He was still partly sacred, partly satanic.

Reflecting this ambiguity, most environmental writing became infused with what philosophers call animism or *panpsychism:* the view that all things possessed a soul — "the truly substantial spiritual world," as Muir said, "whose forms flesh and wood, rock and water, air and sunshine, only veil and conceal." It was this view of universal soul that lay behind Thoreau's belief that a pine tree might be immortal, and behind Nash's claim that rocks have rights. But human beings, although like rocks infused with spirit, were unlike them in possessing the capacity to deny this spirit and thus do evil. We were part of the world but we could destroy it. And this idea — that human beings contained within themselves a germ of the divine but were not identical with it — was not, as we have seen, pantheism. Rather it was the child of another old idea, called Gnosticism.

Gnosticism was the alpha and omega of Judeo-Christian heresies. Theologians called it an eschatological religion, meaning that it described the universe in a story, with a beginning, middle, and end. The story began when an aeon or demi-God stole the divine substance from God. God himself — also called Light, Life, Spirit, Father, the Good, or even the Absolute — was himself merely this same abstracted and transcendent essence, and its theft in effect separated a part of God from himself. This loss started a downward spiral as the archon of the nether powers, the Demiurge, sought to keep the transcendent spirit, or pneuma, from being reunited with God. To keep the pneuma from Light (God), the Demiurge created the universe and man, planting this element in the latter and sometimes in animals for safekeeping.[18]

Adam, primal man, first created as androgynous, only later achieved his sex, exchanging for companionship a fall even further from the light. The universe therefore became for Gnostics a kind of fortress built by the evil Demiurge to keep the divine from reunion with God. Earth and man were the dungeon of this fortress, and their history recorded attempts by the divine in things to achieve reunion with God and counterattempts by the Demiurge to prevent it.

This religion was especially unacceptable to Jews and Christians because, according to Gnostic gospel, the Demiurge was none other than the Old Testament God, Jahweh. Thus, according to Gnosticism, Judeo-Christianity was evil, and whatever the Bible said was a lie. The truth was the opposite of everything said by the Jewish God. For Gnostics, the serpent in the Garden of Eden, rather than being evil, was good. The serpent was attempting to give Adam and Eve knowledge of the Light (the true God), and was thwarted by Jahweh. Thus serpent cults figured prominently in Gnostic history (as well as in other faiths, such as Tantra Yoga), and to understand the Bible, Gnostics stood it on its head.[19]

Spirit, therefore, to Gnostics was one ineffable, abstract reality that surrounded the universe, a portion of which existed in man. But man was kept apart, imprisoned in a body and on earth by Jahweh. Salvation came through union with the Light — a process called gnosis. But because God was an abstract, distant, and otherworldly principle, gnosis could not be achieved through traditional routes to knowledge — such as science — but only through mystical revelation and magical preparation. In fact two extreme kinds of life led to gnosis: the ascetic and the libertine. The ascetic rejected all the world created by Jahweh, trying to avoid further contamination by material things. The libertine rejected traditional morality, which he saw as part of a conspiracy by Jahweh to restrict his freedom. The story ends when all bodies fall to light and are reunited with God.

The parallels between this old heresy and the new environmentalism were uncanny, and for many of the California Cosmologists, quite explicit. The New Philosophy of Nature, like Gnosticism, rejected Christianity and science and embraced mysticism and a union with spirit in nature. Both pictured human beings as doing evil when they were alienated from the Light (nature). Both tolerated diversity of life-styles. Both (as we shall see) preached that the universe was androgynous. And both relied on myths of early American and Eastern cultures, such as those of Tantra Yoga and the Hopi. Therefore, how you felt about Gnosticism said something about how you felt about environmentalism, and vice versa. "We need the gnostic tradition," wrote the German Green ecological activist Rudolf Bahro recently in the leftist publication, *Rot and Grün.* "No doubt about it," wrote Rosak, "any culture that wishes to plunge itself headlong into total psychic alienation and ecological disaster had best begin by treating the Old Gnosis with uncompromising contempt."[20]

The search for new science orbited around the principle that *everything is interconnected.* The earth — even the universe — was one system. "Organic Unity . . . love that, not man apart," the line from Robinson Jeffers that became the motto of Friends of the Earth, contained the main insight of the new science. Harmon wrote, "We recognize our essential connectedness." Modern physics and Eastern mysticism, declared Capra, both recognized the interconnectedness of things: "The most important characteristic of the Eastern world view . . . is the awareness of the unity and mutual interrelation of all things and events, the experience of all phenomena in the world as manifestations of a basic oneness." Likewise, physics "has come to see the universe as an interconnected web of physical and mental relations whose parts are defined only through their connections to the whole." According to the "ecological sensibility," said John Rodman, " 'man' does not stand over against 'his environment' as manager, sight-seer, or do-gooder; he is an integral part of the food chain . . . a microcosm of the cosmos who takes very personally the wounds inflicted on his/her androgynous body."[21]

Feminists, lesbians, and gays, seeing connections between male chauvinism and the domination of nature, viewed the oneness of nature as recognition of an androgynous universe. Forgetting how Taoist Chinese murdered their unwanted little girls by exposure on mountaintops and made cripples of their wives by binding their feet, and overlooking the Hindu habit of bride-burning (still practiced) and purdah (women were secluded from public observation), these environmentalists turned to Eastern cultures as embodying the wisdom of women's liberation.

"The predominance of masculine imagery and the domination by men of Nature and women in the West," wrote Devall, "has been well documented. The feminine principle is associated with birthing, Mother Earth, Gaia [the Greek Earth Goddess] and with yielding as water yields and thus wins its way." The culture of the masculine, sexist West, he suggested, emphasized domination by feminine nature along with females, but the androgynous East, based on "the intermingling of masculine and feminine as in the Yin/Yang," promoted harmony.[22]

Likewise, Berkeley historian Carolyn Merchant, citing the many associations between nature and the feminine, saw the destruction of nature by masculine physical science as a historical outgrowth of sexism. It was, she argued, "the formation of a world view and a science that, by reconceptualizing reality as a machine rather than a living organism, sanctioned the domination of both nature and women."[23]

William Irwin Thomson, founder of the Lindesfarne Association ("a contemplative educational community devoted to the study and realization of a new planetary culture"), emphasized the interconnectedness symbolized by the androgyny of early Gnostic mythology and Hopi religion. Commenting on the ecological insights of the Gnostic Gospel of Saint Thomas, he wrote that "each sex must take on the character of the opposite, before wholeness can be achieved." Similarly, Marilyn Ferguson observed that "self-discovery inevitably involves the awakening of the traits usually associated with the opposite sex."[24]

These ruminations, however, were only ritual bows to a mystical sense of connectedness, not serious scientific undertakings to develop a science of ecology. Traditional scholarship remained suspect. The real ecologists, terribly underfunded, continued to worry about the testability of their theories, while those in more advanced fields such as genetics and the physical sciences moved ever further into the space age, propelled by large governmental and corporate subsidies. Indeed, the more glamorous sciences, by concentrating on topics of global influence and technological import — such as nuclear radiation, satellite mapping, and agricultural eugenics — were, by opening up new areas of controversy, quietly setting the agenda for their less mathematically inclined brethren. But then, this is how it had been for four hundred years: science set the framework for humanistic debate.

The search for new politics rested on one insight: *personal transcendence (or self-realization) is achieved through authentic experience*. Political activism was a kind of authentic experience, not unlike meditation, which led to expanded consciousness. "My contention," said the poet Allen Ginsberg in 1979, "is that if done with a proper human dignity and lightness, meditation can be appropriate to a protest situation." Protest, noted Jerry Rubin, one of the Chicago seven, was "a spiritual movement that's truly revolutionary. Without self-awareness political activism only perpetuates cycles of anger. . . ."[25]

"We stayed in the streets through tear gas and billy clubs," observed activist Lou Krupnik, "and went inside only when holy people whispered Sanskrit mantras into our eager ears."

"A funny thing happened on the way to Revolution," wrote the activist Irving Thomas. "There we were, beating our breasts for social change, when it slowly began to dawn on us that our big-deal social political struggle was only one parochial engagement of a revolution in consciousness. . . ."

An authentic experience was mystical union of self with nature or others. "Self realization," wrote Devall, "for humans means finding a path (or a pathless path in Zen Buddhism) by which the self can be integrated with the Self (Atman), the Great Self of Hindu philosophers of the Tao (the 'way things are') of Taoism. The self is not realized through assertiveness, powerfulness in war or in genetic engineering or in 'seizing the tiller of creation.' The self is realized through humility and modesty." Similarly, suggested John Rodman, we should strive to achieve "ecological sensibility," a way of life "whereby one aligns the self with the ultimate order of things."[26] "Beyond the personal reunification, the inner reconnection, the re-annexing of lost portions of oneself," wrote Ferguson,

> there is connection to an even larger self — this invisible continent on which we make our home. . . . The self is a field within larger fields. When the self joins the Self, there is power. Brotherhood overtakes the individual like an army, not the obligatory ties of family, nation, church, but a living, throbbing connection, the unifying I-Thou of Martin Buber, a spiritual fusion. This discovery transforms strangers into kindred, and we know a new, friendly universe.[27]

Personal transcendence, wrote Harmon, was rejecting the restraints that conventional Western science has put on the power of the mind, and exercising our latent "preternormal knowings" (such as telepathy, clairvoyance, faith healing, retrocognition, precognition, and psychokinesis). In this way, he suggested, we could recognize that "the human mind and the physical universe do not exist independently," that we are all joined together, that a spiritual world exists, and that our capacity for knowledge through unconventional means is nearly limitless.[28]

Just as many roads led to Rome, so too were there many paths to self-transcendence: psychedelic drugs, meditation, contemplation of nature, communal living, or commitment to a political or ecological cause. Indeed, the metaphysics of self-transformation and the politics of confrontation were two sides of the same coin. They were a curious blend of community and communion, sacrilege and sacrament. The sense of comradery in political action enhanced awareness of our connections with others, just as drugs and meditation enhanced our sense of connection with the universe. And through heightened consciousness came the sense of power, that anything was possible. The mind was connected with the universe, and had power over it. *"Let there be transformation, and let it begin with me,"* wrote Ferguson.[29]

"The breakdown of the industrial-era paradigm and its replacement by another," suggested Harmon, ". . . would involve a change in the basic perception of reality." This connection explains, he says,

> why the use of perception-changing techniques has played such a central role. These techniques are of two types. One type is mainly self-chosen and aimed at spiritual or transcendental awareness; it includes meditation, yoga, self-awareness exercises, various group and individual psychotherapies, and psychedelic drugs. The other type is largely imposed on others and is aimed at social and political awareness, that is, the perceiving aspects of the breakdown of the industrial-era paradigm and of the oppressive elements of social institutions; examples are consumer and environmentalist confrontations. . . .[30]

The sense of mystical union with the spiritual world achieved by contemplating nature so extolled by Emerson, Thoreau, Muir, and Krutch came to be seen by this later generation as the kind of raised consciousness also found in psychedelics and politics. One activist at the time said, "LSD gave a whole generation a religious experience." Indeed, Ferguson stated, drugs provided "the entry point experience" to a life dedicated to transcendence and commitment. "It is impossible to overestimate the historic role of psychedelics as an entry point drawing people into other transformative technologies," she wrote. "Drugs were a pass to Xanadu, especially in the 1960s."[31]

And so a revolution that began in consciousness spread to politics. *If we change minds, we change the world,* became their major political insight. Harmon wrote, "We will see the way to a viable global future when we believe it is possible, not the reverse order." All we needed to do to save the world was to want it. "We can begin anywhere — everywhere," wrote Ferguson. " 'Let there be peace,' says a bumper sticker, 'and let it begin with me.' Let there be health, learning, relationship, right uses of power, meaningful work. . . . *Let there be transformation, and let it begin with me.*"[32]

What, then, should this transformation be? The New Philosophers of Nature could not agree. Some, calling themselves "deep ecologists," aimed at a "future primitive." Others, calling themselves members of the "New Age / Aquarian Conspiracy," sought salvation through space-age techniques of planetary engineering.

Deep ecology was conceived by Norwegian philosopher Arne Naess in Oslo in 1973, but was quickly adopted by Californians such as Devall, Sessions, Snyder, and Rodman as the new paradigm. It was based, Naess said, on the principle of "biospherical egalitarianism": that all things have "an *equal right to live and blossom.*" Drawing "from the cosmologies and epistemology of Native Americans, some eastern philosophies such as Mahayanhana Buddhism and Taoism and from pre-Socratic western philosophy," according to Devall, deep ecologists believed all things have equal rights — including the right to "self-realization." It was, according to Snyder, "a thoroughly ethical approach to human relationship with all beings." Its proponents advocated living by "reinhabitation," and practicing "bioregional politics."[33]

Reinhabitation, according to Raymond Dasmann, "means learning to live-in-place in an area that has been disrupted and injured through past exploitation. . . . It means evolving social behavior that will enrich the life of the place, restore its life-support systems, and establish an *ecologically and socially sustainable* pattern of existence with it. . . ."[34]

Bioregionalism was the political counterpart to reinhabitation. It was an idea, according to Jill Engledow, staff writer for the journal *Ecophilosophy,* "in which political organization is derived from the natural boundaries and relationships within a region, rather than being imposed on the region by humanity as an outside force." *The biosphere,* in other words, *should be zoned.* Bioregions themselves, according to Peter Berg, Director of the Planet Drum Foundation (an organization that promotes the idea), were both places on the earth and places in the mind. He wrote, "There is a distinct resonance among living things and the factors which influence them that occurs specifically within each separate place on the planet. Discovering and describing that resonance is a way to describe a bioregion."[35]

Deep ecologists, in short, preached that ecology began at home. The secret to salvation lay in learning to live and act as the ancients did. Let the forces of nature determine our social and political

boundaries! We will do our part to save the earth by organic gardening, cooking by woodstove, reading the *I Ching* by kerosene
lantern, canning our own peaches, studying Hopi mythology, spin-
ning our own cloth, eschewing plastic and polyester, farming with
horse-drawn plows and other "appropriate technologies," and in
general living a life in which small is beautiful.

If to deep ecologists the glass was half empty, to the New Age /
Aquarian Conspirators it was half full. Rather than supposing that
the golden age was in the past, these people — many of them from
Silicon Valley in Santa Clara County — saw it as in the future;
rather than supposing that ecological revolution must begin at home
with a horse-drawn plow, they dreamed of keeping the entire spaceship earth afloat using all technology available; and rather than
believing that intervening in nature was inevitably a sin, they urged
human beings to play the role of benevolent "stewardship."

"Think globally, act locally," became the leading insight of the
New Age / Aquarians. Its advocates — notably Marilyn Ferguson,
Willis Harmon, Fritjof Capra, and Steward Brand (publisher of
the *Whole Earth Catalogue* and the *Co-Evolution Quarterly*) —
saw the earth as sustained only by massive but enlightened human
intervention. The earth was one complete ecosystem that we, for
the first time, had the capacity to understand and control.[36]

Humanity, wrote James Baines in *The Whole Earth Papers,* was
"a product of evolution and an instrument of evolution." Gaia
(Mother Earth), as coevolutionist James Lovelock called it, was a
living organism. It was "a complex entity involving the Earth's
atmosphere, biosphere, oceans, and soil; the totality constituting
a feedback or cybernetic system which seeks an optimal physical
and chemical environment for life on this planet." Yet, thanks to
the technology of coevolution — our capacity to direct evolution
on earth for human beings as well as other species — we had a
chance to save Gaia. Lovelock said, "Cybernetics tells us that we
might safely pass through these turbulent times if . . . we can always control the genie we have let out of the bottle."[37]

Indeed for some coevolutionists anything was possible. J. Peter
Vayk in his *Doomsday Has Been Cancelled* wrote, "Should we find
it desirable, we will be able to turn the Sahara Desert into farms
and forests, or remake the landscape of New England, while we
create the kind of future we dream. . . ." To Willis Harmon a

global future depended on our making an "evolutionary leap to a transindustrial society" that included a new image of man and his capacities. He wrote, "Through a global change of mindset, the world could begin to rebuild human society in a way that will be equitable and sustainable. . . ."[38]

Wrote Ferguson, "For the first time in history, humankind has come upon the control panel of change — an understanding of how transformation occurs. We are living in *the change of change,* the time in which we can intentionally align ourselves with nature for rapid remaking of ourselves and our collapsing institutions."[39]

The division between deep ecologists and New Age / Aquarians was just another battle in the long war between the two cultures of letters and science. Deep ecologists, mostly humanists and social scientists such as Devall, Rodman, and Sessions, looked to the past and the revival of a humane, nonindustrial ethic. The scientifically inclined Aquarians, such as Harmon (an engineer) and Lovelock (a chemist and cyberneticist) looked to a future in which the world would be saved by human ingenuity. Debate between them — and ecological debate among environmentalists nearly everywhere — became increasingly abstracted, far removed from grassroots problems like the fate of beaver in Yellowstone. Deep ecologists remained wedded to the ideal of noninterference without much thought about how this position would affect the fate of species such as the grizzly. New Age / Aquarians remained infatuated with the possibilities of global engineering even while the simplest questions of wildlife ecology remained unasked or unanswered.

But in their continuing split, they demonstrated that the threefold crisis gripping modern environmentalism had not been resolved. Their gnosticism had not found a place for humanity in nature. Their science had not advanced the study of ecology. Their politics had not set an agenda. Only on one thing did they agree: their politics should be revolutionary.

They shared a profoundly egalitarian view of the world. "A great democratic revolution is taking place in our midst," wrote de Tocqueville in 1828, ". . . the most continuous, ancient, and permanent tendency known to history." Yet these ecologists took this old idea several steps further than it had ever been taken before. They not only questioned traditional authority and the rights of industrial nations to take resources from the third world; they also

repudiated the concept of male supremacy, and — most fundamentally — they rejected man's rights over nature. They took egalitarianism to its logical extreme. All *things,* they said, were created
equal.[40]

And through their intense theorizing the California Cosmologists
saw their New Philosophy of Nature assimilated into our national
culture. Spread by seminars, foundation workshops, and the grass-
roots network of public-interest organizations, and propelled by
events — the growing peace movement and a nearly united opposition to President Reagan and his Interior Secretary James Watt —
their ideas became part of mainline thinking. Man, all were agreed,
was the source of environmental evil, nature was sacred, traditional
science suspect; the idea of interconnectedness, though left unex-
plored, earned a ritual bow and was enshrined in the quasi-mystical
idea of the ecosystem.

The national environmental organization Greenpeace, in fusing
this philosophy with political activism, wrote, "Ecology teaches us
that humankind is not the center of life on the planet. Ecology has
taught us that the whole earth is part of our 'body' and that we
must learn to respect it as we respect ourselves . . . life must be
saved by non-violent confrontations and by what the Quakers call
'bearing witness.' " Such ideas were even earlier co-opted by President Nixon. The wilderness and wild animals have, he said in
presenting his environmental message to Congress in 1972, "a higher
right to exist — not granted them by man and not his to take
away."[41]

Activist organizations, following the Greenpeace example and
taking Bookchin's advice, turned left, building an egalitarian, grass-
roots network. And following the star of interconnectedness, they
gravitated from local ecological issues to global ones. They began
to model themselves after Germany's revolutionary "Greens" —
a movement that Fritjof Capra and Charlene Spretnak, in their
recent *Green Politics,* described as dedicated to "ecology, social
responsibility, grassroots democracy, nonviolence, decentralization, postpatriarchal perspectives [that is, feminism] and spiritu-
ality."[42]

Indeed, Capra and Spretnak suggested, this country already had
more than one hundred "green oriented organizations," (including
Greenpeace, Friends of the Earth, Co-Evolution Quarterly, and
many peace groups), which were evolving into an activist network.

"Since the ecological view of reality is one of a network of relationships," they wrote, "the network structure would be especially appropriate . . . they are a necessary first step in the building of a political movement."[43]

The NRAG joined the network, borrowing leaves from the Greens' book of constituency building. In 1984 its staff, inviting Ms. Spretnak to speak, sponsored a seminar on Green politics. It also began to spread the net in the region: establishing, or helping to establish, Western Solidarity (a Peace and Disarmament organization), the Citizens' Alliance for Progressive Action, the Tri-State MX Coalition, the Snake River Alliance, the Wyoming Nuclear Freeze Coalition, the Last Chance Peacemakers Coalition, and Montana Citizens to End the Arms Race. At the same time its staff strengthened their ties with other regional consortia, forming a national patchwork of activists' coalitions. They attended National Networking Seminars for antinuclear groups, participated in the Coalition for a New Foreign and Military Policy conference, and began a program for training peace activists, among other activities.[44]

On the surface, mainline environmental groups did not appear to follow the New Philosophers of Nature. They continued to practice politics as usual, a route many of the California cosmologists disparaged as "Shallow Ecology." It was, deep ecologists thought, too timid; it treated symptoms, rather than causes. It was not sufficiently egalitarian and did not call for a new "paradigm."

"Shallow environmentalism," wrote Sessions, "attempts to curb some of the most glaring environmental abuses such as pollution and resource depletion while calling for no basic change in the anthropocentric urban-industrial social paradigm — what Naess described as a 'fight against pollution and resource depletion. Central objective: the health and affluence of people in the developed countries.' "[45]

Shallow or not, by the mid-1980s mainline environmental groups were cautiously testing the waters of grass-roots politics. While keeping one foot in conventional politics, national environmental organizations took their first steps toward a more radical global activism, establishing their own affinity groups and sponsoring national and regional meetings on the nuclear freeze, nuclear winter, and population growth.

In 1983 the National Audubon Society appropriated the motto of the coevolutionists — "Think globally, act locally" — as the theme of their summer convention and began its own "grass-roots activist program — the Citizen Mobilization Campaign." The Sierra Club established its own "Activist Program." The Audubon Society, Sierra Club, Wilderness Society, Natural Resources Defense Council, Trust for Public Lands, and National Wildlife Federation, among others, sponsored the National Conference on "The World After Nuclear War," held in Washington, D.C. "on the long-term worldwide biological consequences of nuclear war." Likewise, just two weeks earlier, the Environmental Center in Denver had hosted the Rocky Mountain Regional Conference "On the Fate of the Earth," featuring David Brower as speaker and holding sessions on the "aiki of politics," "conflict resolution," and "networking." The conferees urged participants to form "a seventeen-mile chain of human beings encircling the Rocky Flats Nuclear Weapons Plant, to end the arms race."[46]

Just as the Park Service (as we saw in Chapter 14) had redirected its attention from wildlife ecology to global problems such as acid rain, so too were environmentalists drifting toward such issues as peace and disarmament, Green politics, acid rain, population control, and fighting world hunger. Increasingly, their environmental attention became riveted on the ecological effects of Armageddon. And though this issue was crucial to every living individual, it tended to leave other species to their own devices. While worrying about the fate of the earth, they risked losing touch with the fate of places such as Yellowstone.

When they did turn to the plight of natural areas, growing preoccupation with politics encouraged them to search for solutions, not in better biology, but with the right ideology. While some groups on the right — such as the "Sagebrush rebels" or the "New Resource Economists" — argued that government management was the cause of ecocatastrophes and urged that public lands be turned over to private enterprise, the majority of environmentalists thought salvation possible only through their brand of liberalism. As Rick Reese, President of the environmental organization called the Greater Yellowstone Coalition, wrote "The orientation of the present [Reagan] administration toward resource management . . . must be counted as perhaps the greatest threat of all to the wild lands of Greater Yellowstone."[47]

To animals, however, ideological splits among environmentalists were no more meaningful than were the differences between communism and capitalism to the average Cambodian farmer. Death in Yellowstone, as we have seen, was bipartisan.

20

The Land Ethic

THE agenda read: "The Greater Yellowstone Ecosystem — An Ecological Concept." Moderator Franz Camenzind, an independent scientist from Cody, Wyoming, said that this was one of the few times when biologists "will officially have something to say at these meetings. . . ."[1]

Our first task was to define "ecosystem," he suggested, but that would not be easy. He confessed that "ecology perhaps goes beyond science to the realm of an art form and some people might say, a little bit of magic."

He continued, "I am not sure any of us could put a real hardcore definition down on paper or on a map, but what motivated us all is that we wanted to control this unit of ground as an ecosystem, we wanted to look at it as an ecosystem and manage it that way." Instead of defining the word, he showed the audience a map outlining the boundaries of the ecosystem. With suspiciously straight lines, the map encompassed an area of around six million acres — about three times the size of Yellowstone.

They were sitting in the map room of the old and elegant hotel in Mammoth, Yellowstone headquarters. The room was dominated by a large map of the United States, perhaps twenty feet long, which hung along one wall. A gigantic jigsaw puzzle — the shape of each state was cut from a different kind of wood — it was a monument to the skill of artisans who built the early buildings in the park. Now another generation of artisans were at work assembling another puzzle. They were organizing a new environmental group called "The Greater Yellowstone Coalition."

An association of thirty-five groups — including local chapters of the Audubon Society, Sierra Club, Defenders of Wildlife, Wilderness Society, and the National Wildlife Federation — the GYC was the most ambitious effort by environmentalists to protect Yellowstone since the predator-control controversies fifty years earlier. Inspired by the deep ecologists, it was, according to board member William Bryan, a "bioregional approach to identifying and resolving issues in the Greater Yellowstone. . . ."[2]

Calling greater Yellowstone "the largest essentially intact ecosystem remaining in the temperate zones of the earth," its stated purposes were "to promote the scientific concept of the Greater Yellowstone Ecosystem; to create a national public awareness of issues and threats" facing it, and to "utilize the combined effectiveness of the coalition's constituent organizations and individuals to preserve the ecosystem intact."[3]

Discussion in the map room drifted from the semantic and biological to the political: from questions about the nature of the ecosystem to ways in which to protect the region (still undefined) from further encroachment. Many voiced worry that destruction of wilderness around Yellowstone would do irreparable harm to the park's wildlife.

Park boundaries, they said, were not natural, and nearly all its large mammals wandered across them. Yet the national forest areas around the park faced a variety of threats. Geothermal development in Idaho's Island Park area that would possibly destroy the park's fragile system of hot springs and geysers was perhaps in the offing. Oil and gas exploration and hardrock mining in areas critical to wildlife were a threat. The Homestake Mining Company planned to reopen old gold mines at Jardine, Montana, at the very edge of the park and in the heart of the Beartooth-Absaroka Wilderness. Developers from Pennsylvania had received permission from the Forest Service to build a resort called Ski Yellowstone near Hebgen Lake, Montana, in land that biologists knew was prime grizzly habitat. The Forest Service was stubbornly insisting on building many new access roads and trails for hunters and hikers into pristine areas along Yellowstone's northern rim, and it was stupidly accelerating timber sales at prices below cost. As the conferees talked, the list of threats to the integrity of this large island of wilderness continued to grow.

The group agreed to cope with this encroachment by taking a

bold political step. Recognizing that many of the problems facing the area were exacerbated by an incoherent labyrinth of conflicting and overlapping political jurisdictions involving two national parks, five national forests, several federal agencies, and three states, the Coalition resolved to lobby for creation of one administrative entity to replace all those. Under the guidance of its new President, Rick Reese, a former Park Service ranger and member of the Board of Directors of NRAG, the Coalition decided, at this meeting, to ask Congress to enact legislation that would give national recognition to the ecosystem, placing the area under one "coordinated system of management" that would "recognize the importance of Greater Yellowstone's wildlife, scenery, recreation and environmental quality. . . ."[4]

Later the Coalition decided that perhaps Greater Yellowstone would be best protected if administered by the Park Service. Coalition Director Bob Anderson told a meeting in Jackson in the following September that "present park boundaries could be expanded to include the entire ecosystem so that it would be under one agency or everything could be brought within one leadership administratively."[5]

During the months following their meeting in Mammoth, the board continued to press for national recognition of the Greater Yellowstone Ecosystem while it also worked to identify varied developments with which they would take issue: mining, oil and gas exploration, geothermal development, road building, dams, snowmobile trails, Forest Service timber sales, ski resorts, and removal of garbage from grizzlies' access.[6]

Recognizing the need to protect critical wildlife habitat surrounding Yellowstone was a long step in the right direction and quickly enlisted the support of nearly everyone who cared about Yellowstone. Bringing national attention to the declining condition of our first and greatest national park was indeed a vital effort, long overdue. The threats to wilderness in and around the park were immense. And for Yellowstone, the Coalition was the only game in town: if they did not save the park, no one would.

But the Coalition's target was political issues surrounding the park, not biological problems within it. Unlike those who fought to end the killing of wolves and coyotes in the 1930s, the GYC seemed dedicated to extending protection over a larger area of land, not monitoring the condition of those animals in the area already protected. It had no bone to pick with the Park Service. Of the eighty-eight issues listed by the organization as threats to Yellowstone, not one reflected a criticism of Interior Department policies.[7]

Rather than questioning whether the region had sufficient natural food for bears, whether perhaps grizzlies should be declared endangered, or whether the government should comply with the law and designate critical grizzly habitat, they simply supported continuing Interior Department bear management.

Reported GYC President Reese, "We feel the federal agencies are at this time pretty well on target; we feel the grizzly bear recovery plan is a document that is workable and given the proper kind of public support, it will lead to the recovery of the population of grizzly bears in the Greater Yellowstone area." Indeed, he added, "Wild natural food used by grizzlies is available in Yellowstone and the land does provide the bear with the needed food . . . feeding the bear is not necessary and we hope it will never become necessary at any point in the future."[8]

Although supporting closure of Fishing Bridge, they did not mention development of Grant Village or planned expansion of the John D. Rockefeller, Jr., Memorial Parkway.

The problem with elk, they suggested, was not that they were too many, but that their range was too small. They reported that "the southern herd is below the average." And in the northern range, "the availability of adequate food during winter has been, and still is, the most important factor limiting the distribution and abundance of ungulates. . . ." The major threat to elk, they concluded, was "development pressure" surrounding Yellowstone.[9]

Although they cited the dangers of geothermal energy development outside Yellowstone, they did not question the wisdom of deep drilling in it (a project announced by the Park Service at the June 1984 meeting).[10]

These well-intentioned people, so aware of the threats outside the park, remained blind to dangers within it, because the Interior Department's philosophy of wildlife management was theirs as well.

"In modern times," wrote GYC President Reese recently, "the National Park Service has taken seriously its mission of managing Yellowstone as a natural area. . . . Today enlightened park managers prefer a continually evolving policy of non-interference with the park's life forms and their natural environment." Although he stated that man may "reinterfere" to right past wrongs (such as weaning "garbage-hooked bears"), Reese suggested that these actions would seldom be necessary:

> Greater Yellowstone is large enough, remote enough, diverse enough, and with the exception of wildlife poaching, protected enough to provide most, and in some cases all, of the necessary ingredients for a smooth-functioning, self-regulating ecosystem. Presumably if man pulled out of the area today, totally absenting himself from five or six million acres, native plant and animal life forms would continue to survive, even flourish. That is so because sufficient area remains in a natural condition for self-sustaining, self-balancing natural forces to function.[11]

Indeed, the government policy of natural regulation was the *raison d'être* of this environmental group. By establishing as its purpose preservation of the "Greater Yellowstone Ecosystem," it had adopted Park Service "ecosystems management" as an argument against development of lands around the park.

Although dressed as a new scientific theory, this "bioregional approach" was the old preservationism underneath. *True* ecosystems management — based on adequately studying the past and the ecological role of aboriginal people, not resting on the illusion that the "original ecosystem" was still "intact," and devoted to sustaining or, where necessary, compensating for, processes no longer operating — was indeed a necessary approach to preservation of wilderness. But this expression was used by these activists, as it had been by the government itself, as another name for noninterference.

The primary goal was still keeping people away from nature, a nature too complex and too sacred to be defiled by manipulation or analysis. Just as California deep ecologist George Sessions thought that "nature is not only more complex than we think, but it is more complex than we can ever think," and Master Plan architect Olson believed that "mystery and the unknown" remained in wilderness, so too did Hank Fischer, a founding board member of the GYC, reject "those who would manipulate wilderness," because (quoting Barry Commoner) "nature knows best." We can never know enough to manage wilderness, these environmentalists still insisted; therefore we should not try.[12]

And because "Save the Yellowstone ecosystem!" had a ring of scientific objectivity and spiritual mystery that "Save wildlife habitat!" did not have, this expression was used, not as a scientific concept for evaluating wildlife-management practices, but as an emotive slogan for saving wilderness. Whereas scholars saw ecosystems as fictional models conceived so as to further understanding, activists drew them on maps, endowing them with identifiable boundaries. Although scientists supposed that human activities were part of ecosystems, conservationists defined the idea so as to exclude them.

They were also carrying the rationale for ecosystems management one step beyond official theory. Whereas park managers had argued that, except for the grizzly and portions of the elk herd, the park was an "intact" ecosystem, spokesmen for the GYC suggested that only "Greater Yellowstone" was of sufficient size to sustain all the life it contained. The ecosystems approach, they implied, when limited to Yellowstone, did not work; it would succeed only if the ecosystem were enlarged. In this way they endorsed official policy as a rationale for preservation even while they admitted it had so far failed.

And so with a curious reverse logic, the failure of Park Service policy became the rationale for Yellowstone's expansion. If we refused to intervene to save wildlife, animals could survive only if we "protected the ecosystem," and if a species or the range was in trouble, this problem was a further argument to extend habitat. The more ecosystems management failed in the park, the poorer the habitat and more threatened the fauna, the stronger the argument in favor of extending control of surrounding lands. Whether a species (such as the elk) was too abundant or (like the grizzly) becoming rare, the solution was the same: expand wilderness. Wildlife became hostage to a political war over land use.

Yet in taking this course they had paradoxically made a mistake that Aldo Leopold warned against. He said, "Land is not merely soil." By Land Ethic he meant an ethic of all living things, including human beings. The principle that was needed, he implied, was a Life Ethic — rules by which all living things could live together in peace. And yet by supposing that protection of land was a solution to all wildlife problems, environmentalists had, in effect, taken Leopold literally. Once the idea of preservation was enshrined, the Land Ethic became simply a rationale for protecting land.

Conclusion

We has met the enemy, and it is us.
— Walt Kelly

21

Yellowstone Elegy

THE Roosevelt arch stands alone at the edge of an empty field, facing a Park Service gravel quarry. Arriving by automobile, most visitors now bypass it, taking the cutoff road at the other end of town. The tree-lined park, hay meadow, trout pond, and train station are gone. A heating plant and large water-storage tank, intended to provide hot water for the concessionaire's laundry, is under construction on the hill above.

Here, as throughout the park, the slow rate of change — and the short span of our lives — help obscure Yellowstone's fate. In a place so beautiful, those without the perspective of time may not notice what is missing. They probably will not recognize the high-lining, the sparse vegetation, the spread of exotic grasses, the absence of willow, and the aging stands of aspen on the northern range. They may not miss the animals that are no longer there, such as the wolf and white-tail deer, nor will they realize that the animals Roosevelt saw in numbers are now scarce: the mule deer, antelope, bighorn, and, of course, the beaver. They will not realize that many of these species are common in the regions surrounding the park. They may remember the black bears that once begged along the road, but the plight of the grizzly will — as it had with the wolf — pass unnoticed.

Instead, visitors will continue to rejoice at all they do see: the geysers, the tame elk and bison, the occasional moose, the fabulous fishing, and the green vistas in places such as Hayden Valley or the upper Gibbon.

They probably will not miss all that has been lost, for it is difficult to mourn the absence of something one never knew, something that disappeared, perhaps, in one's grandparents' day. Instead, the transformation of Yellowstone, for many visitors, would be a subtle one: not a change that they would see, but a gradual, and perhaps unconscious, impoverishment of their Yellowstone experience.

But some day visitors may realize that the experience, once so special, that the park once offered, is gone. And then they will ask: who — or what — destroyed Yellowstone?

We know what the Park Service answer would be, for we hear it today. Their policies were successful, we will be told. Yellowstone was destroyed by outside forces beyond their control.

Former Park Service naturalist-historian Paul Schullery observed in 1984:

> Yellowstone is in trouble. The immensity of the threats and already active destroyers is staggering: geothermal energy exploration up against the west boundary; ever larger recreational vehicles and ever larger crowds; reckless, wholesale, and shortsighted development in surrounding national forests; water-, mineral-, and energy-starved cities hundreds of miles away; an avowedly antiwilderness — antinature really — government administration; and most dangerous of all, a diffident and parsimonious citizenry. Few believe there's much hope in the long run.[1]

In the park, by contrast, he said, "you find a wild setting that is in better shape ecologically than it has been since before 1900. Internally it is robust. It has the flaws . . . too few wolves, a lack of influence by pre-Columbian cultures . . . too many meadows growing to timothy and other domestic grasses. . . ."[2]

Blame for the destruction of Yellowstone, in short, will be placed, not on Park Service policy, but on outside pressures and public indifference. And indeed, that would be partly true. These things on the periphery hurt some wildlife such as the grizzly and may threaten Yellowstone's thermal features. But the destruction within the park would be the product of something else. It would be testimony to the bewitching power of a false idea.

In late winter 1982, a buffalo, walking across the frozen Yellowstone River near Fishing Bridge in the park, fell through the ice. Four men on snowmobiles, seeing the struggling animal, threw a rope over its horns and began towing it to shore. Just before it reached safety, the rope slipped off its horns and it drowned.[3]

The men, after their effort, were visibly upset. What disturbed them even more than the death of the animal, was the behavior of the Park Service official who watched the drama from the bridge. The official, Bobbi Seaquist, did not help. She just stood there, saying the drowning "was nature's way."

Seaquist was following orders. Her superior, ranger Ron Sprinkle, had told her to tell the men "to stop tinkering with nature's progression."

But what was nature's progression?

Lying behind Sprinkle's admonition was an attractive idea that occupied the core of contemporary environmentalism. It was a flower of the seed planted by the New Philosophy of Nature, the religion founded by Emerson, Thoreau, and Muir. And it was this philosophy that guided Park Service managers and shaped American attitudes toward nature.

Actions of environmentalists in Yellowstone reflected the religious, scientific, and political insights that emerged from the redwood think tanks of California during the last two decades. In their unquestioned acceptance of the principle of noninterference, environmental activists of the northern Rockies embodied the major insight of the deep ecologists. In their ritual use of the word "ecosystem" and explicit adoption of "bioregionalism," they joined with the California Cosmologists in expressing belief in the interconnectedness of things, at the same time remaining ambivalent about traditional science. By insisting that nature always knows best, by wrapping the world in ineffability, they provided a rationale for doing nothing. In their preference for land-use issues over wildlife problems, they sought political solutions to biological questions. Attracted to activism, coalition building and co-option, they created an alliance with bureaucrats that stifled debate on wildlife issues.

Yellowstone had become a casualty of the environmental crisis that the California Cosmologists had sought to resolve. Yet the deepening difficulties of the park only made it clear that the problem had resisted solution. Environmentalists, together with the Park Service, still supposed they could save Yellowstone, as Lynn White had said, "by deep-freezing an ecology . . . as it was before the first Kleenex was dropped." But, as White had implied, this was a Quixotic act of people caught up in a crisis they did not understand.

Yellowstone was caught squarely in the ecologic crisis of our time. This land, which evoked images of a wonderland to Langford and his contemporaries, and was made a park in the year that Lewis Carroll's *Through the Looking-Glass* appeared, was indeed a wonderland; but it was also a looking-glass world, a place that mirrored our national ideals. As a symbol of our hopes for natural preservation it reflected what we wished to see.

In this way the park became, like other national symbols, infused with myth. Rather than recognizing how Yellowstone had changed, we saw it as a picture of the primeval scene. Rather than recognizing the ecological role of Indians, we continued to portray them as noble and ineffective savages. Rather than recognize that visitation to the park had remained roughly stable for the last twenty-five years, we insisted on placing blame for its problems on (Schullery's words) the "ever larger crowds."

Behind this wishful thinking lay our ambivalence about the role of man in nature. What did Yellowstone represent to us, and what did we want it to be? Was it a natural area or a cultural institution? Whatever we decided, we were convinced it could not be both. From the highly compartmentalized world of science, in which anthropologists seldom talked to biologists, came no inkling that culture and nature might have intertwined in Yellowstone. From the ranger corps, trained to the requisites of "visitor safety and protection," preservation came to mean keeping people away from "the resource." From environmentalists seeking a new holistic paradigm wherein man might be part of nature, came the message: the two must be kept apart.

Yet we were mistaken. Natural areas were not made less natural by human presence. The worlds of nature and culture overlapped. We had tried to draw too fine a line between the two.

Yellowstone was not just a natural area. It was not natural to see elk grazing on the mown bluegrass lawns of the administration buildings at Mammoth, to wake in the morning at the Lake Hotel to find a bison asleep under one's window, to meet a moose walking through the parking lot, to see a bear along the road. It was not natural, but it was nice. The park was a place where nature and culture mixed, a community of animals and people, a laboratory where we could learn how to coexist with other creatures.

It demonstrated for the rest of the country how a community that was visited by more than two million people a year could remain beautiful; how a settlement that generated five million gallons of sewage a day could still have pure water; how streams that received hundreds of thousands of fishermen a year could still be natural fisheries.

It was in this way, as a model society side by side with nature, that the park should have served as a symbol for our hopes for preservation. It could have shown that what people touch need not be made ugly by the touching, and that there was still a place where we might live in peace with other creatures.

Instead, it reflected our own misperceptions — of ourselves and of nature. If Yellowstone dies its epitaph will be: "Victim of an Environmental Ideal."

Epilogue

How to Save Our National Parks

It might be of some comfort to believe that the story of Yellowstone is unique. After all, the national park system is extremely diverse; what happens in one place need not occur in another.

Unfortunately, Yellowstone is not unique. The entire park system is in trouble. In those parks classified as natural zones (such as Yellowstone), wildlife and their habitat are disappearing even as some of these species continue to thrive outside the park system.

In Everglades, the future of the panther and wood stork remain in doubt, even as the latter is rebounding elsewhere; the peregrine falcon is declining in Guadalupe Mountains National Park while staging a comeback in many other places; and humpback whales, according to Park Service records, dwindle in Glacier Bay, Alaska. Mount Rainier and other parks of the West may, like Yellowstone, have far too many elk. Exotic species are a ubiquitous threat. Goats, rabbits, pigs, burros, and mongoose are displacing native animals in Channel Islands, Haleakala, Hawaii Volcanoes, Olympic, Death Valley, and Virgin Islands. Altogether, according to a 1987 Park Service study, "at least 157 units of the System contained at least one exotic vertebrate species."[1]

A 1987 study by University of Michigan researcher William Newmark, published in the British journal *Nature*, found that fourteen of our western national parks have already lost forty-two populations of mammals and are in danger of losing more. "Without active intervention by park managers," Newmark said, "it is quite likely that a loss of mammalian species will continue."[2]

Indeed, while these and similar problems have not been widely publicized, they have caused mounting concern among wildlife professionals. The National Park Service, in its first state of the parks report in 1980, identified 4,345 specific threats to park resources. In 1984 Destry Jarvis, Vice President of the National Parks and Conservation Association, told a group of park professionals, "Our national parks system exists in a state of crisis."[3]

To be sure, much of this attention is focused on "external" threats: growing population pressures and industrial development in areas surrounding parks. But while encroachment does present a clear and present danger, it is not the only cause of worry.[4]

Rather, there is a rising consensus among park professionals that a major source of our parks' decline is the National Park Service itself.

"While the highest level of scrutiny and concern understandably focuses on the composition and condition of the resource base itself, the National Park *System*," noted a joint working group of the Audubon Society, Wilderness Society, Conservation Foundation, and Sierra Club in 1983, "the welfare of that System is in major ways influenced by the mechanism designed to protect and administer national park units, the National Park *Service*" (italics in original).[5]

"The National Park Service today," the working group continued, "is an agency confronted with serious problems. Some existing management and personnel policies have compromised the effectiveness of the Service and thus endangered the quality of the System. There are a number of actions which need to be initiated to improve morale, professionalism, and effectiveness of the National Park Service."[6]

The biggest threat to the national parks, former Park Service chief scientist Robert Linn told me recently, is that "strong leadership at NPS's top echelon combined with the original idealism is lacking and has been lacking for well over a decade. Professionalizing the Service must be top priority."[7]

The greatest dangers to the national parks, according to William Burch, professor of social ecology at Yale University, "come from within the policies, practices, traditions of the Service. NPS has spent so long [two decades or so] pointing outside itself." Among the problems now facing the Service, Burch told me, include a "loss of morale among park employees . . . lack of systematic training for the professional performance of park rangers," and ". . . no direct research [and] experimentation" on the best ways to achieve natural preservation.[8]

"From an internal perspective, two serious problems in the park system," says John Reed, former chief of the Biological Resources Division in Washington, "are the lack of current, well-articulated guidelines for science and natural resource management; and a serious lack of real accountability in actions and decision making pertaining to many natural resource management activities.[9]

"Accountability is a problem," Reed explained, "when our organization permits individuals seriously lacking in professional skills and relevant experience in natural resources to occupy key natural resource management and advisory positions and to make major policy, funding and programmatic decisions without truly being held accountable for their actions. I have reviewed scores of funding proposals from parks, many addressing the same kinds of issues, and I saw that the wheel was being reinvented over and over again, and that there was nothing in place to correct this in any organized fashion service-wide. Each region and in many cases each park often appears to operate almost like an island, focusing all its energies inwardly with little awareness or involvement in the generic, service-wide issues."[10]

The story of Yellowstone, in short, is a state of the parks report in microcosm. If we do not act soon, this immense national resource may be imperiled. What may we do to save it?

The language of the National Park Service Act has always bothered Park Service officials. Commanding the agency both to prescrve the parks and to provide for the "enjoyment of the same" by the public, it appears to give the agency two mutually inconsistent missions. How can officials carry out the mandate of preservation while promoting public use of our parks? More people visiting parks will do more damage to them. In attempting to balance use versus preservation, many rangers feel they are playing a zero-sum game they cannot win.

This ambivalence is aggravated by the fact that the Park Service has, for much of its history, served two constituencies with very distinct interests: environmentalists and the recreation industry. This dual service has only aggravated the agency's ambivalence.

The environmental community naturally wishes to see preservation take precedence over use, while the recreation industry wants to reverse this order of priorities.[11]

Not surprisingly, therefore, coincident with these two conceptions of how our parks ought to be used are competing ideas of what parks ought to be. One—which we might call the "landscape philosophy"—envisions national parks as larger versions of city parks. They are places of great natural beauty, to be sure, but more importantly, they are places for *people*. Rather than presenting true wilderness to the visitor, they offer pleasing, if contrived, vistas, pleasant accommodations, and recreational opportunities. According to the recreationist philosophy, in short, parks are cultural entities, the products of *artifice*.

The second view of what national parks ought to be might be called the "wilderness vision." This is the perception of parks as little pieces of undisturbed, primeval wilderness.

These two ideals entail quite different techniques of park management. Landscaping requires the services of landscape architects and maintenance personnel. The wilderness vision, by contrast, requires no management at all. Instead, if national parks are true wilderness, they only need *protection* from the impact of people and civilization.

Over its history, the Park Service equivocated between these two models, emphasizing sometimes use, sometimes preservation—shifts that reflected to some degree the relative strengths of the competing interest groups' lobbying power.

During periods when the Service was more anxious to please recreationists, it pushed developments in the parks, such as planting game fish in streams, building hotels, resorts, and even golf courses.

This emphasis affected wildlife. The campaign of predator extermination, carried out by the Park Service between 1916 and 1935, was, as we saw, an attempt to please the public by eliminating "bad" animals—predators such as wolves and mountain lions—to increase the number of "good" animals—such as elk and bison—which the new agency believed brought more people to the parks.

When, on the other hand, conservationism was ascendent in America, the Service tended to put a higher value on preservation. During the 1930s, for example, when the railroads were declining and environmentalism was on the rise, the Park Service terminated its predator-control program and established the Wild Life Division.

During the last two decades, a time of rising environmental consciousness among Americans, the Service has again become more sensitive to the pleas of conservationists. As public-land policy specialist Ronald A. Foresta noted in 1984, this agency continues to "move its base of support away from the park-using public at large and toward the environmental public-interest groups." And an offspring of this renewed relationship, as we have seen, is the policy of natural regulation.[12]

Park policy, then, is a willow bending in the prevailing political wind, swinging between development and a passive form of preservationism—even though both policies have proved disastrous for the parks. And as long as the Park Service believes the choice is *between* preservation and public use, there is no way it can formulate a consistent mission for itself.

Nevertheless, it has found a formula for management that serves both recreationism and the wilderness vision. For despite the incompatibility of these two philosophies, there is one critical point on which they are in agreement: that the role of the Park Service, in managing the park system, is principally *protective*. If parks are managed for people—as the recreationist philosophy dictates—then the Park Service has a duty to serve and protect those people. On the other hand, if parks are managed to protect wilderness, then the Park Service has a duty to protect the park *from* people.

Either way, the role of rangers is one of protection. The mission of the Park Service, in short, is interpreted not as managing the resource, but managing *people*. Consequently, whether the Park Service emphasizes visitation or protection, it sees the need to train rangers in law enforcement.

This emphasis has not changed since publication of *Playing God in Yellowstone*. *Protection* remains the byword—a lion's share of monies still go to "visitor safety and protection." Rangers receive insufficient training in ecology and history. Superintendents and regional directors still control research and resource management, even though, according to a recent poll by the Association of Park Rangers, only 2.2 percent of superintendents' prior training and experience has been in resource management (as compared with 5.8 percent in maintenance and 24.5 percent in "visitor protection").[13]

Resource management, Richard Briceland told me last year, remains "an underground activity in the Park Service." There is no "career ladder" for resource management, so that rangers who choose this field find little room for advancement. The total fiscal 1987 budget for protecting natural resources (including such items as fire suppression management and backcountry patrols) is around $57 million—just six percent of total Park Service spending.[14]

Meanwhile, the Service's effort in scientific research remains miniscule. Today there are only around seventy researchers for the 337 units of the national park system. Research spending is around $16 million (less, Briceland explained to me, than the budget of one Forest Service Experiment Station).[15]

And there are no immediate signs things are about to get better. Last year, for example, the regional directors vetoed proposals to professionalize the Service—including establishing a career ladder in resource management and taking control of research away from superintendents.[16]

Unfortunately, as the story of Yellowstone demonstrates, protection alone is an inappropriate model for managing our national parks. Western civilization has radically altered these places, which were never complete ecosystems, and they remain as tiny islands surrounded by technological society. The eviction of the Indians, elimination of predators, introduction of exotic species of plants and animals, and a century of fire control have thrown even the "wildest" parks into ecologic disequilibrium.

And once an ecosystem has been truncated and thrown out of balance, it no longer has the capacity to cure itself. Like a seriously ill person whose vital organs are no longer functioning, these places, if left alone, will die. A policy of protection, therefore, will neither arrest further change nor ensure that all that happens is "natural." Rather, over time it will produce historically unprecedented conditions—an entirely new regime of fauna and flora—which we will be unable to predict and which we probably will not like.

What our national parks need, therefore, is not only protection, but *restoration*. If the parks are to be preserved, they first must be restored to conditions where some semblance of ecological balance prevails.

Restoration of the land, moreover, is not a utopian ideal, but a developing science. The task of restoration ecology is like searching for, and then assembling, parts of a puzzle to make a picture. One finds isolated communities of native genetic types and carefully transplants these species to preserves reconstructed to replicate their original habitats.[17]

There are today many examples of successful ecological restoration. Throughout the Midwest and West, grasslands prairies and wetlands have been nursed back to life. The University of Wisconsin Arboretum, for example, was founded in 1934 for this specific purpose. "Our idea," wrote its first director, Aldo Leopold, "is to reconstruct . . . a sample of original Wisconsin—a sample of what Dane County looked like when our ancestors arrived here during the 1840s." Today the Arboretum has successfully restored over 400 acres of wetlands, forest, and prairie, and plans to restore several hundred more.[18]

Another ambitious project now under way is the reclamation of 430 square miles of dryland tropical forest in Guanacaste National Park, Costa Rica. Considered by wildlife experts to be perhaps the most important and unusual biological experiment undertaken in several decades, the Guanacaste project is headed by a team of experts under the supervision of Daniel H. Janzen, professor of biology at the University of Pennsylvania.[19]

Restoration ecology, according to William R. Jordan III, director of public programs for the University of Wisconsin's Arboretum and a leader in the field, "represents both agriculture and medicine transformed by the ecological consciousness.[20]

"The techniques and methods of restoration are those of agriculture—the cultivation of the land. . . . [Yet] unlike agriculture, restoration is not a mode of production, but like medicine, a healing art."[21]

The analogy with medicine is, I think, particularly apt as a description of the task now facing park managers. Our parks today are seriously ill. Keeping a seriously ill patient alive and nursing him back to health requires elaborate life-support mechanisms and constant attendance.

So reclamation can never be entirely complete. Even after a relatively successful restoration, we will not have the luxury of supposing that we can then leave parks to their own devices. We cannot bring extinct species back to life, nor can we reproduce all the conditions that prevailed before the coming of the white man. We cannot expect animals that evolved in ecosystems the size of half a continent to survive unaided in the relatively tiny areas we call national parks, any more than we can expect Indians to live as hunter-gatherers in the postage-stamp-sized reservations to which they are now consigned.

The habitat of scavengers such as condors and grizzly bears, for example, cannot be completely reclaimed. In pre-Columbian times, these species depended heavily on dieoffs of abundant and ubiquitous animal species such as bison or spawning salmon, as well as on carrion left by Indians and predators. Yet the animal world will never be as fecund or dispersed as it once was, nor will the Indians and predators play their ecological role to the extent they once did. So if we wish to preserve such species, we may need to find substitutes for the food sources on which they once depended.

Restoration, therefore, leads to a kind of management we might call "sustenance ecology," a process that would take four steps. First, collecting what scientists call "baseline data": information—gathered by historians, anthropologists, archaeologists, and biologists—that would tell us what the parks were like in pre-Columbian times. Second, taking an inventory to see how the park had changed: what species have been lost, and what exotic species have been introduced. Third, removal of exotic species and reintroduction of native plants and animals now missing. And finally, devising strategies to compensate for those conditions that prevailed in pre-Columbian times but cannot be recovered. Such strategies might include, for example, providing "carrion" for scavengers, culling game herds, and burning forests and grasslands to replace lost Indian hunting and fire practices.

How would sustenance ecology work in practice? It would mean that in Yellowstone, for example, the first work should be archaeological studies to determine whether elk and bison were native to the park, and anthropological and ethnological investigations to learn the ecological roles of aboriginal Americans. This research should be followed by multidisciplinary efforts to determine just how European civilization had changed the park. If these studies confirmed the evidence adduced in this book, then restoration would require culling the elk and bison herds, reintroducing wolves and mountain lions, replacing exotic grasses with native genetic types, embarking on an aggressive program of broadcast burning, and leaving food—probably killed elk—for grizzlies.

While such a regime will not eliminate the conflict between public use and preservation, it could mitigate—and compensate for—the unfortunate effects of people on wildlife.

Protectionism contains no contingency plan to cope with the problems arising when human use damages animal habitats. It is, therefore, a pure ideal that can be no more successful than King Canute's command for the sea not to rise. For while it is inevitable that at times human activities will affect wilderness, the protectionist has no way to repair the damage caused by this encroachment. Rather, he must simply wring his hands at the prospect that another bit of "undisturbed wilderness" is "lost forever." Once a land is thrown out of equilibrium, he offers nothing that will make it right again.

Protectionism, therefore, magnifies the dangers that public use poses to wildlife, not only because it permits ecological imbalances to persist or worsen, but also because it does not offer any formulas to cope with habitat destruction. Sustenance ecology, by contrast, does not depend exclusively on excluding people from wilderness in order to protect it. It is a philosophy specifically for parks, dedicated to reestablishing and sustaining ecological equilibrium to lands that receive a reasonable amount of public use.

The challenge facing the Park Service, therefore, is not merely protection, but the far more ambitious task of restoring and sustaining our national parks.

This, of course, is not a new idea. As we saw, George M. Wright and his colleagues Joseph S. Dixon and Ben H. Thompson urged that "the need to supplement protection with more constructive wildlife management has become manifest with a steady increase of problems both as to number and intensity."[22]

The Leopold Report, in suggesting the Park Service recreate "a reasonable illusion of primitive America," made a plea for restoration. Yet surprisingly, although the Leopold Report was made official policy, little restoration was attempted in any of our national parks. With few exceptions, baseline studies were never undertaken; almost no historical research—including archaeology and anthropology—was done; native-species restoration was stalled, and proliferating exotic species, rather than being removed, were, for many years, ignored.[23]

In truth, the Park Service was unable to implement the Leopold Report. Restoring our parks required scientific resources the Park Service did not have. And the only way reformers in the sixties could gain acceptance of an increase in the science budget was to put research and resource management under the supervision of the rangers. This is, as we saw in Chapter 14, just what they did.

Further, each of the ten regional headquarters was given control of research and resource management in their areas. And within each region, the major parks were themselves almost completely autonomous. In these ways superintendents and regional directors acquired nearly total control over scientific activity.

The delegation of such powers to the regions and superintendents decentralized the Service, preventing any coordinated scientific undertaking. The flow of information and chain of accountability between the parks and Washington were broken. A superintendent could—and many did—prevent results of research that might reflect badly on a park administration from leaving the park.

The Balkanization of the Park Service was further encouraged by the national park system reorganization of 1964, which divided parks into three zones: natural, historical, and recreational. These divisions effectively prevented the kind of sustained interdisciplinary research that true restoration ecology required. They meant that while biological research might continue in natural zones such as Yellowstone, "mission-oriented" historical research was considered inappropriate in such places.[24]

Through these steps, the national park system has come to resemble a feudal society, looking not unlike France before the reign of Louis XI. Just as fifteenth-century France was a country with independent duchies and a weak king, so the national park system is one of independent regions and a weak director. Each region is nearly autonomous; Washington has no so-called line authority cover them. Given the isolation of major parks, superintendents of these areas control largely independent fiefdoms and have nearly total control over scientific activity.[25]

There are several consequences of this arrangement. First, superintendents can prohibit studies that might make their decisions look bad. Second, they can influence the results of these studies.

Third, bad news does not travel to Washington, and therefore is not subject to the scrutiny of top officials within Interior and the administration, the national press corps, or the mainline environmental groups. Fourth, in the absence of sound scientific information reaching Washington, the director cannot make intelligent decisions. And fifth, the absence of sound research that could serve as the basis for planning encourages the politicization of decision-making, where the National Park Service has no data with which to counter the competing claims of pressure groups.

Thus while the Park Service has expanded its research effort in terms of dollars spent, it has not yet developed the strong research and management it would need to replace protectionist policies with more constructive ones. Instead, natural regulation remains intrenched, even though few biologists believe this regime alone will save our parks.

"A laissez-faire approach to management is simply untenable," Bruce A. Wilcox, director for the Center for Conservation Biology at Stanford University told a *Newsweek* reporter last year.[26]

"The overall management of resources philosophy needs to be changed," Peter Dangermond, former director of the California park system and now a national parks consultant, told me recently. "We need to admit that man has been in the modification and management business for a long time, and that we have made some changes to the ecosystems which have thrown nature out of balance. A program to correct past mistakes should be made public in a big way."[27]

How then do we get the Park Service back on track?

Faced with these and other challenges to the park system, last year Park Service director William Penn Mott, Jr. introduced a twelve-point plan aimed at "shaping a new vision and stimulating a new enthusiasm for protecting, preserving, and perpetuating all units of the national park system." Among the aims of the plan is to establish a "blue-ribbon panel" of "outside" experts to reexamine the policies of wildlife management that have guided the Service for the last twenty-five years.[28]

"Current NPS policies for wildlife management," the plan noted, "are largely based on a 1963 report by an advisory board headed by A. Starker Leopold. . . . It is time to reexamine the principles of ecological management propounded in the Leopold Report. The relationships between science, research, and resource management need to be examined and clarified."[29]

Although the blue-ribbon panel has not yet been named, it faces a daunting challenge. Regardless of the value of the advice offered by the panel, it is unlikely the Park Service will be any more able to heed that advice any better than it did the prescriptions of the Leopold Committee. As the problems facing the park system reflect institutional inadequacies, saving our parks requires rescuing the Park Service first. How can this be done?

Some believe it can be done by larger congressional appropriations for such things as baseline research. "If this administration has hurt us," said Paul Pritchard, president of the National Parks and Conservation Association last year, "it is in denying us the raw data. The Park Service simply doesn't have the money to do an inventory of the animals."[30]

By contrast, those who call themselves "new resource economists" assert the Park Service is incapable of reform. The agency's shortcomings, according to this view, are an inevitable consequence of public management: "When everyone owns a resource, such as air and water, no one actually owns it—and no one is accountable for treating it properly," said Jane Shaw, an associate of the Political Economy Research Center (PERC), a think tank studying privatization of public lands. As a consequence, some at PERC suggest the parks be turned over to profit-making corporations.[31]

"I have no problem with the idea of the Disney people running Yellowstone," said Terry Anderson, a professor of economics at Montana State University and a colleague of Shaw.[32]

Others—for example, John Baden, PERC founder and now director for the Foundation for Research on Economics and Environment (FREE)— advocate turning management of the national parks over to private, nonprofit wilderness endowment boards, organized like art or history museums, but accountable to Congress.[33]

Neither money nor privatization, however, is, in my opinion, the magic formula that will turn the Park Service into a professional, ecologically minded agency.

Resource management and research, to be sure, are starved for funds, and have been during the entire history of the agency. But the Scrooge in the system is neither Congress nor the President, but the Park Service itself. This agency failed consistently to submit appropriations requests for what it should spend on resource management and research. The Robbins Committee urged that spending on natural-history research be at least ten percent of the budget. Twenty-four years later, in fiscal 1987, it remains under two percent.[34]

Americans clearly desire natural preservation, but there is not enough money in it to ensure that entrepreneurs working at preservation will earn a profit. As ecological change is very slow, sometimes taking decades to become apparent, management requires a very long time horizon—far longer than is usually perceived by the heads of private corporations, who usually are concerned with the next quarterly statement.

The challenge of preservation is unique—quite different from the goals of managing lands for farming or hunting, for example— and offers an intrinsic value which is difficult to measure. Many of the mistakes made by the Park Service are consequences of the tendency of that agency to be entirely *too* entrepreneurial, when it promotes recreation at the expense of preservation. And while wilderness endowment boards may not be entrepreneurial, putting the parks in their control would further feudalize the system, making it even more difficult to keep management accountable to the public.

Saving our parks, therefore, cannot be achieved by infusions of money alone or by privatization of the service. Rather, this goal can only be met by a reform of the National Park Service. Such a process would entail, I believe, the following twelve steps:

1. The President should appoint a commission to review the purposes, policies, and performance of the Park Service. This commission should be aided by the National Science Foundation and should include prominent scholars who have not been associated with the Park Service in the past.
2. The Park Service should make clear its own interpretation of its Congressional mandate—something which, surprisingly, it has never done.[35] In particular, the agency would address the relationship of its role in preservation to its responsibilities to provide parks for public use. It should replace its present park management with a more holistic one, which would preserve the parks while keeping them as places designed for appropriate public use. The Park Service should make clear that it is

more than a custodian, that its responsibilities include restoration and nurturative management and, as a corollary, sponsorship of intensive research.

3. The present division between natural and historical areas should be abolished. Although the purpose of the parks set forth in the Leopold Report—preserving vignettes of primitive America—is a fine one, it implies that all parks—even socalled natural areas—are historic places, established to preserve various features of our natural and cultural history. And just as no park is purely a natural area without any historical relevance, so all historical parks have some natural values.

4. To preserve the past, our parks should reflect, where appropriate, the role Native Americans played in these lands. Unfortunately, both the Leopold Committee and the Park Service almost entirely overlooked the contributions of the Indians to the early culture and ecology of primitive ecosystems. Where appropriate, the Indian heritage of our parks should be emphasized, and Indian techniques for modifying the environment should be considered by resource managers.

5. To give the Park Service more independence and to make it more accountable to the American people, it should be made an independent agency, apart from the Department of the Interior, which is almost entirely devoted to the development of natural resources. As long as the Park Service—with its mission of preservation—remains in Interior, it will be treated as a stepchild.

6. To reduce the power of special-interest groups and enhance the professional stature of the agency, anyone appointed director of the Natural Park Service should be subject to Senate confirmation, and his or her credentials clearly established.

7. To make management more sensitive to the contributions of science, the ranger corps should be completely professionalized. That is, the Service should create a cadre of resource managers who would be required to have graduate training in relevant academic disciplines to advance through the ranks. Resource management should be the primary career ladder that reaches all the way to the top. Qualifications for superintendents and their staffs should be similar to those we expect of college administrators or museum directors.

8. Scientific research should be made the core of all Park Service

activities. The number of researchers within the Service should be multiplied at least five-fold. A quarter of all researchers should be recruited from university ranks. The Park Service should establish clear guidelines for public access to the fruits of government research.

9. To de-Balkanize the Service and ensure accountability for resource managers in the individual parks, the chief of research in each national park should report not to his superintendent, but to the district chief scientist, who in turn should report to Washington. Resource management should be a separate division within each national park, rather than — as today — an activity supervised by the park's chief ranger; and the chief of this division should be responsible to a regional chief of resource management.

10. To ensure national coordination of resource management, an Office of Ecological Studies should be established in Washington to direct research throughout the park system and serve as a research facility for natural and historic preservation.

11. To promote field research and provide the best possible information for park resource managers, multidisciplinary research stations should be established in or near the major parks, or those in critical condition.

12. To minimize co-optation of independent scholars doing research in the parks, the process by which these people participate in mission-oriented research should be revised. Most needed are long-term studies collecting baseline data on the state of the parks, and equally long-term, basic research on ecosystem dynamics. Direction and support for such research should not come from short-term contracts awarded by each park or by Service contracting agents as it does now, but from a funding program administered by peer-review panels composed of nongovernmental scholars, following the pattern of the National Endowment for the Humanities and National Science Foundation.

Are these reforms likely to happen?

To answer this, I think we must take an ecological point of view. Ecology is the science of interconnections. Similarly, the problems of the parks are interrelated, a consequence of faulty policies and institutional inadequacies. But they have a cultural origin as well.

Our parks represent our past. Very much a part of that past was the frontier. This magic place, as Frederick Jackson Turner noted in 1920, bound the nation together. It represented the qualities we have come to think set us apart as a people: youth, innocence, energy, possibility, opportunity, and equality.[36]

Yet to many Americans the frontier also represents the negative side of the American character: the desire to conquer nature. For three hundred years, the land lying beyond the frontier was a place which our forefathers believed was their manifest destiny to control. This attitude eventually led to the disappearance of wilderness altogether. The lesson many Americans draw from this sad history is that if we want wild lands, we must keep people out of them.

Both the positive and negative sides of the frontier mentality are still very much alive in the American psyche. We want to think there are still places where wilderness exists, where the land, as artist Charley Russell once put it, belongs "only to God." And we still fear that any human intervention in such places, no matter how well intended, can only lead to harm.

The parks—particularly those classified as natural zones—have been represented to us as such places. They are, we have been told by the Park Service, the last vestiges of wild America. And while the Leopold Committee might have called them "illusions of primitive America," to the Park Service and to many of our countrymen they are not illusions, but the real thing.

In suggesting our parks remain intact ecosystems, therefore, Park Service policy has reinforced this peculiarly American myth. Saving our parks, however, can only begin when we reject this myth. We must embark on a program of restoration that treats our parks as places to be nurtured, and we must recognize that the frontier—and true wilderness—is gone forever.

Hence, preservation involves two paradoxes: First, we can restore and sustain the appearance of undisturbed wilderness only by admitting that undisturbed wilderness no longer exists. And second, recreating the illusion that parks remain untouched by modern civilization can be done only by using all that technology and science have to offer.

I, for one, believe Americans will experience these epiphanies, at least eventually. Older societies have reached similar understandings at some point of their maturation. Unfortunately, too often they did not realize their wilderness was gone until they had destroyed much of the wildlife in it. Let's not make the same mistake.

Paradise Valley, Montana
March 1, 1987

Notes and Sources

1. THE STATE OF NATURE IN YELLOWSTONE

1. The town park was built in 1903. See Hiram M. Chittendon and John Millis, *Annual Report of the Chief of Engineers, Construction, Repair, and Maintenance of Roads and Bridges in Yellowstone National Park* (Washington, D.C.: U.S. Government Printing Office, 1904), Appendix FFF.
2. Theodore Roosevelt, "Wilderness Reserves," reprinted in Paul Schullery, ed., *Old Yellowstone Days* (Boulder, Colo.: Associated University Press, 1979), pp. 185–205.
3. Ibid.
4. Ibid., p. 189.
5. Osborne Russell, *Journal of a Trapper* (Lincoln: University of Nebraska Press, 1955), pp. 27–28.
6. Aubrey L. Haines, *The Yellowstone Story* (Yellowstone: Yellowstone Library and Museum Association in cooperation with Colorado Associated University Press, 1977), p. xix.
7. Alan W. Cundall and Herbert T. Lystrup, *Hamilton's Guide to Yellowstone* (West Yellowstone, Montana: Hamilton Stores, 1981), p. 11; Bryan Harry and Willard E. Dilley, *Wildlife of Yellowstone and Grand Teton National Parks* (Salt Lake City: Wheelwright Press, 1972), p. 3; Glen F. Cole, "Nature and Man in Yellowstone National Park," *Information Paper No. 28,* June 1975, Yellowstone National Park; U.S. Department of the Interior, National Park Service, "Population Status of Large Mammal Species in Yellowstone National Park," June 1982, Yellowstone Park files.
8. The Yellowstone Park Act (U.S. Statutes at Large, vol. 17, ch. 24, pp. 32–33), signed into law by President Ulysses S. Grant, March 1, 1872.
9. John Muir, *Our National Parks* (Madison: University of Wisconsin Press, 1981), p. 39.
10. Yellowstone Park Act, second paragraph.
11. The Lacey Act (U.S. Statutes at Large, vol. 28, p. 73).
12. The National Park Service Act (U.S. Statutes at Large, vol. 39, p. 535), signed into law by President Wilson, August 25, 1916.
13. William J. Robbins, *A Report by the Advisory Committee to the National Park Service on Research of the National Academy of Sciences–National Research Council* (Washington, D.C.: National Academy of Sciences–National Research Council, August 1, 1963), p. 32; interview with A. Starker Leopold, 1982. Some Americans credit this country with having first conceived the "national park concept." (See Stewart Udall, *The Quiet Crisis* [New York: Holt, Rinehart & Winston, 1963]). But saying so requires forgetting that the preservation of game began at least 4,000 years ago. For a good history of the "national park concept," see Alistair Graham, *The Gardeners of Eden* (London: George Allen & Unwin, 1973), esp. ch. 2, "In the Pride of his Grease."

2. JONAS AND THE BEAVER

1. Based on interviews with David de L. Condon and Robert J. Jonas.
2. Robert J. Jonas, "A Population and Ecological Study of the Beaver (*Castor Canadensis*) of Yellowstone National Park," M.S. thesis, University of Idaho, 1955; Robert J. Jonas, "Beaver," *Naturalist*, vol. 10, no. 2, 1959, pp. 60–61.
3. Aubrey L. Haines, *The Yellowstone Story* (Yellowstone: Yellowstone Library and Museum Association, 1977), ch. 3, "In Pursuit of Peltry"; Osborne Russell, *Journal of a Trapper* (Lincoln: University of Nebraska Press, 1977); Walter W. deLacy, "A Trip up the South Snake River in 1863," *Contributions to the Historical Society of Montana*, vol. 1 (Helena: Rocky Mountain, 1876), p. 119; Earl of Dunraven, *The Great Divide: Travels in the Upper Yellowstone in the Summer of 1874* (Lincoln: University of Nebraska Press, 1967), p. 72; John Muir, *Our National Parks* (Madison: University of Wisconsin Press, 1981), p. 37; Milton P. Skinner, "The Predatory and Fur-Bearing Animals of the Yellowstone National Park," *Roosevelt Wild Life Bulletin*, vol. 4, no. 2, June 1927, p. 205.
4. Robert J. Jonas, "Northern Yellowstone's Changing Water Conditions and Their Effect upon Beaver," *Yellowstone Nature Notes*, vol. 30, no. 1, January–February, 1956; M. Call, "A Study of the Pole Mountain Beaver and Their Relation to the Brook Trout Fishery," M.S. thesis, University of Wyoming; Thomas C. Collins, "Stream Flow Effects on Beaver Populations in Grand Teton National Park," undated, Department of Zoology, University of Wyoming; Durward L. Allen, *Wolves of Minong: Their Vital Role in a Wild Community* (Boston: Houghton Mifflin, 1979), pp. 253–257; L. David Mech, *The Wolf: The Ecology and Behavior of an Endangered Species* (Garden City, N.Y.: Natural History Press, 1970), pp. 172–175.
5. Jack R. Nelson and Thomas A. Leege, "Nutritional Requirements and Food Habits," *Elk of North America: Ecology and Management*, Jack Ward Thomas and Dale E. Toweill, eds. (Harrisburg, Pa.: Stackpole, 1982), pp. 323–367; N. T. Hobbs et al., "Composition and Quality of Elk Diets During Winter and Summer: A Preliminary Analysis," *North American Elk: Ecology, Behavior and Management*, Mark S. Boyce and Larry D. Hayden-Wing, eds. (Laramie: University of Wyoming Press, 1979), pp. 47–53; C. E. Olmsted, "The Ecology of Aspen with Reference to Utilization by Large Herbivores in Rocky Mountain National Park," *North American Elk*, pp. 89–97.
6. Other shrubs, such as serviceberry and chokecherry, also provide nitrogen, but they are not as readily available to ungulates in winter as aspen and willow.
7. W. J. Rudersdorf, "The Coactions of Beaver and Moose on a Joint Food Supply in the Buffalo River Meadows and Surrounding Area in Jackson Hole, Wyoming," M.S. thesis, Utah State University, 1952.
8. Jonas, "A Population and Ecological Study"; Edward R. Warren, "A Study of the Beaver in the Yancey Region of Yellowstone National Park," *Roosevelt Wildlife Annals*, vol. 1, no. 1, October 1926, pp. 13–191; Edward R. Warren, "The Life of the Yellowstone Beaver," *Roosevelt Wild Life Bulletin*, vol. 1, no. 2, August 1922.

3. THE PERILS OF PLAYING GOD

1. Hiram Martin Chittenden, *The Yellowstone National Park* (Norman: University of Oklahoma Press, 1961), p. 17.
2. Elers Koch, "Big Game in Montana from Early Historical Records," *Journal of Wildlife Management*, vol. 5, no. 4, October 1941, pp. 357–370.
3. Chittenden, *Yellowstone*, pp. 16–18.
4. For a good summary of this historical record, see Milton P. Skinner, "The Predatory and Fur-Bearing Animals of the Yellowstone National Park," *Roosevelt Wild Life Bulletin*, vol. 4, no. 2, June 1927. especially pp. 165–179.
5. Aubrey L. Haines, *The Yellowstone Story* (Yellowstone: Yellowstone Library and Museum Association, 1977), vol. 1, p. 87.

6. Nathaniel P. Langford, *The Discovery of Yellowstone Park* (Lincoln: University of Nebraska Press, 1971), especially pp. 72, 83–84, 98, 104, 116, 120; F. V. Hayden, *Preliminary Report of the United States Geological Survey of Montana and Portions of Adjacent Territories* (Washington, D.C.: U.S. Government Printing Office, 1872), p. 131; J. W. Barlow, *Letter from the Secretary of War, Accompanying an Engineer Report of a Reconnaissance of the Yellowstone River in 1871*, Senate Ex. Doc. No. 66, 42nd Congress, 2nd Session, April 18, 1872, pp. 11, 14, 35, 37, 40–41; William A. Jones, *A Report upon the Reconnaissance of Northwestern Wyoming, Made in the Summer of 1873 by Captain William A. Jones, Corps of Engineers* (Washington, D.C.: U.S. Government Printing Office, 1873), pp. 19, 27.

7. Earl of Dunraven, *The Great Divide: Travels in the Upper Yellowstone in the Summer of 1874* (Lincoln: University of Nebraska Press, 1967), pp., 6, 10, 252–253, 296, 336, 346–348, 373; William Ludlow, *Report of a Reconnaissance from Carroll, Montana Territory, on the Upper Missouri to the Yellowstone National Park, and Return, Made in the Summer of 1875* (Washington, D.C.: U.S. Government Printing Office, 1876), pp. 30, 63, 69–71; Theodore B. Comstock, "The Yellowstone National Park," *American Naturalist*, vol. 8, 1874, pp. 65–79, 155–166.

8. Dunraven, *The Great Divide*, p. 5.

9. W. E. Strong, *A Trip to the Yellowstone Park in July, August and September, 1875* (Washington, D.C.: Government Printing Office, 1876), pp. 80, 92–93; Haines, *The Yellowstone Story*, vol. 1, pp. 204–205.

10. Robert Easton and Mackenzie Brown, *Lord of Beasts: The Saga of Buffalo Jones* (Tucson: University of Arizona Press, 1972), p. 10; Curtis K. Skinner et al., "History of the Bison in Yellowstone Park," Yellowstone National Park Library, 1942; Paul Schullery, "Another Look at Buffalo Jones in Yellowstone Park," Yellowstone Library, 1976.

11. Department of the Interior, *Report upon the Yellowstone National Park to the Secretary of the Interior for the Year 1877* (Washington, D.C.: U.S. Government Printing Office, 1877), p. 842; anonymous, "The Destruction of Elk for Their Teeth," undated, Yellowstone National Park archives; Philip H. Sheridan, J. F. Gregory, and W. W. Forwood, *Report of an Exploration of Wyoming, Idaho, and Montana, in August and September, 1882* (Washington, D.C.: U.S. Government Printing Office, 1882).

12. Margaret M. Meagher, *The Bison of Yellowstone Park*, National Park Service Scientific Monograph Series, no. 1 (Washington, D.C.: U.S. Government Printing Office, 1973).

13. Aubrey L. Haines, *The Yellowstone Story* (Yellowstone: Yellowstone Library and Museum Association, 1977), vol. 1, p. 269.

14. H. Duane Hampton, *How the U.S. Cavalry Saved Our National Parks* (Bloomington: Indiana University Press, 1971).

15. Milton Skinner, "Predatory and Fur-Bearing Animals"; *Report of the Acting Superintendent of the Yellowstone National Park, 1899* (Washington, D.C.: U.S. Government Printing Office, 1900) (See also Acting Superintendents' reports for the years 1903, 1904, and 1914); Haines, *The Yellowstone Story*, vol. 2, p. 83; letter to Secretary of Interior from Chester Lindsley, Acting Superintendent of Yellowstone Park, December 28, 1916.

16. *Acting Superintendent's Report*, 1890.

17. *Acting Superintendent's Reports*, 1891, 1894, 1895, 1899; Schullery, "Old Yellowstone Days," p. 194.

18. *Acting Superintendent's Reports*, 1912, 1914; Douglas B. Houston, *The Northern Yellowstone Elk: Ecology and Management* (New York: Macmillan, 1982), pp. 12–15.

19. George Shiras, 3rd, "Wild Animals That Took Their Own Pictures by Day and by Night," *National Geographic*, vol. 24, no. 7, July 1913; E. W. Nelson, "Description of a New Subspecies of Moose from Wyoming," *Proceedings of the Biological Society of Washington*, vol. 27, April 25, 1914, pp. 71–74; Houston, *Northern Yellowstone Elk*, p. 158; Douglas B. Houston, "The Shiras Moose in Jackson Hole, Wyoming,"

Technical Bulletin No. 1, Grand Teton National History Association, 1968; *Acting Superintendent's Report,* 1895.

20. John D. Varley, "A History of Fish Stocking Activities in Yellowstone National Park Between 1881 and 1980," Yellowstone Information Paper No. 35, 1980.
21. Ibid.; B. B. Arnold, "A Ninety-seven Year History of Fishery Activities in Yellowstone National Park, Wyoming," (Yellowstone: Yellowstone Library March 19, 1967); F. Sheldon Dart, "History of the Fish Planting in Yellowstone National Park," (Yellowstone: Yellowstone Library, 1936); R. J. Fromm, "An Open History of Fish and Fish Planting in Yellowstone National Park," (Yellowstone: Yellowstone Library, 1926); Hugh M. Smith and William C. Kendall, "Fishes of the Yellowstone National Park," Bureau of Fisheries Document No. 504 (Washington, D.C.: U.S. Government Printing Office, 1921).
22. Varley, "History of Fish Stocking," pp. 1–4.
23. In 1901 Congress appropriated $15,000 for saving the Yellowstone bison. See Curtis K. Skinner et al., "History of the Bison"; Schullery, "Another Look at Buffalo Jones"; Haines, *Yellowstone Story,* vol. 2, pp. 67–74.
24. Curtis K. Skinner et al., "History of the Bison"; Schullery, "Another Look at Buffalo Jones."
25. *Acting Superintendent's Report,* 1912.
26. Department of the Interior, National Park Service, *Hearings Before a Subcommittee of the Committee on Appropriations, United States Senate, 90th Congress, First Session, On Control of Elk Population, Yellowstone National Park* (Washington, D.C.: U.S. Government Printing Office, 1967), p. 13.
27. See Reports of Acting Superintendents for this period; also, cf. Vernon Bailey, "Plan for the Control and Management of the Elk Herds of the Yellowstone Region of Wyoming, Montana, & Idaho," U.S. Biological Survey and Forest Service, Yellowstone Library, 1916.
28. Henry S. Graves and E. W. Nelson, *Our National Elk Herds,* U.S. Department of Agriculture Circular 51 (Washington, D.C.: U.S. Government Printing Office, 1919).
29. Ibid.
30. Ibid.
31. See *Superintendent's Monthly Reports,* September 1919 through June 1920; *Livingston Enterprise* articles for October 23, 24, 25, November 9, 11, 12, 25, 30, 1919; Edgar H. Fletcher, *Annual Meteorological Summary, with Comparative Data, Yellowstone Park, Wyoming* (Salt Lake City: U.S. Department of Agriculture, Weather Bureau, 1928).
32. *Livingston Enterprise* articles for December 3, 9, 13, 14, 19, 31, 1919, March 24, 31, April 14, 18, May 12, 14, 18, June 10, 1920.
33. See *Superintendent's Monthly Reports* for January, March, April, 1920; Horace Albright, *Superintendent's Annual Report,* 1920.
34. Stephen D. Mather, Director of the Park Service, to Chester A. Lindsley, Acting Superintendent of Yellowstone, February 13, 1918.
35. Horace M. Albright, Superintendent of Yellowstone, to Arno Cammerer, Director of the Park Service, October 18, 1937; "Antelope Soon Extinct," *In the Open,* January 1914; William J. Barmore, Jr., "Population Characteristics, Distribution, and Habitat Relationships of Six Ungulate Species on Winter Range in Yellowstone National Park," Yellowstone Library, 1980; Milton P. Skinner, "White-tailed Deer Formerly in the Yellowstone Park," *Journal of Mammalogy,* vol. 10, no. 2, May 1929, pp. 101–115.
36. Barmore, *Population Characteristics,* Tables 103 and 109; Edmund Heller, "The Big Game Animals of Yellowstone National Park," *Roosevelt Wild Life Bulletin,* vol. 2, no. 4, February 1925.
37. Lt. Col. Lloyd Brett, Acting Superintendent of Yellowstone, to W. T. Hornaday, Director, New York Zoological Society, July 29, 1912; Milton Skinner, "White-tailed Deer."
38. W. M. Rush, *Northern Yellowstone Elk Study,* 1932, Montana Fish and Game Com-

mission. Conducted under the auspices of the state of Montana, Rush's study was at first privately financed. Later expenses were borne jointly by the U.S. Forest Service, the Park Service, and the Montana Department of Fish and Game.

39. Ibid., pp. 64–65.
40. Ibid.
41. George M. Wright, Joseph S. Dixon, and Ben H. Thompson, *Fauna of the National Parks of the United States: A Preliminary Survey of Faunal Relations in National Parks, Fauna Series No. 1*, May 1932 (Washington, D.C.: U.S. Government Printing Office, 1933); George M. Wright and Ben H. Thompson, *Fauna of the National Parks of the United States: Wildlife Management in the National Parks, Fauna Series No. 2*, July 1924 (Washington, D.C.: U.S. Government Printing Office, 1935).
42. Wright and Dixon, *Fauna Series No. 1*, p. 85.
43. Ibid., p. 124.
44. Adolph Murie, *Ecology of the Coyote in Yellowstone, Fauna Series No. 4*, 1940 (Washington, D.C.: U.S. Government Printing Office, 1940), p. ix.
45. Ibid., pp. 57, 147.
46. James B. Trefethen, "The Terrible Lesson of the Kaibab," *National Wildlife*, vol. 5, no. 4, 1977; D. I. Rasmussen, "Biotic Communities of Kaibab Plateau, Arizona," *Ecological Monographs*, vol. 11, no. 3. Cf. C. John Burk, "The Kaibab Deer Incident: A Long-Persisting Myth," *Bioscience*, vol. 23, no. 2.
47. Trefethen, "Terrible Lesson."
48. Ibid.
49. No one mentioned that another predator had been eliminated as well: Indians had been evicted from the Kaibab not long before it had been made a national wildlife refuge. See I. P. Kelley, "Southern Paiute Ethnography," University of Utah Anthropological Paper no. 69, 1964.
50. Rasmussen, "Biotic Communities."
51. Trefethen, "The Terrible Lesson."
52. Rasmussen, "Biotic Communities," pp. 268-269.
53. Aldo Leopold, *A Sand County Almanac* (Oxford: Oxford University Press, 1949); J. H. Feil and Harvey Gillette, "The Fox, Wolf and Deer," quoted in Susan L. Flader, *Thinking Like a Mountain* (Lincoln: University of Nebraska Press, 1974).
54. Flader, *Thinking Like a Mountain*, p. 2.
55. Ibid., p. 209.
56. Ibid., p. 184. Although Leopold used the word "irruption," many today use "eruption."
57. Aldo Leopold, Lyle K. Sowls, and David L. Spencer, "A Survey of Over-Populated Deer Ranges in the United States," *Journal of Wildlife Management*, vol. 11, no. 2, April 1947, pp. 162–177; letter from Aldo Leopold to Charles J. Kraebel, Superintendent of Glacier National Park, January 18, 1927.
58. U.S. Department of the Interior, National Park Service, "Wildlife Management Background Information," Yellowstone Library, 1964; Department of the Interior, National Park Service, "Management Plan for Northern Elk Herd," Yellowstone Library, 1958.
59. National Park Service, "Wildlife Management Background Information."
60. Rudolf L. Grimm, "Northern Yellowstone Winter Range Studies, *Journal of Wildlife Management*, vol. 3, no. 4, October 1939, pp. 295–306; Victor Cahalane, "Wildlife Surpluses in National Parks," *Transactions of Sixth Annual North American Wildlife Conference*, vol. 6, 1941; Walter H. Kittams, "Management of the Northern Yellowstone Range and Elk," Yellowstone Library, undated; Curtis K. Skinner, "Problems of Surplus Elk in Yellowstone Park," *Wyoming Wild Life*, vol. 14, no. 6, 1950; Walter H. Kittams, "Future of the Yellowstone Wapiti," *Naturalist*, vol. 10, no. 2, 1956.
61. Kittams, "Future of the Yellowstone Wapiti," p. 34.
62. Ibid., p. 31.
63. Ibid., p. 35.

64. Ibid., pp. 34–35.
65. Paul E. Packer, "Soil Stability Requirements for the Gallatin Elk Winter Range," *Journal of Wildlife Management,* vol. 27, no. 3, July 1963; Paul E. Packer, "Effects of Ground Cover, Soil Bulk Density, and Summer Rainstorm Intensity on Soil Stability in the Gallatin Winter Range," Yellowstone Library, 1963; Paul E. Packer to W. W. Dresskell, August 25, 1960.
66. Jonas, "A Population and Ecological Study of the Beaver (*Castor Canadensis*) of Yellowstone National Park," M.S. thesis, University of Idaho, 1955, p. 166.
67. Department of the Interior, "Wildlife Management Background Information," Record of Elk Reductions, 1964; Robert E. Howe, "Yellowstone River and Gallatin River Elk Herds," Yellowstone Library, July 17, 1961.
68. Department of the Interior, National Park Service, "Management Plan for Northern Elk Herd, Yellowstone National Park," Yellowstone Library, November 21, 1958.
69. Ibid.
70. U.S. Department of the Interior, National Park Service press release, November 14, 1961; Department of the Interior, National Park Service, "Long-Range Management Plan for the Northern Yellowstone Elk Herd," November 1961; Department of the Interior, National Park Service, press release, November 26, 1961.
71. Robert E. Howe, "Final Reduction Report, 1961–1962 Northern Yellowstone Elk Herd," Yellowstone Library, May 17, 1962.
72. National Park Service press release, January 15, 1962; W. Leslie Pengelly et al., "Statement Regarding the Management of the Northern Yellowstone Elk Herd," Yellowstone Library, 1961; John J. Craighead et al., "Elk Migration in and near Yellowstone National Park," *Wildlife Monographs,* no. 29, August 1972.

4. KILLING ANIMALS TO SAVE THEM

1. Robert E. Howe, "Final Reduction Report, 1961–1962 Northern Yellowstone Elk Herd," Yellowstone Library, May 17, 1962.
2. Ibid.
3. Ibid.
4. Ibid.
5. Ibid.
6. From an eyewitness account.
7. Arnold Olsen, "Yellowstone's Great Elk Slaughter," *Sports Afield,* October 1962.
8. See Foreword by Garrison in Howe, "Final Reduction Report."
9. A. S. Leopold et al., "Wildlife Management in the National Parks," Report of the Advisory Board on Wildlife Management to Secretary of Interior Udall. March 4, 1963.
10. Interview.
11. Leopold, "Wildlife Management."
12. Ibid.
13. Ibid.
14. Ibid.
15. Ibid.
16. Interview.
17. Neil J. Reid, "Ecosystem Management in the National Parks," *Transactions of the 33rd North American Wildlife and Natural Resources Conference,* March 11–13, 1968 (Washington, D.C.: Wildlife Management Institute, 1968); Secretary of Interior to Director, National Park Service, July 10, 1964.
18. Leopold, "Wildlife Management."
19. Glen Cole, "Wildlife Management Plan for Yellowstone National Park," Yellowstone Library, August 1967; Robert E. Howe and William B. Barmore, "Research and Investigations Relative to Proposed Artificial Revegetation of the Northern Yellowstone Winter Range," Yellowstone Library, November 1962; George B. Harzog, Jr., "Management Program, Northern Yellowstone Elk Herd," Yellowstone Library,

March 1, 1967; According to Maurice Hornocker, then Team Leader of the Idaho Cooperative Wildlife Research Unit, University of Idaho, he discussed reintroduction of the mountain lion to Yellowstone with Superintendent John S. McLaughlin in 1966.

20. William J. Barmore, "Pronghorn–Mule Deer-Range Relationships on the Northern Yellowstone Winter Range," Yellowstone Library, undated; John S. McLaughlin, "1965–1966 Pronghorn (Antelope) and Habitat Management Plan for Yellowstone National Park," Yellowstone Library, September 1965; Barmore, "Population Characteristics of Six Ungulate Species on Winter Range in Yellowstone National Park," Yellowstone Library, 1980, vol. 3, Table 105.

21. Bart W. O'Gara and Kenneth R. Greer, "Food Habits in Relation to Physical Condition in Two Populations of Pronghorns," *Proceedings,* 4th Antelope States Workshop, 1970; Bart W. O'Gara, "A Study of the Reproductive Cycle of the Female Pronghorn," Ph.D. thesis, University of Montana, 1968.

22. Jack K. Anderson, Superintendent of Yellowstone, to Wynn Freeman, Montana Department of Fish and Game, January 15, 1974.

23. Hartzog, "Management Program, Northern Elk Herd," p. 16.

24. Ibid.; Glenn L. Erickson, "Northern Yellowstone Elk Herd Numbers and Removals, 1922–1981," Table 13, February 21, 1984, Montana Department of Fish, Wildlife and Parks; Meagher, *The Bison of Yellowstone,* p. 147.

25. Interview.

26. Department of the Interior, National Park Service, Hearings Before a Subcommittee of the Committee on Appropriations, United States Senate, 90th Congress, First Session, On Control of Elk Population, Yellowstone National Park (Washington, D.C.: U.S. Government Printing Office, 1967), National Park Service, "Distribution of Live Elk Shipments from Yellowstone National Park," Yellowstone Library, 1967; National Park Service, "Summary of a Public Meeting to Discuss Management of the Northern Yellowstone Elk Herd," Yellowstone Library, 1967.

5. AN ENVIRONMENTAL IDEAL AND THE BIOLOGY OF DESPERATION

1. Interview with Cole; National Park Service, "Natural Control of Elk," Yellowstone Library, December 5, 1967; Glen F. Cole, *The Elk of Grand Teton and Southern Yellowstone National Park,* Research Report GRTE N-1 (Yellowstone: National Park Service, 1969).

2. National Park Service, "Natural Control of Elk"; National Park Service, "Information Paper — Northern Yellowstone Elk Herd," Yellowstone Library, September 17, 1968; National Park Service and Montana Department of Fish and Game, "Summary, Interagency Elk Management Meeting, Northern Yellowstone Herd," Bozeman, Montana, June 15, 1967, Yellowstone Park administrative files (obtained through the Freedom of Information Act); Superintendent of Yellowstone to NPS Regional Director, September 25, 1968, Yellowstone administrative files (obtained through the Freedom of Information Act); Glen F. Cole, "Elk and the Yellowstone Ecosystem," Yellowstone Library, 1969; National Park Service, "Administrative Policy for the Management of Ungulates," Yellowstone Library, September 1, 1971.

3. Glen F. Cole, "Where Do We Go from Here?" *Montana Deer Management,* Montana Fish and Game Department Information Bulletin no. 1, January 1958.

4. Ronald F. Lee, *Family Tree of the National Park System* (Philadelphia: Eastern National Park & Monument Association, 1972), p. 38; Udall to Harzog, July 10, 1964.

5. National Park Service, *Administrative Policies for Natural Areas of the National Park System* (Washington, D.C.: U.S. Government Printing Office, 1968), p. 17.

6. Ibid. p. 16.

7. Ibid., p. 25.

8. Robert M. Linn, "The Ecosystem Concept and the National Parks," in *Proceedings of the Meeting of Research Scientists and Management Biologists of the National Park Service*, pp. 14–15; Cole, "Elk and the Yellowstone Ecosystem"; Douglas B. Houston, "Ecosystems of National Parks," *Science*, vol. 172, May 14, 1971.
9. See Chapters 11 and 12.
10. Leopold, "Wildlife Management."
11. Linn, "The Ecosystem Concept"; *Management of Natural Areas;* National Park Service, *Management Policies* (Washington, D.C.: U.S. Government Printing Office, 1978), p. iv–1.
12. Lemuel A. Garrison, *The Making of a Ranger: Forty Years with the National Parks* (Chicago: Institute of the American West, 1983), p. 286.
13. Lee, *Family Tree*, pp. 64–89.
14. Among environmentalists brought into government at this time were A. Starker Leopold, Ansel Adams, Sigurd Olson, and Nathaniel Reed. See Chapter 18.
15. Rachel Carson, *Silent Spring* (Boston: Houghton Mifflin, 1962); Barry Commoner, *The Closing Circle* (New York: Alfred A. Knopf, 1971), p. 37; Paul R. Ehrlich, *The Population Bomb* (New York: Sierra Club / Ballantine, 1968).
16. Carson, *Silent Spring*, p. 126; Ellsworth Hastings et al., "Repopulation by Aquatic Insects in Streams Sprayed with DDT," *Annals of the Entomological Society of America*, vol. 54, no. 3, May 1961, pp. 436–437; Ellsworth Hastings et al., "Distribution and Repopulation of Terrestrial Insects in Sprayed Areas," *Annals of the Entomological Society of America*, vol. 54, no. 3, May 1961, pp. 433–435. David de L. Condon took motion pictures of this ecological disaster, but somehow the film disappeared shortly afterward.
17. Commoner, *The Closing Circle*, p. 37.
18. Lee, *Family Tree*, p. 69.
19. Diseases such as smallpox reached the West at least by the eighteenth century, if not earlier; the "backpacker's disease," *giardia lamblia*, was found early throughout the West (Owen R. Willilams, *Giardia and the Water Borne Transmission of Giardia: A General Review*, Forest Service Monograph, 1981).
20. *Management of Natural Areas*, Appendixes I and J; David Wallechinsky and Irving Wallace, *The People's Almanac* (Garden City: Doubleday, 1975); Dee Brown, *Bury My Heart at Wounded Knee* (New York: Holt, Rinehart and Winston, 1970); Vine Deloria, Jr., *Custer Died for Your Sins* (New York: Macmillan, 1969).
21. National Park Service, *Master Plan for Yellowstone National Park* (Denver: Midwest Regional Office, National Park Service, 1973), p. 2.
22. Ibid., p. 24.
23. Glen F. Cole, "Nature and Man in Yellowstone National Park," Information Paper No. 28, Yellowstone Library, June 1975; Glen F. Cole, "An Ecological Rationale for the Natural or Artificial Regulation of Native Ungulates in Parks," *Transactions of the Thirty-sixth North American Wildlife and Natural Resources Conference*, March 7–10, 1971 (Washington, D.C.: Wildlife Management Institute, 1971).
24. *Management of Natural Areas*, pp. 17–18.
25. See Chapters 6, 7, 10, and 12.
26. National Park Service and Montana Department of Fish and Game, "Memorandum of Understanding on Northern Yellowstone Elk Herd Management," August 27, 1968.

6. A SOLUTION TO THE ELK PROBLEM

1. A. L. Olson, "The 'Firing Line' in the Management of the Northern Elk Herd," *University of Idaho Bulletin*, 1939; Russell Chatham, "Shooting Elk in a Barrel," *Sports Illustrated*, vol. 44(5) 1976, pp. 62–63.
2. Chatham, "Shooting Elk."
3. Eyewitness report.
4. Eyewitness report.

5. W. B. Shore, "An Incident of the Annual Elk Migration," *Outdoor Life*, February 1912; "Game Notes" (anonymous), *Outdoor Life*, February 1912.
6. *Billings Gazette*, October 6, 1972.
7. Jon Swenson to Charles Kay, August 3, 1984; Swenson to LeRoy Ellig, February 21, 1984. From 1981 through 1984, 5,730 elk were killed in the special hunt; between 1962 and 1968, 1,709 elk were shot by rangers.
8. *Livingston Enterprise*, December 8, 1983 and February 4, 1984. See also Chapter 7.
9. Anderson to Harzog, October 23, 1967 (obtained through the Freedom of Information Act). Harzog's handwritten comment: "Looks good to me — you run it with Glen [Cole] and Starker [Leopold]."
10. Barmore to Cole, March 13, 1968.
11. Interview with Barmore, 1983.
12. Interviews with LeRoy Ellig and Arnold Foss of Montana Department of Fish, Wildlife and Parks; cf. Ellig's comments, "Minutes of Northern Yellowstone Elk Herd Management Meeting," Mammoth, Wyoming, September 27, 1968 (obtained through the Freedom of Information Act).
13. For the story of this "comeback," see Chapter 10.
14. The winter census of 1971 reported 7,281 elk (Glenn L. Erickson, "The Northern Yellowstone Elk Herd — A Conflict of Policies," address to the Western Association of Fish and Wildlife Agencies, July 13–17, 1981, Table 3).
15. The words "eruption" and "irruption" are synonymous. See Graeme Caughley, "Eruption of Ungulate Populations with Emphasis on Himalayan Thar in New Zealand," *Ecology*, vol. 51, no. 1, Winter 1970, pp. 53–72; Graeme Caughley, *Analysis of Vertebrate Populations* (New York: John Wiley, 1977); Graeme Caughley, "What Is This Thing Called Carrying Capacity?" *Northern American Elk: Ecology, Behavior and Management*, Mark S. Boyce and Larry D. Hayden-Wing, eds. (Laramie: University of Wyoming Press, 1977), pp. 2–9.
16. M. M. Douglas, "The Warning Whistle of Thar, *Journal of the Tussock Grasslands*, no. 35, May 1977.
17. Caughley, "Eruption of Ungulate Populations."
18. Ibid.
19. Ibid.
20. Douglas, "Warning Whistle."
21. A. R. E. Sinclair, *The African Buffalo: A Study of Resource Limitation of Populations* (Chicago: University of Chicago Press, 1977); A. R. E. Sinclair and M. Norton-Griffiths, eds., *Serengeti: Dynamics of an Ecosystem* (Chicago: University of Chicago Press, 1979).
22. Sinclair, *African Buffalo*, pp. 258–259; cf. George B. Schaller, *The Serengeti Lion: A Study of Predator-Prey Relations* (Chicago: University of Chicago Press, 1972).
23. Sinclair, *African Buffalo*, pp. 261 ff.
24. In Mount Cook National Park in New Zealand, 20,800 thar were shot between 1956 and 1977; between 1974 and 1977, 25,000 thar were shot for export game meat. See Douglas, "Warning Whistle."
25. Paul S. Martin and H. E. Wright, Jr., *Pleistocene Extinctions: The Search for a Cause* (New Haven: Yale University Press, 1967); Charles A. Reed, "Extinction of Mammalian Megafauna in the Old World Late Quaternary," *Bioscience*, vol. 20, no. 5, March 1, 1970, pp. 284–288; Richard G. Klein, "Middle Stone Age Man–Animal Relationships in Southern Africa: Evidence from Die Kelders and Klasier River Mouth," *Science*, vol. 190, October 17, 1975.
26. Sinclair and Griffiths, *Serengeti*, p. 284.
27. Interview with Peek; James M. Peek, "Natural Regulation of Ungulates (What Constitutes a Real Wilderness?)," *Wildlife Society Bulletin*, vol. 8, no. 3, Fall 1980; James M. Peek, "Comments on Caughley's Comment," *Wildlife Society Bulletin*, vol. 9, no. 3, Fall 1981.
28. Plato, "The Meno," *Collected Dialogues of Plato*, Edith Hamilton and Huntington Cairns, eds. (New York: Pantheon Books, 1963), p. 381.
29. Interview.
30. Interview with Frank Golley; Schaller, *Serengeti Lion*, p. 404.

31. Barmore, "Population Characteristics of Six Ungulate Species," pp. 448–456.
32. Interview with Peek.
33. Frank L. Miller, "Wolf-Related Caribou Mortalities on a Calving Ground in North-Central Canada," *Wolves in Canada and Alaska: Their Status, Biology and Management,* Ludwig N. Carbyn, ed. (Edmonton: Canadian Wildlife Service, 1983), pp. 100–101; Douglas H. Pimlott, "Wolf Predation and Ungulate Populations," *American Zoology,* vol. 7, 1967, pp. 267–278.
34. Allen, *Wolves of Minong;* L. David Mech, *The Wolf: The Ecology and Behavior of an Endangered Species* (Garden City, N.Y.: Natural History, 1970); Arthur T. Bergerud, "Prey Switching in a Simple Ecosystem," *Scientific American,* vol. 249, no. 6, 1983; L. David Mech and Patrick D. Karns, "Role of the Wolf in a Deer Decline in the Superior National Forest," United States Forest Service Research Paper No. 148, 1977; Ludwig N. Carbyn, "Wolf Predation on Elk in Riding Mountain National Park, Manitoba, *Journal of Wildlife Management,* vol. 47, no. 4, 1983, pp. 963–976; John L. Weaver, "Wolf Predation upon Elk in the Rocky Mountain Parks of North America: A Review," in Boyce and Hayden-Wing, *North American Elk,* pp. 29–33; A. T. Bergerud, W. Wyett, B. Snider, "The Role of Wolf Predation in Limiting a Moose Population," *Journal of Wildlife Management,* vol. 47, no. 4, 1983, pp. 977–988.
35. Glen F. Cole, "An Ecological Rationale," p. 419.
36. Mary Meagher, "Yellowstone's Bison: A Unique Wild Heritage," *National Parks & Conservation Magazine,* May 1974; Meagher, *Bison of Yellowstone,* pp. 68, 110–113.
37. Meagher, "Yellowstone's Bison."
38. Ibid.
39. Douglas B. Houston, "The Status of Research on Ungulates in Northern Yellowstone National Park," presented at the American Association for the Advancement of Science Symposium on Research in National Parks, December 28, 1971; Douglas B. Houston, "Ecosystems of National Parks," *Science,* vol. 172, May 14, 1971; Douglas B. Houston, "Yellowstone Elk: Some Thoughts on Experimental Management," *Pacific Park Science,* vol. 1, no. 3, Spring 1981.
40. Pengelly to Professor Alan Beetle, Department of Range Science, University of Wyoming, August 19, 1974.
41. See Chapter 14.
42. Of the few pieces that were published, very few appeared in "refereed" journals — scholarly publications in which articles submitted for publication are evaluated by experts in the same field. Much of Meagher's work remains "in preparation" and is still unavailable. Houston has been the most prolific, but much of his most important work was not published until the late 1970s. His work *The Northern Yellowstone Elk,* for example, was substantially completed in 1974, but did not appear in published form until 1982 (cf. Douglas B. Houston, "The Northern Yellowstone Elk," Parts I and II, Yellowstone National Park Research Office, 1974, and Parts III and IV, 1976, Yellowstone Park Research Office).
43. Houston, "The Status of Research on Ungulates," 1971; Douglas B. Houston, "A Commentary on the History of the Northern Yellowstone Elk," *BioScience,* vol. 25, no. 9, September 1975.
44. Houston, "The Northern Yellowstone Elk," 1974, p. 6.
45. Houston, *The Northern Yellowstone Elk,* 1982, pp. 193–195.
46. Ibid., pp. 23–25, 34–35; Douglas B. Houston, "The Northern Yellowstone Elk — Winter Distribution and Management," presented at a symposium on Elk Ecology and Management, Laramie, Wyoming, April 1978.
47. Ibid., p. 194.
48. Glen F. Cole, "A Naturally Regulated Elk Population," presented at a Symposium on Natural Regulation, Northwest Section of the Wildlife Society, Vancouver, British Columbia, March 1978; Cole, "An Ecological Rationale for Natural Regulation, 1971"; Cole, "Elk and the Yellowstone Ecosystem," 1969.
49. Glen Cole, "Population Regulation in Relation to *K,*" presented at the annual meeting of the Montana Chapter of the Wildlife Society, Bozeman, Montana, February 22, 1974.

50. Douglas B. Houston, "Ecosystem Concept Applied to National Parks," abridgment and modification of "Ecosystems of National Parks," *Naturalist,* vol. 21, no. 1, 1971; Houston, "The Status of Research on Ungulates," 1971.

51. National Park Service, "Some Considerations in the Use of Exclosures to Assess the Biotic Effects of Herbivores and Departures from Natural Conditions in Yellowstone National Park," Information Paper No. 13, May 15, 1971.

52. R. Daubenmixe, *Plant Communities* (New York: Harper and Row, 1968), pp. 236–237; Houston, "The Status of Research on Ungulates"; Alan A. Beetle, "The Zootic Disclimax Concept," *Journal of Wildlife Management,* vol. 27, no. 1, January 1974; Linda J. Cayot et al., "Zootic Climax Vegetation and Natural Regulation of Elk in Yellowstone National Park," *Wildlife Society Bulletin,* vol. 7, no. 3, 1979.

53. Houston, *Northern Yellowstone Elk,* pp. 182, 184.

54. Douglas B. Houston, "Wildfires in Northern Yellowstone National Park," *Ecology,* vol. 54, no. 5, Summer 1973; Houston, *Northern Yellowstone Elk,* pp. 101–106, 182; Robert E. Sellers and Don G. Despain, "Fire Management in Yellowstone National Park," *Proceedings,* Tall Timbers Fire Ecology Conference and Intermountain Fire Research Council Fire and Land Management Symposium, vol. 14, 1976; Don G. Despain and Robert E. Sellers, "Natural Fire in Yellowstone National Park," *Western Wildlands,* Summer 1977; Dale L. Taylor, "Some Ecological Implications of Forest Fire Control in Yellowstone National Park, Wyoming," *Ecology,* vol. 54, no. 6, Autumn 1973.

55. Houston, "Wildfires."

56. Ibid.

57. Based on interviews and review of the literature.

58. The most glaring example of citing data not publicly available is the use to which Barmore's dissertation has been put. Although never submitted to his thesis committee, it has nevertheless been repeatedly cited by Houston and Yellowstone resource managers even though independent researchers — including this writer — have not been permitted to quote from it.

59. Houston, *Northern Yellowstone Elk,* p. 194.

60. Charles W. Fowler and William J. Barmore, "A Population Model of the Northern Yellowstone Elk Herd," *Proceedings of Scientific Research in the National Parks,* R. M. Linn, ed., National Park Service Proceedings Series, no. 5.

61. Houston, *Northern Yellowstone Elk,* p. 194; Larry A. Lahren, "The Myers-Hindman Site: An Exploratory Study of Human Occupation Patterns in the Upper Yellowstone Valley from 7000 B.C. to A.D. 1200," Anthropologos Researchers International, Livingston, Montana, April 1976; Larry A. Lahren, "Archeological Investigations in the Upper Yellowstone Valley, Montana: A Preliminary Synthesis and Discussion"; *Aboriginal Man and Environments on the Plateau of Northwest America,* A. H. Stryd and R. A. Smith, eds. (Calgary: Student Press, 1971).

62. Lahren, "The Myers-Hindman Site"; Larry A. Lahren, "Archaeological Salvage Excavations at a Prehistoric Indian Campsite (24YE344) in Yellowstone National Park," 1973 (obtained from author). Although Lahren did turn up one "ungulate bone" near the Gardiner River in Yellowstone, neither he — nor any other archaeologist — has ever identified an elk bone in the park.

63. Houston, *Northern Yellowstone Elk,* p. 24, *passim.* I questioned Dr. Meagher about this matter in February 1984, and was told it represented work in progress. Yet it had been cited by Houston as early as 1974 (in Houston, "Northern Yellowstone Elk").

64. Barbee to Pengelly, April 11, 1984, in response to the latter's Freedom of Information Act request.

65. Colonel Lloyd Brett, Acting Superintendent of Yellowstone, to the Secretary of the Interior, September 18, 1915.

66. U.S. Department of Agriculture, Biological Survey, "Report on Investigations of the Elk Herds in the Yellowstone Region of Wyoming, Montana, and Idaho," Yellowstone Library, December 14, 1915; Vernon Bailey and Alva A. Simpson, "Plan for the Control and Management of the Elk Herds of the Yellowstone Region of Wyoming, Montana, and Idaho," Yellowstone Library, January 1916.

67. Captain F. T. Arnold, Acting Superintendent, Yellowstone National Park, to Secretary of the Interior, April 5, 1916, Yellowstone Archives, box no. 89; Report of Elk Census by Vernon Bailey, March 2–14, 1916, reporting 11,564 elk, Yellowstone Archives, box no. 89; letter from Colonel L. R. Brett, Acting Superintendent of Yellowstone, to the Secretary of the Interior, May 5, 1916, Yellowstone Archives, box no. 89.
68. Houston, *Northern Yellowstone Elk,* pp. 12–15. In 1933 Aldo Leopold cited these early counts as examples of "a dependable census" (Aldo Leopold, *Game Management* [New York: Charles Scribner's Sons, 1933], p. 144).
69. Interview.
70. James M. Peek, "On Counting Elk," *Bugle,* vol. 2, no. 1, 1985. In an interview with Yellowstone Park biologists in February 1984, I was told that they could not estimate the efficiency of their elk counts.
71. Houston, *Northern Yellowstone Elk,* pp. 12–15.
72. Houston, *Northern Yellowstone Elk,* p. 14; *Superintendent's Monthly Reports,* March and April, 1920; U.S. Weather Bureau, "Annual Meteorological Summary"; Edgar H. Fletcher, "Climatic Features of Yellowstone National Park," *Scientific Monthly,* vol. 25, October 1927, pp. 329–336.
73. Interview; Ian McTaggert Cowan, "Range Competition Between Mule Deer, Bighorn Sheep and Elk in Jasper National Park, Alberta," *Twelfth North American Wildlife Conference;* Edward P. Cliff, "Relationships Between Elk and Mule Deer in the Blue Mountains of Oregon," *Fourth North American Wildlife Conference.*
74. Richard A. Dirks, "Climatological Studies in Yellowstone and Grand Teton National Parks," 1974, Department of Atmospheric Science, University of Wyoming; Arthur V. Douglas and Charles W. Stockton, "Long-Term Reconstruction of Seasonal Temperature and Precipitation in the Yellowstone National Park Region Using Dendroclimatic Techniques," prepared for the National Park Service, Yellowstone Library, June 1975; Charles W. Stockton, "A Dendroclimatic Analysis of the Yellowstone National Park Region, Wyoming-Montana," prepared for the National Park Service, Yellowstone Library, 1973. Results of the tree-ring analysis were not consistent with the historical record, and showed that summers since 1850 were on the average drier throughout the northern Rockies than those earlier. This drying, however, would not explain the decline of vegetation that has occurred in Yellowstone since 1930.
75. J. R. Habeck, comments on Douglas B. Houston's "Wildfires in Northern Yellowstone National Park," reviewed for *Ecology,* October 13, 1972. Quoted with permission of the author.
76. Houston, "A Comment on the History of the Northern Yellowstone Elk"; Caughley, "What Is This Thing Called Carrying Capacity?"
77. National Park Service, *Final Environmental Statement, Yellowstone National Park Master Plan* (Denver: Midwest Regional Office, 1974), p. 30.
78. Federal Trade Commission survey, 1982; Consumer Preference Corporation survey, 1977; *Livingston Enterprise,* October 14, 1982.
79. National Park Service, "Wilderness Recommendation, Yellowstone National Park," August 1972; Bruce M. Kilgore, "Fire Management Programs in National Parks and Wilderness," presented at the Symposium on the Field Effects of Fire, October 19–21, 1982, Jackson, Wyoming; Bruce M. Kilgore, "Restoring Fire to National Park Wilderness," *American Forests,* March 1975; Bruce M. Kilgore, "Fire Management in the National Parks: An Overview," *Proceedings,* Tall Timbers Fire Ecology Conference No. 14, 1976; National Park Service, "The Natural Role of Fire — A Fire Management Plan for Yellowstone National Park," 1983, Yellowstone Park files; National Park Service, "Final Environmental Assessment, Natural Fire Management Plan for Yellowstone National Park," 1981, Yellowstone Park files.

7. NATURE TAKES ITS COURSE

1. Interviews with Swenson and Nelson.
2. Douglas B. Houston, *The Northern Yellowstone Elk* (New York: Macmillan, 1982),

pp. 23–24. Park biologists had postulated that in primeval times a kind of faunal gridlock prevented elk from leaving Yellowstone. The lower valleys had their own large populations of elk, they argued, and these animals would block the path of any Yellowstone elk that might try to migrate off the plateau.

3. The elk are still wearing collars. Some have now wandered a considerable distance into Yellowstone. Eventually, this research might determine whether Cole and Houston were correct in hypothesizing the existence of two distinct herds of elk, one that leaves the park in winter and the other that does not.

4. Based on interviews with Swenson, 1984–1985.

5. Erickson, "A Conflict of Policies."

6. Cole to Anderson, October 26, 1967 (obtained through the Freedom of Information Act); National Park Service, "Information Paper — Northern Yellowstone Elk Herd," September 17, 1968; Cole, "Elk and the Yellowstone Ecosystem"; Houston, "The Status of Research on Ungulates."

7. Houston, "The Northern Yellowstone Elk," pp. 72–73.

8. Erickson, "A Conflict of Policies."

9. Western Association of Game and Fish Commissioners, "Resolution on National Park Service Wildlife Management Policy," December 3, 1975; *Missoulian,* June 2, 1975.

10. Interview, 1983.

11. National Park Service, "World Heritage Designation," Yellowstone Library, 1978; National Park Service, "Yellowstone National Park Is a Biosphere Reserve," Yellowstone Library, 1972.

12. On bears, see Chapters 11, 12; eventually collars were removed from alligators (James A. Kushlan and Terri Jacobsen, "Management of Alligator Nest Flooding," January 1984, South Florida Research Center, Everglades National Park).

13. Curtis Skinner et al., "History of the Bison in Yellowstone."

14. Reed to Dingell, January 24, 1973 (obtained through the Freedom of Information Act).

15. Department of the Interior, Fish and Wildlife Service, *Yellowstone Fishery Investigations, Annual Project Report — Calendar Year 1975,* Yellowstone National Park, 1976.

16. *Yellowstone Fishery Investigations,* 1982.

17. *Yellowstone Fishery Investigations,* 1975.

18. Paul Schullery, *Mountain Time* (New York: Nick Lyon Books, 1984), p. 112; Paul Schullery, "A Reasonable Illusion," *Rod and Reel,* November/December 1979.

19. Schullery, *Mountain Time,* p. 112.

20. Flynn to Barbee, February 8, 1983; Montana Department of Fish, Wildlife and Parks, "Comments on the Natural Resources Management Plan for Yellowstone National Park," 1983.

21. Montana, "Comments on Natural Resources Management Plan."

22. *Livingston Enterprise,* February 4, 1985.

23. Interview.

24. Interview.

25. Interview.

26. Based on my own inspection. See Chapter 8.

27. Interview.

28. Gene F. Payne, Professor of Range Science, Montana State University, to Gary E. Everhardt, Superintendent, Grand Teton National Park, October 20, 1972; Robert G. Haraden, Acting Superintendent, Yellowstone, to Payne, October 30, 1972 (both obtained through the Freedom of Information Act). The more recent removal of exclosures was confirmed by Yellowstone's Research Office in February 1984.

29. George E. Gruell, "Fire's Influence on Wildlife Habitat on the Bridger-Teton National Forest, Wyoming, vol. 2, Changes and Causes, Management Implications," Forest Service Research Paper INT-252 (Ogden, Utah: Department of Agriculture, 1980), p. 20.

30. Ibid.

31. For years, repeated offers by park geologists to study this watershed were rejected. In 1985, John Bailey, president of the local chapter of *Trout Unlimited,* persuaded

Yellowstone authorities to permit a joint study of the Yellowstone watershed by the Park Service, Montana Department of Fish, Wildlife and Parks, and the U.S. Geological Survey.

32. Wright to Pengelly, February 2, 1984; Gary A. Wright, *People of the High Country: Jackson Hole Before the Settlers* (New York: Peter Lang, 1984).

33. Charles E. Kay, "Aspen Reproduction in the Yellowstone Park–Jackson Hole Area and Its Relationship to the Natural Regulation of Ungulates," paper presented at the Western Elk Management Symposium, Utah State University, Logan, Utah, April 19–20, 1984.

34. Bruce M. Kilgore, "Fire Management Programs in National Parks and Wilderness," presented at the Symposium on the Field Effects of Fire, October 19–21 1982, Jackson, Wyoming; National Park Service, "Chronological Natural Fire History, Yellowstone National Park," 1984, Yellowstone National Park Resource Manager's Office.

35. The inhibiting effect of ungulate grazing on the spread of fires has been noticed elsewhere, paradoxically, among other places, in the Serengeti (M. Norton-Griffiths, "Influence of Grazing, Browsing and Fire on Vegetation Dynamics," *Serengeti*, ed. by Sinclair and Norton-Griffiths, p. 333).

36. Interview.

37. Interview.

38. Interview.

39. The most recent elk census found one moose.

40. Interview with Erickson; Glenn L. Erickson, "Statewide Wildlife Survey and Inventory," Job No. W-130-R-10, July 1, 1978–June 30, 1979, Montana Department of Fish, Wildlife and Parks; Glenn L. Erickson, "Summary of Helicopter Survey of Mule Deer Winter Range in Hunting Districts 313, 314, and 317, on March 30 and April 3, 1979."

41. Interview with Keating.

42. Kimberly Alan Keating, "Population Ecology of Rocky Mountain Bighorn Sheep in the Upper Yellowstone River Drainage, Montana/Wyoming," M.S. thesis, Montana State University, March 1982. Houston, using "admittedly precarious manipulations," discounted this evidence, suggesting instead that the censuses became increasingly efficient. But this argument is circular.

43. Jim Ackerman, "The Majestic Bighorn Falls to Pinkeye," *Montana Magazine*, May/June, 1983; Mary Meagher, "An Outbreak of Pinkeye in Bighorn Sheep, Yellowstone National Park: A Preliminary Report," *Biennial Symposium, North American Wild Sheep and Goat Council*, vol. 3, 1983; Mary Meagher, "An Outbreak of Pinkeye in Bighorn Sheep, Yellowstone National Park — Preliminary Report," 1983, Yellowstone National Park Research Office.

44. I witnessed the death throes of many rams at this time. See also Jim Robbins, "A Cruel Mother Nature Rules the Parks, *High Country News*, vol. 16, no. 7, April 16, 1984.

45. The extent of the die-off was confirmed for me by Meagher in February 1984.

46. Howe, "Yellowstone River and Gallatin Elk Herds," Yellowstone Library, July 17, 1961.

47. Chief Ranger to Superintendent of Yellowstone, October 17, 1984.

48. M. Meagher, "Elk Distribution Flight, Northern Winter Range, Yellowstone National Park, 1981," Yellowstone Research Office.

49. Ibid.

50. Bison figures given to me by Meagher on February 6, 1984.

51. *Missoulian*, April 1, 1984. That buffalo wandered should not have surprised anyone. In 1911, four bulls roamed forty-three miles from the park before the cavalry sent scouts to drive them back (Lt. Col. Lloyd Brett to Secretary of the Interior, June 30, 1911).

52. *Billings Gazette,* March 10, 1984.

53. *Livingston Enterprise,* March 1 and 21, 1984.

54. Associated Press wire dispatch, February 9, 1984; *Livingston Enterprise,* February 25 and 26 and March 1, 1985.

55. For Montana ranchers this is a serious matter. State law requires that, if any cow is found to be infected with brucellosis, the entire herd must be destroyed.
56. Houston, *Northern Yellowstone Elk,* p. 49.
57. Interview with park biologists, February 6, 1984.

8. THE SECRET OF YANCEY'S HOLE

1. Aubrey L. Haines, *The Yellowstone Story* (Yellowstone: Yellowstone Library and Museum Association, 1977), pp. 442–443.
2. Ibid., pp. 238, 241–242.
3. Ernest Thompson Seton, *Life-Histories of Northern Animals,* vol. 1, Grass-eaters (New York: Charles Scribner's Sons, 1909), pp. 455–458.
4. Edward R. Warren, "A Study of the Beaver in the Yancey Region of Yellowstone National Park," *Roosevelt Wildlife Annals,* vol. 1, no. 1, October 1926, pp. 13–191; Warren, "The Life of the Yellowstone Beaver," *Roosevelt Wildlife Bulletin,* vol. 1, no. 2, August 1922.
5. This is range plot number 25, established in 1936.
6. Photograph taken May 4, 1953.
7. As an example of this kind of defense, see Paul Schullery, "The Yellowstone Dilemma," Part 2, *Bugle,* Spring 1985.
8. Houston, *Northern Yellowstone Elk,* pp. 157–159, 167–169, 184–185.
9. In 1981 Meagher reported 2,006 bison and 16,019 elk.
10. Houston, *Northern Yellowstone Elk,* p. 159.
11. Schullery, "The Yellowstone Dilemma," Part II, *The Bugle,* vol. 2, Issue 1, Winter 1985.
12. Wright to Pengelly, February 2, 1984.
13. A June 1982 publication of the Yellowstone Research Office states, "natural regulation of all ungulate species is still a topic of intensive research" ("Population Status of Large Mammal Species in Yellowstone National Park"). Yet although Houston was employed in elk research until 1978, many of the most crucial studies remained undone. See Chapter 14.
14. It was, in fact, substantially the same book as the monograph, "Northern Yellowstone Elk," completed in 1976.
15. A. R. E. Sinclair, Foreword to *Northern Yellowstone Elk.*
16. Gary Blonson, "Where Nature Takes Its Course," *Science* 83, November 1983.
17. See Chapter 17.
18. Interview, 1984.

9. GROWING APPLES IN EDEN

1. *Letter from the Secretary of War Communicating the Report of Lt. Gustavus C. Doane upon the So-called Yellowstone Expedition of 1870,* 41st Congress, 3rd Session, Ex. Doc. No. 51, pp. 5–6.
2. Jefferson to John Adams, May 27, 1813, quoted in Stephen J. Pyne, *Fire in America* (Princeton: Princeton University Press, 1982), p. 75.
3. Kenneth L. Pierce, *History and Dynamics of Glaciation in the Northern Yellowstone National Park Area* (Washington, D.C.: U.S. Government Printing Office, 1979); Ake Hultkrantz, "The Indians of Yellowstone Park," *Shoshone Indians* (New York: Garland, 1974); George C. Frison, *Prehistoric Hunters of the High Plains* (New York: Academic Press, 1978); George C. Frison, "Prehistoric Occupations of the Grand Teton National Park," *Naturalist,* vol. 21, no. 1, 1971; Gary A. Wright, *People of the High Country: Jackson Hole Before the Settlers* (New York: Peter Lang, 1984); J. Jacob Hoffman, "A Preliminary Archaeological Survey of Yellowstone National Park," 1961, Montana State University; Gary A. Wright et al., "Regional Assessment, Archaeological Report," Department of the Interior, *Greater Yellowstone Cooperative Regional Transportation Study* (Lincoln, Nebraska: Midwest Archaeological Center, National Park Service, June 1978).

4. Pyne, *Fire in America*, chapter 1; Omer C. Stewart, "Fire as the First Great Force Employed by Man," *Man's Role in Changing the Face of the Earth* (Chicago: University of Chicago Press, 1955); Omer C. Stewart, "Barriers to Understanding the Influence of Use of Fire by Aborigines on Vegetation," *Proceedings,* Tall Timbers Fire Ecology Conference, vol. 2.

5. Pyne, *Fire in America*, p. 71; Henry T. Lewis, "Maskuta: The Ecology of Indian Fires in Northern Alberta," *Western Canadian Journal of Anthropology*, vol. 7, no. 1, 1977; Stephen W. Barrett, "Relationship of Indian-Caused Fires to the Ecology of Western Montana Forests," M.S. thesis, June 1981, University of Montana; Stephen W. Barrett, "Indians and Fire," *Western Wildlands*, Spring 1980; Dean A. Shinn, "Historical Perspectives on Range Burning in the Inland Pacific Northwest," *Journal of Range Management*, vol. 33, no. 6, November 1980.

6. Pyne, *Fire in America*, pp. 20–45; Earle F. Layser, "Forestry and Climatic Change," *Journal of Forestry*, November 1980.

7. Don G. Despain and Robert E. Sellers, "Fire Management in Yellowstone National Park," *Proceedings,* Tall Timbers Fire Ecology Conference and Intermountain Fire Research Council Fire and Land Management Symposium, vol. 14, 1976; Despain and Sellers, "Natural Fire in Yellowstone National Park," *Western Wildlands*, Summer 1977; George E. Gruell, "Fire's Influence on Wildlife Habitat in the Bridger-Teton National Forest, Wyoming." In Yellowstone other factors, such as the different volcanic soils, also help to determine the vegetation (see Chapter 15). The succession of lodgepole pine in particular seems not to be reversed by fire (see Don G. Despain, "Nonpyrogenous Climax Lodgepole Pine Communities in Yellowstone National Park," *Ecology*, vol. 64, no. 2, 1983).

8. Lewis, "Ecology of Indian Fires"; James M. Peek et al., "Evaluation of Fall Burning on Bighorn Sheep Winter Range," *Journal of Range Management*, vol. 32, no. 6, November 1979.

9. Pyne, *Fire in America*, pp. 20–45; Dale L. Taylor, "Some Ecological Implications of Forest Fire Control in Yellowstone National Park, Wyoming," *Ecology*, vol. 54, no. 6, Autumn 1973.

10. Thomas Morton, *New English Canaan: or, New Canaan*, 1637 rpt. (New York: Arno Press, 1972), pp. 52–54; Carol O. Sauer, *Seventeenth-Century Northern America* (Berkeley: Turtle Island, 1980); Pyne, *Fire in America*, p. 76; Barrett, "Indians and Fire."

11. Jefferson to Adams, May 27, 1813; Pyne, *Fire in America*, p. 46.

12. Ibid., p. 48.

13. Ibid., pp. 84–85.

14. Shinn, "Historical Perspectives on Range Burning"; Barrett, "Indians and Fire"; Pyne, *Fire in America*, pp. 71 ff., p. 252; Russell, *Journal of a Trapper*, pp. 30 ff. For reference to early burning near Yellowstone, see Meriwether Lewis and William Clark, *History of the Expedition Under the Commands of the Captains Lewis and Clark* (New York: Atherton, 1922), vol. 2, p. 107.

15. F. V. Hayden, *Sixth Annual Report of the United States Ecological Survey of the Territories* (Washington, D.C.: U.S. Government Printing Office, 1873), pp. 22, 81, 229, 244, 255, 265, 267; *Superintendent's Report for 1885.* DeLacy ran into Indian fire throughout his trip. See deLacy, "A Trip up the South Snake River in 1863," pp. 103, 113, 115, 122–123.

16. Shinn, "Historical Perspectives on Range Burning"; Barrett, "Indian-Caused Fires"; Gary A. Wright, "Homage to Gustavus Cheney Doane: Science in Our National Parks," *Proceedings,* First Conference on Scientific Research in the National Parks, American Biological Institute, 1979; Barrett, "Indians and Fire"; P. Mehringer, S. Arnold, and K. Peterra, "Post Glacial History of Lost Trail Pass Bog, Bitterroot Mountains, Montana," *Arctic and Alpine Research*, vol. 9, no. 4, 1977.

17. Gary A. Wright and Susanne J. Miller, "Prehistoric Hunting of New World Wild Sheep: Implications for the Study of Sheep Domestication," presented to the Symposium on Recent Research in Anatolian Prehistory at the 72nd General Meeting of the American Anthropological Association, December 1, 1973; Harold McCracken

et al., "The Mummy Cave Project in Northwestern Wyoming," Buffalo Bill Historical Center, Cody, Wyoming, 1978; George C. Frison, "Shoshonean Antelope Procurement in the Upper Green River Basin, Wyoming," *Plains Anthropologist*, vol. 16, no. 64; Frison, *Prehistoric Hunters of the High Plains*.

18. Frison, *Prehistoric Hunters of the High Plains;* David Dominick, "The Sheepeaters," *Annals of Wyoming*, vol. 36, 1964; Julian H. Steward, *Basin-Plateau Aboriginal Sociopolitical Groups*, Smithsonian Institution Bureau of American Ethnology, Bulletin 120 (Washington, D.C.: U.S. Government Printing Office, 1938); B. Robert Butler, "A Guide to Understanding Idaho Archaeology, 3rd ed.: The Upper Snake and Salmon River Country," 1978, a special publication of the Idaho Museum of Natural History, Pocatello, Idaho.

19. Aubrey L. Haines, "The Bannock Indian Trails of Yellowstone National Park," Yellowstone National Park Library, undated; Wayne Replogle, "Great Bannock Trail," *Naturalist*, vol. 10, no. 2, 1959; Wayne Replogle, *Yellowstone's Bannock Trails* (Yellowstone: Yellowstone Library and Museum Association, 1956); Frison, "Prehistoric Occupations."

20. Replogle, *Yellowstone's Bannock Trails;* Hoffman, "A Preliminary Archaeological Survey"; Wright et al., "Archaeological Report"; Larry A. Lahren, "Archaeological Salvage Operations at a Prehistoric Indian Campsite in Yellowstone National Park," Yellowstone Library, March 22, 1973.

21. McCracken et al., "The Mummy Cave Project."

22. Robert H. Lowie, *The Crow Indians* (Lincoln: University of Nebraska Press, 1935), pp. 12, 54, 48, 82, 96, 103, 184, 190, 214; Dominick, "The Sheepeaters."

23. Frison, "Shoshonean Antelope Procurement"; Frison, *Prehistoric Hunters*, pp. 254–256.

24. Frison, *Prehistoric Hunters*, especially pp. 239–244; Leslie B. Davis, "The 20th Century Commercial Mining of Northern Plains Bison Kills, Bison Procurement and Utilization — A Symposium," *Plains Anthropologist*, Memoir 14, 1978, Leslie B. Davis and Michael Wilson, eds., pp. 254–286.

25. Wright, "Homage to Doane"; Frison, *Prehistoric Hunters;* R. Robert Butler et al., "The Wasden Site Bison: Sources of Morphological Variation," *Aboriginal Man and Environments on the Plateau of Northwest America*, Arnoud H. Stryd and Rachel A. Smith, eds. (Calgary: Archaeological Association of Calgary, 1971); Barry W. Nimmo, "Population Dynamics of a Wyoming Pronghorn Cohort from the Eden-Farson Site, 48SW304," *Plains Anthropologist*, vol. 35, 1971.

26. Charles A. Reed, "Extinction of Mammalian Megafauna in the Old World Late Quarternary, *BioScience*, vol. 20, no. 5, 1970; Charles A. Reed, "They Never Found the Ark," *Ecology*, vol. 50, no. 2, Early Spring 1969; Martin and Wright, *Pleistocene Extinctions;* Paul S. Martin, "The Discovery of America," *Science*, vol. 179, March 1973; James E. Mosimaon and Paul S. Martin, "Simulating Overkill by Paleoindians," *American Scientist*, vol. 63, May–June, 1975; Austin Long and Paul S. Martin, "Death of American Ground Sloths," *Science*, vol. 186, November 15, 1974.

27. Paul S. Martin, "Pleistocene Overkill"; Frances Haines, "The Northward Spread of Horses among the Plains Indians," *American Anthropologist*, vol. 40, 1938.

28. Long and Martin, "Death of American Ground Sloths."

29. Martin, *Pleistocene Extinctions*, p. vi.

30. Mark N. Cohen, *The Food Crisis in Prehistory: Overpopulation and the Origins of Agriculture* (New Haven: Yale University Press, 1977), pp. 157–223.

31. Cohen, *Food Crisis in Prehistory*, pp. 188 ff.; J. E. Guilday, "Differential Extinction During Late Pleistocene and Recent Times," in Martin and Wright, *Pleistocene Extinctions*.

32. Cohen, *Food Crisis in Prehistory*, p. 189.

33. Haines, "The Northward Spread of Horses"; Gary A. Wright, "Notes on Chronological Problems on the Northwestern Plains and Adjacent High Country," *Plains Anthropologist*, vol. 27, 1982; Gary A. Wright, "High Country Adaptations," *Plains Anthropologist*, vol. 25, no. 89, August 1980; Gary A. Wright, "The Shoshonean Migration Problem," *Plains Anthropologist*, vol. 23, 1978.

34. McCracken et al., "The Mummy Cave Project"; Barry W. Nimmo, "Population

Dynamics of a Wyoming Pronghorn Cohort"; Butler et al., "The Wasden Site Bison." Analysis of the age structure of animals found at most of these sites shows that the age profile of the wildlife populations in prehistory was quite young. A young population suggests heavy predation — that few animals die of old age.

35. Wright to Pengelly, February 2, 1984.
36. Ibid.
37. Gary Wright, interview. This relative abundance was also reflected in Mummy Cave excavations. At lower elevations than Yellowstone, excavations have unearthed remains of elk, but even here they were not as numerous as other species. The Myers-Hindman site, for instance, the excavation Houston cites to show the abundance of elk, produced 28 bison bones, 25 deer, 25 mountain sheep, and only 12 elk.
38. Lewis and Clark, *History of the Expedition*, vol. 2, pp. 55 ff.; B. Robert Butler, "A Guide to Understanding Idaho Archaeology"; Replogle, *Yellowstone's Bannock Indian Trails*.
39. Meriwether Lewis and William Clark, *The History of the Lewis and Clark Expedition*, ed. by Elliott Coues (New York: Dover, 1893), vol. 3, p. 1197; C. M. Gates, ed., *Five Fur Traders of the Northwest: Being the Narrative of Peter Pond and the Diaries of John Macdonnell, Archibald N. McLeod, Hugh Aries, and Thomas Conner* (Minneapolis: University of Minnesota Press, 1933); G. B. Grinnell, *The Fighting Cheyenne* (Norman: University of Oklahoma Press, 1956); F. G. Roe, *The North American Buffalo: A Critical Study of the Species in Its Wild State* (Toronto: University of Toronto Press, 1951).
40. Harold Hickerson, "The Virginia Deer and Intertribal Buffer Zones in the Upper Mississippi Valley," A. Leeds and A. P. Vayda, eds., *Man and Culture in America: The Role of Animals in Human Ecological Adjustments* (Washington, D.C.: American Association for the Advancement of Science, 1978), pp. 43–65.
41. Sherburne F. Cook, "The Significance of Disease in the Extinction of the New England Indians," *Human Biology*, vol. 45, no. 3, September 1973.
42. E. Wagner Stearn and Allen E. Stearn, *The Effect of Smallpox on the Destiny of the Amerindian* (Boston: Bruce Humphries, 1945), p. 47.
43. Ibid.
44. Ibid., Dominick, "The Sheepeaters," pp. 141–142; Robert F. and Yolanda Murphy, "Shoshone-Bannock Subsistence and Society," *Anthropological Records*, vol. 16, no. 7, 1959.
45. Langford, *Discovery of Yellowstone Park*, p. 122.
46. Philitus W. Norris, *Superintendent's Annual Report for the Year Ended June 30, 1880*, vol. 2 (Washington, D.C.: U.S. Government Printing Office, 1880), p. 605; Hoffman, "A Preliminary Archaeological Survey." Russell ran into Blackfoot throughout the park and deLacy saw signs of them regularly (Russell, *Journal of a Trapper, passim;* deLacy, "A Trip up the Snake River").
47. I have visited these wickiups. For a description, see Hoffman, "A Preliminary Archaeological Survey."
48. This story was told me by David Spirtes, then North District ranger in Yellowstone.
49. Michael D. Yandell, ed., *National Parkways Comprehensive Guide to Yellowstone National Park*, pp. 7, 70; TW Services, *Yellowstone National Park* (brochure), March 1, 1982.
50. Meagher to Dwain W. Cummings, *Buffalo Magazine*, December 1975.
51. Schullery, *Mountain Time* (New York: Nick Lyons Books, 1984), p. 75.
52. Taylor, "Forest Fires in Yellowstone National Park."
53. Houston, *Northern Yellowstone Elk*, p. 189.
54. National Park Service, "Cultural Conflict, Eyewitness Accounts of the Nez Perce War," Yellowstone Library, undated; Aubrey L. Haines, *The Yellowstone Story* (Yellowstone: Yellowstone Library and Museum Association, 1977), vol. 2, ch. 8, "Warfare in Wonderland"; Mark H. Brown, *The Flight of the Nez Perce* (New York: G. P. Putnam's Sons, 1983); Alvin M. Josephy, Jr., *Chief Joseph's People and Their War* (Yellowstone: Yellowstone Library and Museum Association, 1964).
55. Haines, *Yellowstone Story*, pp. 237–238; Hutzkrantz, "The Indians of Yellowstone Park," p. 140.

56. Hoffman, "A Preliminary Archaeological Survey"; Norris, *Superintendent's Annual Report, 1880,* p. 605; Hultkrantz, "The Indians of Yellowstone Park," p. 145.
57. Hoffman, "A Preliminary Archaeological Survey"; Haines, *Yellowstone Story,* vol. 1, pp. 237–238; Aubrey L. Haines, "The Indians in Our Past," *A Manual of General Information on Yellowstone National Park,* Yellowstone, 1963; Hultkrantz, "The Indians of Yellowstone Park," p. 137.
58. Peter Steinhart, "Ecological Saints," *Audubon,* vol. 86, no. 4, July 1984.
59. Steinhart, "Ecological Saints"; Peter Matthiessen, *Indian Country* (New York: Viking Press, 1979), p. 3.
60. John Locke, *Two Treatises of Government,* Thomas I. Cook, ed. (New York: Hafner, 1947), p. 145.
61. Pyne, *Fire in America,* p. 81; C. B. Macpherson, *The Political Theory of Possessive Individualism: Hobbes to Locke* (Oxford: Clarendon Press, 1962).
62. Pyne, *Fire in America,* pp. 51 ff.
63. Stewart, "Barriers to Understanding the Influence of Use of Fire by Aborigines."
64. Ibid.
65. Lewis, "Ecology of Indian Fires."
66. Pyne, *Fire in America,* p. 81.
67. All quotations from Stephen Fox, *John Muir and His Legacy: The American Conservation Movement* (Boston: Little, Brown, 1981), p. 350.
68. Today, when there is little game over which the white man can fight with the Indian, the principal issue is use of water. It is in this province that great legal battles with the Indian are now being fought — and generally won — by the white man.
69. F. V. Hayden, *Preliminary Report,* 1872, p. 174.

10. THE WOLF MYSTERY

1. Interview with Marshall Gates.
2. Based on an interview with more than forty wolf experts, in Yellowstone and elsewhere. See also John Weaver, *The Wolves of Yellowstone,* U.S. Department of the Interior, National Park Service, National Resources Report No. 14 (Washington, D.C.: U.S. Government Printing Office, 1978).
3. Glen F. Cole, "Yellowstone Wolves *(Canis lupus irremotus),*" Research Note No. 4, April 1971, Yellowstone National Park.
4. Based on extensive interviews with biologists, park rangers, seasonal rangers, officials of the Montana Department of Fish, Wildlife and Parks, local residents, and park maintenance staff.
5. L. David Mech, "The Status of the Wolf in the United States, 1973," *Wolves,* Douglas H. Pimlott, ed. (Morges, Switzerland: International Union for Conservation and Natural Resources, 1975).
6. John Ise, *Our National Park Policy: A Critical History* (Baltimore: Johns Hopkins University Press, 1961), pp. 592 ff.
7. Stanley P. Young and Edward A. Goldman, *The Wolves of North America* (Washington, D.C.: American Wildlife Institute, 1944), p. 382.
8. Ibid., pp. 377–381, 383.
9. According to the historical record, wolves lived entirely unmolested in Yellowstone until 1914. Indeed, only four wolf reports were recorded in the entire history of the park up to 1912. In 1912, in a letter to the Director of the New York Zoological Society, the Acting Superintendent wrote that "[Scout] McBride has been in the park for many years, and is not convinced that there have ever been any gray wolves here. Statements have been made that they have been seen, but none have ever been killed or captured inside the park . . ." (Brett to Hornaday, July 29, 1912). The status of mountain lions, however, was quite different. Nearly every early explorer reported that these animals were numerous, and substantial unofficial hunting of these animals began long before the government's predator-control program began. See Paul Schullery, "Another Look at Buffalo Jones in Yellowstone Park," Yellowstone Library, 1976.

10. *Superintendent's Report for 1887.*
11. Victor H. Cahalane, "The Evolution of Predator Control Policy in the National Parks," *Journal of Wildlife Management,* vol. 3, no. 3, July 1939; Aubrey L. Haines, *The Yellowstone Story* (Yellowstone: Yellowstone Library and Museum Association: 1977), vol. 2, p. 82; letter from Acting Superintendent to Secretary of the Interior, March 14, 1911; letter from Acting Superintendent to Commissioner of Dominion Parks Branch, Ottawa, Canada, March 11, 1912.
12. Young and Goldman, *Wolves of North America,* pp. 383–385.
13. *The National Park Service Act.*
14. *The National Park Service Act;* Arno B. Cammerer, Director of the National Park Service to J. N. Dymond, Chairman of the Ecological Society of America, undated, Yellowstone National Park Archives.
15. Vernon Bailey to "Elmer" (probably Chester) Lindsley, October 8, 1915; Bailey, "Report on the Investigations of Elk Herds," December 14, 1915; Bailey and Simpson, "Plan for the Control and Management of Elk," January 1916; Acting Superintendent, Yellowstone, to Secretary of the Interior, September 18, 1915; Vernon Bailey, "Destruction of Deer by the Northern Timber Wolf," U.S. Department of Agriculture, Bureau of Biological Survey Circular No. 58, May 4, 1907.
16. Bailey to Lindsley, October 8, 1915. On October 7, 1915, Cruse Black and Donald Stevenson were appointed as the first rangers ever to specialize in predator control. "Your special work," read their letters of appointment, "will be hunting and killing mountain lions, wolves and coyotes in the park by trapping, shooting and poison."
17. Mather to Lindsley, January 13, 1918; Mather to M. W. Nelson, Chief, Biological Survey, January 13, 1918; Lindsley to Kerb Garner (Biological Survey predator control agent), January 25, 1918; Mather to Lindsley, February 9, 1918; Lindsley to Mather, February 15, 1918; Lindsley to Mather, February 18, 1918; Acting Chief, Biological Survey, to Mather, March 18, 1918; Charles J. Bayer, Predatory Animal Inspector, Biological Survey, to Lindsley, April 12, 1918; Lindsley to Bayer, April 18, 1918; National Park Service, "Condensed Chronology of Service Predator Policy," Yellowstone Library, undated.
18. Cahalane, "The Evolution of Predator Control Policy," p. 235.
19. National Park Service, "Condensed Chronology of Predator Policy."
20. Ibid.
21. Ibid., Schullery, *Bears of Yellowstone,* pp. 66–67; George M. Wright and Ben H. Thompson, *Fauna of the National Parks of the United States, Wildlife Management in the National Parks, Fauna Series Number 2* (Washington, D.C.: U.S. Government Printing Office, 1934), pp. 32–33.
22. National Park Service, "Condensed Chronology of Predator Policy."
23. Cahalane, "Evolution of Predator Control Policy," p. 235.
24. Minutes of the 12th Superintendents' Conference on Predator Control, 1932.
25. National Park Service, "Chronological Record of General Correspondence Relative to Predator Control," Yellowstone Library, undated; National Park Service, "Organizations and Individuals Opposing Service Predator Policy," Yellowstone Library, undated.
26. Minutes of 10th Superintendents' Conference on Predator Control, 1928.
27. National Park Service, "Chronological Record"; George M. Wright, Ben H. Thompson, and Joseph S. Dixon, *Fauna of the National Parks of the United States: A Preliminary Survey of Faunal Relations in National Parks, Fauna Number 1* (Washington, D.C.: U.S. Government Printing Office, 1933).
28. National Park Service, "Chronological Record."
29. Albright to Society of Mammalogists, October 13, 1930; Horace Albright, "National Park Service Predator Control Policy," *Journal of Mammalogy,* May 1931.
30. Minutes of 10th Superintendents' Conference, 1928.
31. Cammerer to Dymond, undated, but circa 1938.
32. National Park Service, "Condensed Chronology of Predator Policy."
33. Interviews with Cahalane, 1983–1984.
34. Minutes of 12th Superintendents' Conference, 1932. The best-known "sighting" of wolves in Yellowstone occurred during the height of the predator-control controversy.

In 1934, Marguerite L. Arnold, wife of the Chief Ranger and daughter of Chester Lindsley (the Acting Superintendent of Yellowstone who initiated widespread predator control in Yellowstone), reported seeing four wolves near Tower Creek (Marguerite L. Arnold, "Yellowstone Wolves," *Nature,* August 1937).

35. Order from Director of the Park Service, prohibiting any control without his specific written authority, November 12, 1934; Director to Superintendent of Region 3, assigning Murie to "a thorough Coyote study in Yellowstone," March 25, 1937; Albright to Rogers, October 18, 1937.

36. Rogers to Director, May 9, 1939; Albright to Cammerer, October 18, 1937; Cammerer to A. E. Demaray, Associate Director of the Park Service, November 4, 1937.

37. Murie submitted a progress report on his studies on January 7, 1938, and his final administrative report on July 16, 1938; National Park Service, "Chronological Record of Correspondence."

38. Albright to Cahalane, November 24, 1937; Rogers to Cammerer, July 17, 1939; John R. White, Chief of Operations of Yellowstone, to Cammerer, June 27, 1939.

39. Albright to Cahalane, November 24, 1937.

40. David R. Madson to Cammerer, May 20, 1939.

41. Grinnell to Director, August 16, 1938; Adolph Murie, *Ecology of the Coyote in Yellowstone, Fauna Series No. 4* (Washington, D.C.: U.S. Government Printing Office, 1940).

42. The correspondence during the decade following the Murie study clearly indicates that Yellowstone authorities were chafing at the restrictions against predator control, and occasionally followed their inclination to do away with coyotes. (Rogers to Director, July 8, 1939; David de L. Condon, "Special Incident Report," November 11, 1942; Rogers to the Director, November 13, 1942; Memorandum for the Files, November 25, 1942; Director to the Regional Director, December 1, 1942; District Ranger Hugh Ebert to Chief Ranger, January 5, 1943.) The plague scare occurred before the Murie study, but after the Park Service predator-control policy was in effect (Rogers to Director, July 11, 1936; Surgeon General, U.S. Public Health Service, to Superintendent, Yellowstone, "Confidential Report," August 19, 1936; Rogers to Demaray (telegram), August 20, 1936; Acting Superintendent, Yellowstone, to Superintendent of Custer Battlefield National Monument, December 19, 1951; Acting Superintendent, Yellowstone, to Regional Director, National Park Service, November 22, 1949; National Park Service, "Mouse Count," Yellowstone National Park Archives.

43. National Park Service, Advisory Board on Wildlife Management, "Predator and Rodent Control in the United States," Yellowstone Library, 1964; Warren F. Hamilton, Acting Superintendent, Yellowstone, to E. G. Grund, District Agent, U.S. Fish and Wildlife Service, February 3, 1953.

44. Interviews with Hornocker and John Craighead.

45. National Park Service, "Manual of General Information on Yellowstone National Park," 1963, Yellowstone National Park.

46. These political pressures had existed in the twenties, and had provided motivation for the predator-control policy in the first place (National Park Service, "Organizations and Individuals in Favor of Service Predator Policy," Yellowstone Library, 1939).

47. National Park Service, *Administrative Policies for Natural Areas* (Washington, D.C.: U.S. Government Printing Office, 1968), p. 25.

48. Interviews with various park personnel and with Gordon Eastman.

49. Interview with Sideler.

50. National Park Service, "Natural Control of Elk," December 5, 1967; Glen F. Cole, "Yellowstone Wolves," April 1971; interview with Gates; "Wolves Returning to Yellowstone," *Travel Log,* Wyoming Travel Commission, 1968.

51. National Park Service, "Yellowstone Wolves," Yellowstone Library, November 8, 1968; "Yellowstone Wolves," *Travel Log.*

52. Ibid.; Cole, "Yellowstone Wolves," 1971. The "situation room" was described to me by many persons.

53. Interviews with Shoesmith.

54. Interviews with Sumner, Gaab, and others.
55. Interviews with Jonas, and several maintenance personnel.
56. Several interviews with Weaver.
57. Cole, "Yellowstone Wolves," 1971.
58. Ibid.; Department of the Interior, "Yellowstone Wolves," November 8, 1968; Cole, "Gray Wolf," April 1969.
59. In interviews with several personnel working in the park at this time, I also discussed this reconstruction of the record with John Weaver, who had researched it thoroughly, and, through the Freedom of Information Act, obtained a complete record of these reconstructed "sightings." Weaver also gave me a copy of his original tabulations of wolf sightings (John Weaver, "Yellowstone National Park Wolf Reports, 1926–1966"). See also Weaver, *The Wolves of Yellowstone.* For Cole's explanation of his method, see Cole, "Yellowstone Wolves," 1971.
60. Interview with Kittams.
61. "Endangered Species Act Oversight," *Hearings before the Subcommittee on Resource Protection of the Committee on Environment and Public Works, United States Senate,* 95th Congress, 1st Session, July 20, 21, 22, and 28, 1977, Serial No. 95-H33 (Washington, D.C.: U.S. Government Printing Office), p. 385.
62. Mech to Cole, September 9, 1975.
63. Weaver, *Yellowstone Wolves,* p. 15.
64. Interview with Eastman.
65. Of more than three dozen wolf experts whom I interviewed, nearly all admitted suspicions that these wolves had been secretly planted. I spent a year investigating the feasibility of such a transplant, and found that it would, indeed, have been easy to accomplish. Wolves could have been trapped privately in Minnesota and transported west with little difficulty; and at that time they could be taken across the international boundary between Canada and the United States without special permit. Eastman, for instance, regularly took wolves across the border during this period without being asked to show documentation.
66. The quotations of Jonas, Novakowski, Reynolds, and Ward were taken from interviews conducted from 1983 to 1985.
67. Dick Randall, "Returning to the Wild," *Defenders,* vol. 54, no. 4, August 1979.
68. Linn to Raymond F. Dasmann, January 21, 1971.
69. I tracked down many of these sightings, interviewing the rancher near Sheridan who shot the two wolves, the taxidermist who cured the hide of one, and the member of the Wyoming Game and Fish Department who inspected it (see Weaver, *Wolves of Yellowstone,* p. 15, and Gary L. Day, "The Status and Distribution of Wolves in the Northern Rocky Mountains of the United States," M.S. thesis, University of Montana, 1981).
70. Weaver, *Wolves of Yellowstone,* p. 5.
71. Ibid., p. 21.
72. Interview with Weaver.
73. Interview with O'Gara.
74. Reed to Anthony Wayne Smith, President and General Counsel, National Parks and Conservation Association, March 19, 1973.
75. Dunmire to Craighead, July 24, 1972.
76. Interview with Karen Craighead Haynam; Karen Craighead, *Large Mammals of Yellowstone and Grand Teton National Parks,* copyright 1978, Karen Craighead.
77. National Park Service, *Final Environmental Statement, Yellowstone National Park Master Plan* 1974, p. 20; National Park Service, *Natural Resources Management Plan,* 1983, Yellowstone National Park; Haines, *The Yellowstone Story,* p. xix; *National Parkways Guide to Yellowstone,* p. 73; National Park Service, "Population Status of Large Mammal Species in Yellowstone," June 1982; National Park Service, "Checklist — Mammals, Yellowstone National Park," YELL-170, undated (but still distributed), Yellowstone National Park; Fran Hubbard, *Animal Friends of Yellowstone* (Fredericksburg, Texas: Awani Press, 1971).
78. Douglas B. Houston, "Cougar and Wolverine in Yellowstone National Park," Research Note no. 5, Yellowstone Research Office, November 1, 1978.

79. Reed to Director, National Park Service, October 2, 1973.
80. National Park Service, "Population Status of Large Mammal Species in Yellowstone," June 1982; National Park Service, "Checklist — Mammals"; National Park Service, *1983 Natural Resources Management Guide*.
81. Interview, 1985.
82. Office of Endangered Species and International Activities, Bureau of Sport Fisheries and Wildlife, U.S. Department of the Interior, "Northern Rocky Mountain Wolf, *Canis lupus irremotus* (Goldman, 1937)," *Threatened Wildlife of the United States*, Resource Publication 114 (Washington, D.C.: U.S. Government Printing Office, March 1973), p. 235.
83. Ibid.
84. Townsley to Regional Director, National Park Service, Rocky Mountain Region, March 16, 1978; Northern Rocky Mountain Wolf Recovery Team, *Northern Rocky Mountain Wolf Recovery Plan*, U.S. Fish and Wildlife Service, May 28, 1980; Dick Randall, "Returning to the Wild"; Jim Robbins, "Crying Wolf — Restoring the 'Rapacious Predator' to the Rockies," *High Country News*, vol. 13, no. 18, September 18, 1981; Hank Fischer, "Can Western Wolves Make a Comeback?" *Defenders*, vol. 59, no. 3, May/June 1984.
85. The Minnesota plan employs a system with which the Fish and Wildlife Service reimburses farmers for losses suffered from wolf predation (Department of the Interior, Fish and Wildlife Service, "Summary of Basic Data from FWS Livestock Depredation Control Program, 1979–1983," U.S. Fish and Wildlife Service, University of Minnesota, Minneapolis, Minnesota; John Weaver, "Of Wolves and Livestock," *Western Wildlands*, Winter 1983).
86. Glen Cole, "Restoring a Viable Wolf Population in Yellowstone National Park," draft, August 7, 1975 (obtained through the Freedom of Information Act); Estey to Dean Bobzean, August 25, 1976; Mech to Cole, September 9, 1975.
87. Townsley to Hershler, July 30, 1982; Hershler to John Crowell, Jr., Assistant Secretary of Agriculture, May 24, 1982 (both obtained through the Freedom of Information Act).
88. Northern Rocky Mountain Wolf Recovery Team, "Draft Wolf Management Guidelines and Control Plan," October 1982, U.S. Fish and Wildlife Service; Northern Rocky Mountain Wolf Recovery Team, "Northern Rocky Mountain Wolf Recovery Plan," 1985; Chris Cauble, "Yellowstone Won't Get Wolves This Year," *Livingston Enterprise*, January 19, 1984.
89. Northern Rocky Mountain Wolf Recovery Team, "Northern Rocky Mountain Wolf Recovery Plan," 1985, p. iii; Fischer, "Can Western Wolves Make a Comeback?"
90. *Oversight Hearing before the Subcommittee on Public Lands and National Parks of the Committee on Interior and Insular Affairs, House of Representatives*, "Public Land Management Policy," 98th Congress, 1st Session, on Fiscal Year 1984 Budget Request for the National Park Service, February 21, 1983, pp. 14–15.

11. RENDEZVOUS AT DEATH GULCH

1. Richard R. Knight et al., *Yellowstone Grizzly Bear Investigations*, 1981 (Bozeman: U.S. Department of the Interior, National Park Service, July 1982), pp.46–51; Richard R. Knight and Bonnie M. Blanchard, *Yellowstone Grizzly Bear Investigations*, 1982 (Bozeman: U.S. Department of the Interior, July 1983), pp. 28–30; National Park Service, "Bear Mortalities, Greater Yellowstone Area, 1981," November 1, 1981, Yellowstone Park files; National Park Service, "Bear Management Summary," September 2, 1982, Yellowstone Park files. Bears numbers 8, 69, and 90 were sent to British Columbia; numbers 5 and 81 were destroyed by the state of Montana; an adult female was killed with an accidental dose of tranquilizer.
2. Roland Wauer to members of the Interagency Grizzly Bear Steering Committee, August 17, 1982. Knight and Eberhardt were then working on a paper, "Population Dynamics of the Yellowstone Grizzly Bear," which was completed in the following

summer and published in the April 1985 issue of *Ecology,* vol. 66, no. 2, pp. 323–334.

3. "Yellowstone Grizzly Reported Imperiled," *New York Times,* September 1, 1982.
4. National Park Service, "Information Paper No. 24," April 1977, Yellowstone Park files; National Park Service, "Black Bear: *Ursus americanus*" (YELL-466) January 1979, Yellowstone Park; National Park Service, "The Bears of Yellowstone Fact Sheet: Black Bear," Yellowstone Park files; National Park Service, "Interagency Grizzly Bear Study Team Observation Reports, 1975–83."
5. Leakage of the Wauer Memorandum was followed by a spate of publicity on the plight of the grizzly, including a CBS Special in August 1983. See "Grizzlies Scarcer in Yellowstone," *Washington Post,* August 31, 1982; "Grizzlies Seen as Imperiled in Wyoming," *New York Times,* October 10, 1982; Tom McNamee, "Breath-holding in Grizzly Country," *Audubon,* November 1982; and Alston Chase, "The Last Bears of Yellowstone," *Atlantic,* February 1983.
6. National Park Service, *Proceedings of the Meeting of Research Scientists and Management Biologists of the National Park Service* (Grand Canyon National Park: Horace M. Albright Training Center, April 6, 7, and 8, 1968). My account of this meeting is based on this report and on interviews of several individuals who participated.
7. It was at this meeting that the term "ecosystem concept" was first applied to the management of our national parks. The Leopold Report did not mention the word "ecosystem," and the Green Book used it only once, in this sentence: "National Parks, preserved as natural, comparatively self-contained ecosystems, have immense and increasing value to civilization as laboratories for basic research" (National Park Service, *Administrative Policies for Natural Areas,* p. 43).
8. A. Starker Leopold, "The View from Berkeley and Madison," in *Proceedings,* pp. 8–9.
9. Robert M. Linn, "The Ecosystem Concept and the National Parks," in *Proceedings,* ibid., pp. 14–15.
10. Although the degree of worry about bears was emphasized to me by several of the participants, it was not reflected in the *Proceedings.* See Clifford J. Martinka, "Grizzly Research Program of Glacier National Park," and Glen Cole, "Research Goals, Yellowstone National Park," in *Proceedings,* pp. 21, 47–49.
11. F. Dumont Smith, *Book of a Hundred Bears* (Chicago: Rand McNally, 1909), pp. 39–40; Paul Schullery, *The Bears of Yellowstone,* especially pp. 133–136.
12. The largest grizzlies found in Yellowstone weighed in excess of 1,000 pounds, although the average weight, for males, was around 500 pounds and the average weight for females around 350 (John J. Craighead and John A. Mitchell, "Grizzly Bear," *Wild Animals of North America,* J. A. Chapman and G. A. Feldhamer, eds. (Baltimore: Johns Hopkins University Press, 1982), pp. 515–555; Frank C. Craighead, Jr., *Track of the Grizzly* (San Francisco: Sierra Club, 1979); Bill Schneider, *Where the Grizzly Walks* (Missoula: Mountain Press, 1977); Thomas McNamee, *The Grizzly Bear* (New York: Alfred A. Knopf, 1984), and William H. Wright, *The Grizzly Bear* (Lincoln: University of Nebraska Press, 1909, reprinted 1977).
13. Nearly all those visiting the Yellowstone plateau in the 1860s and 1870s reported seeing bear; however, they often did not identify the species they saw. When a species was mentioned, it was most often the grizzly. Whether this frequency meant that these animals were more common than blacks, or whether it meant only that these animals impressed the explorers more, we cannot know. Lewis and Clark encountered their first grizzly on the lower Missouri, during spring 1805; their encounters became more frequent — and more threatening — as they approached the great falls of the Missouri. (Meriwether Lewis and William Clark, *History of the Expedition under the Command of Captains Lewis and Clark to the Sources of the Missouri,* 3 vols. [New York: American Explorers Press, 1973], pp. 268–271).
14. Ibid., pp. 306–310, and pp. 365–366. All along the Missouri in Montana, the explorers found great numbers of dead buffalo, and (inevitably accompanying the buffalo) grizzlies as well, for example: "we found a number of carcases of the Buffaloe lying along shore, which had been drowned by falling through the ice in winter . . . we saw also many tracks of the white bear of enormous size, along the river shore and

about the carcases of the Buffaloe, on which I presume they feed." *The Journals of Lewis and Clark,* Bernard De Voto, ed. (Boston: Houghton Mifflin, 1953), p. 95.

15. John D. Varley, "A History of Fish Stocking Activities in Yellowstone National Park Between 1881 and 1980," Yellowstone Information Paper No. 35, 1980; John D. Varley and Paul Schullery, *Yellowstone Fishes and Their World* (Yellowstone: Yellowstone Library and Museum Association, 1983); L. P. Glenn et al., "Reproductive Biology of Female Brown Bears, *Ursus arctos,* McNeil River, Alaska, Paper 39, *Third International Conference on Bears: Their Biology and Management* (Binghamton, N.Y., and Moscow, U.S.S.R., June 1974), pp. 381–390.

16. Theodore Roosevelt, "Wilderness Reserves," reprinted in Paul Schullery, ed., *Old Yellowstone Days* (Boulder, Colo.: Associated University Press, 1979), p. 202.

17. Paul Schullery, *The Bears of Yellowstone* (Yellowstone: Yellowstone Library and Museum Association, 1980), p. 133.

18. Ibid., especially p. 67. In 1912, the Acting Superintendent estimated the grizzly population in the park to be fifty animals (Brett to Hornaday, July 29, 1912); Albright to Superintendent of Yellowstone, June 18, 1926; press release, Superintendent's Office, July 13, 1938; Olaus J. Murie to Newton B. Drury, Director of the National Park Service, March 24, 1944, and Lewis Gannett, *Sweet Land* (Garden City, N.Y.: Doubleday, Doran, 1934), pp. 169–177.

19. John J. Craighead and Frank C. Craighead, Jr., "Management of Bears in Yellowstone National Park," July 1967, National Park Service files, YELL-67, pp. 17–24; John J. Craighead and Frank C. Craighead, Jr., "Grizzly Bear–Man Relationships in Yellowstone National Park," *BioScience,* vol. 21, no. 16, August 15, 1971; and John J. Craighead et al., "A Population Analysis of the Yellowstone Grizzly Bears," Missoula: Montana Cooperative Wildlife Research Unit, 1974.

20. On July 31, 1983, for instance, a young Frenchman visiting Yellowstone was gored by a bison and later died from the wounds (National Park Service press release, August 1, 1983).

21. Smith, *Hundred Bears,* pp. 44–46.

22. Schullery, *Bears of Yellowstone,* p. 70–72; and Craighead and Craighead, "Management of Bears in Yellowstone," pp. 15, 93 ff.

23. Craighead and Craighead, "Management of Bears in Yellowstone," p. 19, Table 4, and "National Park Service, Bear Incidents, 1931–1968," Yellowstone files.

24. The lawsuit in Sequoia was confirmed by the National Park Service Public Information Office in November 1982.

25. The "data base" accumulated by the Craigheads was more than twice as great as that later gathered by the Interagency Grizzly Bear Study Team. See Chapter 12.

26. Craighead and Craighead, "Management of Bears in Yellowstone," Table 7, pp. 38, 72.

27. Victor G. Barnes, Jr., and Olin E. Bray, "Population Characteristics and Activities of Black Bears in Yellowstone National Park" (Colorado State University: Colorado Cooperative Wildlife Research Unit, June 1967); National Park Service, "Yellowstone Black Bears"; Glen F. Cole, "Management Involving Grizzly and Black Bears in Yellowstone National Park, 1970–1975," 1976, Yellowstone Park files; Mary Meagher and Jerry R. Phillips, "Restoration of Natural Populations of Grizzly and Black Bears in Yellowstone National Park," address delivered to Fifth International Conference on Bear Research and Management, International Association for Bear Research and Management, Madison, Wisconsin, February 1980.

28. Stephen Herrero, *Bear Attacks: Their Causes and Avoidance* (New York: Nick Lyons, 1985), pp. 56–63; Jack Olsen, *Night of the Grizzlies* (New York: New American Library, 1969).

29. Linn to Joseph P. Linduska, Associate Director, Bureau of Sport Fisheries and Wildlife, May 21, 1968; John Craighead to James M. Coutts, Acting Chief, Division of Wildlife Research, May 29, 1968.

30. Linn to Linduska, May 21, 1968; Glen F. Cole, "Management Involving Grizzly Bears and Humans in Yellowstone National Park, 1970–1973 — An Interim Report," reprinted in *BioScience,* vol. 24, no. 6, June 1974, pp. 335–338.

31. Craighead and Craighead, "Management of Bears in Yellowstone," p. 6. The Craig-

heads mistakenly supposed that the grizzly had been put on the list in 1966. In fact, it did not appear there until March 11, 1967 (*Federal Register* vol. 32, no. 8).

32. Craighead and Craighead, "Management of Bears in Yellowstone," p. 72.
33. Ibid., p. 85.
34. Ibid., pp. 96–98.
35. Ibid., p. 94–96.
36. Ibid., pp. 98–99.
37. Linn to Linduska, May 21, 1968.
38. Schullery, *Bears of Yellowstone,* p. 88; John J. Craighead and Frank C. Craighead, Jr., "Turning in on the Grizzly," *Science Year Special Reports,* pp. 35–49.
39. Craighead and Craighead, "Bear Management in Yellowstone"; John Craighead, "The Ecocenter," address delivered to Murie Chapter of the Audubon Society, April 28, 1984.
40. National Park Service, *Final Environmental Statement, Yellowstone Master Plan,* June 11, 1974, p. 32.
41. Craighead et al., "A population analysis of the Yellowstone Grizzly Bears," p. 6; National Academy of Sciences, "Report of Committee on the Yellowstone Grizzlies" (Washington, D.C.: National Academy Press, 1974), pp. 15–21.
42. National Academy of Sciences, "Report," pp. 15–21; Frank C. Craighead, Jr., *Track of the Grizzly,* p. 217.
43. National Academy of Sciences, "Report," pp. 33–35; Cole, "Management Involving Grizzly Bears and Humans in Yellowstone National Park–An Interim Report."
44. Barnes and Bray, *Population Characteristics of Black Bears.*
45. Linn to John Craighead, October 13, 1969.
46. Curtis Bohlen, Deputy Assistant Secretary of the Interior, to Senator Robert F. Griffin, November 27, 1973.
47. A. Starker Leopold, Stanley A. Cain, and Charles E. Olmsted, "A Bear Management Policy and Program for Yellowstone National Park," meeting of the Natural Sciences Advisory Committee of the National Park Service, Yellowstone National Park, September 6, 1969.
48. A. Starker Leopold and Durward L. Allen, "Memorandum to Director Whalen," National Park Service, August 2, 1977.
49. National Park Service, press release, March 17, 1969.
50. Schullery, *Bears of Yellowstone,* pp. 80–81; Frank C. Craighead, Jr., *Track of the Grizzly,* pp. 196 ff.
51. This figure represents my own calculations from the "Lake Bear Logs" and varies but slightly from those presented by Frank Craighead in *Track of the Grizzly* (p. 196).
52. National Academy of Sciences, "Report," pp. 26–33.
53. Harry V. Reynolds, Jr., to Dr. Ian McT. Cowan, Chairman of the Committee on the Yellowstone Grizzlies, January 19, 1974.
54. Interview, 1982.
55. Interview, 1982; notes of interview Morris had with Dick Randall, in May 1973.
56. Hunting of the Yellowstone grizzly in Montana did not end until 1975.
57. Minutes of Grizzly Bear Meeting, Bozeman, Montana, March 29, 1972, morning session.
58. Ibid.
59. In the period 1968 to 1973, I was told, Yellowstone authorities were encouraged to dispose of old records. The "incident reports" that I was shown referred to ranger logs, but the original logs are nowhere to be found.
60. U.S. Department of the Interior, "Draft Environmental Statement, Proposed Grizzly Bear Management Program," September 18, 1974. By this time the Environmental Policy Act had been law for four years and the "proposed" management plan had been in effect for six.
61. Deputy Associate Director, National Park Service, to regional directors, March 1, 1972; Director of the Midwest Region, National Park Service, to superintendents in region, March 14, 1972; Regional Director to Superintendent, Yellowstone, March

27, 1972; Harold J. Estey, Chief Park Ranger, Yellowstone, to the Bayer Company, August 29, 1972.

62. Frank C. Craighead, Jr., *Track of the Grizzly,* p. 208.
63. Ibid., pp. 201–202; Schullery, *Bears of Yellowstone,* p. 86. A ban on further marking of bears was initiated by Superintendent Anderson in 1969.
64. Cole to Linn, October 10, 1972.
65. See, for instance, Robert C. Haraden, Acting Superintendent, Yellowstone, to Yellowstone Park personnel, March 31, 1975: "We are asking your cooperation in reporting all sightings of bears — both black and grizzly — to the Communications Center. . . . Please supply as much information as you can, using the enclosed sample reporting sheet as a guide."
66. Reed to Senator Philip A. Hart, November 1, 1973.
67. Reed to Representative James Haley, Chairman, Committee on Interior and Insular Affairs, October 17, 1973.
68. The Fund for Animals, Inc., news release, October 1973; John J. Craighead, "A Briefing on the Grizzly Bear Situation in Yellowstone," Mammoth, Yellowstone National Park, September 19, 1972.
69. Regenstein to Bohlen, December 28, 1973.
70. Seater to Mike Tharp of the *Wall Street Journal,* February 14, 1974.
71. Seater to Regenstein, September 5, 1973.
72. Seater to Tharp, February 14, 1974.
73. Reed to Director of the National Park Service, November 19, 1973; Reed to Regenstein, November 6, 1973.
74. Morton to Handler, February 6, 1973.
75. National Academy of Science, "Report," pp. 13–33.
76. Ibid., pp. 33–35.
77. Ibid., pp. 39–43.
78. Watts to Director, National Park Service, December 19, 1974. By the time this letter was written, the Steering Committee was already under the direction of Knight, a Park Service employee.
79. Knight's office was moved from Mammoth to Bozeman, Montana only after the earlier arrangement had received criticism from the press.
80. Letter from Finley to the editor of *Wildlifer,* March 22, 1983.
81. Letter from Anderson to James B. White, Director, Wyoming Game and Fish Department, March 20, 1975.
82. These actions were pursuant to a petition, filed by the Fund for Animals, February 14, 1974, with the Park Service, "urging certain steps to protect the grizzly bear *Ursus arctos horribilis* as an 'endangered species.' " Yet today grizzlies are still legally hunted in northern Montana.
83. Russell Peterson to Rogers C. B. Morton, February 3, 1975.
84. Interagency Grizzly Bear Study Team, "Individual Bear Histories"; Knight to Dr. W. Leslie Pengelly, Director, Wildlife Biology Program, University of Montana, April 12, 1984, in response to Pengelly's Freedom of Information Act request of January 24, 1984.
85. One can tell a great deal from the age profile of a wildlife population. That a population is young usually suggests that it is either suffering from, or recovering from, excessive mortality.
86. Interview with John Craighead, 1985; J. Craighead, "The Ecocenter."
87. Interagency Grizzly Bear Study Team, "Observation Reports." Yellowstone officials, however, point to the large number of "sightings" of black bears reported by visitors and park personnel. Because everyone is encouraged to report bears sighted, however, and because one bear near a road can be sighted by scores of people in one afternoon, these figures are literally meaningless.
88. Ibid.
89. Richard R. Knight, Bonnie M. Blanchard, and Katherine C. Kendall, *Yellowstone Grizzly Bear Investigations: Report of the Interagency Study Team, 1980* (Bozeman: U.S. Department of the Interior, National Park Service, July 1981), p. 27.

90. Ibid., p. 29.
91. Interagency Grizzly Bear Study Team, "Individual Bear Histories"; Trevor Povah to Robert C. Haraden, Assistant Superintendent, Yellowstone, September 15, 1975.
92. *Yellowstone Grizzly Bear Investigations,* 1978–1981.
93. The story of garbage in these towns is one of successive attempts to make the dumps inaccessible to grizzlies. The Cooke City dump, for instance, was first moved to a new location; then a chain-link fence was put around it. Because careless villagers sometimes left the gate to the dump open, the dump was locked and refuse could be deposited there only at specified times each week. This precaution required villagers to keep garbage behind their houses longer, attracting bears into town.
94. Craighead and Mitchell, "Grizzly Bear," pp. 546–548.
95. Herbst to Senator Richard Stone, November 9, 1977.
96. Handwritten memorandum from Associate Director for Fish and Wildlife Resources to Herbst, November 2, 1977.
97. Glen F. Cole, "Management Involving Grizzly Bears in Yellowstone National Park, 1970–1974," contributing paper, 26th Annual AIBS Meeting of Biological Sciences, Oregon State University, Corvallis, Oregon, August 1975.
98. National Park Service, "Yellowstone Black Bears"; National Park Service, "The Bears of Yellowstone Fact Sheet"; Glen F. Cole and Mary M. Meagher, "Restoration and Maintenance of Natural Grizzly and Black Bear Populations in Yellowstone National Park," Information Paper No. 15, revised March 1979; National Park Service, "Summary of Yellowstone Bear Protection and Management Policies," revised April 1979, Yellowstone Park files; Mary Meagher, "Evaluation of Bear Management in Yellowstone National Park, 1977," Research Note No. 8, February 10, 1978, Yellowstone Park files.
99. Cole and Meagher, "Restoration of Natural Populations of Grizzly and Black Bears."
100. National Park Service, "The Bears of Yellowstone," undated, but currently available at visitor centers in the park.
101. A. Starker Leopold, Foreword to Schullery, *The Bears of Yellowstone,* p. 1.
102. *The Bears of Yellowstone,* p. 97.
103. National Park Service, "Bear Management Summary," Yellowstone National Park, 1969–1983, Yellowstone files.
104. Leopold and Allen, "Memorandum to Director Whalen," August 2, 1977, p. 13.
105. From extensive interviews, I discovered profound disillusionment in the ranks of those who worked in bear management a decade ago. But few wished to discuss the matter, and absolutely no one would talk to me about it on the record.
106. T. A. Jaager, "Death Gulch a Natural Bear Trap," *Appleton's Popular Science Monthly,* vol. 54, February 1899, p. 475. The first white men to visit the gulch were probably Frank H. Knowlton and Walter H. Weed of the United States Geological Survey, who were exploring the park in 1888 (Frank H. Knowlton, "A Gulch of Death," *Popular Science News,* from an undated clipping in the Yellowstone archives).
107. Ernest Thompson Seton, *The Biography of a Grizzly* (New York: Century, 1899).
108. *Hearing before the Subcommittee on Environmental Pollution of the Committee on Environment and Public Works, U.S. Senate, 98th Congress, "Grizzly Bear Management in the Yellowstone Ecosystem, August 11, 1983,"* Cody, Wyoming (Washington, D.C.: U.S. Government Printing Office, 1984), p. 32.
109. Letter from Maurice Hornocker to the editor of the *Wildlifer,* No. 195, November–December 1982.

12. THE GRIZZLY AND THE JUGGERNAUT

1. Draft of a report to the Secretary of the Interior, William Clark, from Wildlife Committee of the National Park System Advisory Board, John Turner, Chairman, December 16, 1984.
2. Yellowstone Park press release, July 12,1984.
3. Richard R. Knight and Lee L. Eberhardt, "Population Dynamics of Yellowstone

Grizzly Bears"; *Ecology*, vol. 66, no. 2, April 1985, pp. 323–334; Richard R. Knight, "Projected Future Abundance of the Yellowstone Grizzly Bear," *Journal of Wildlife Management*, vol. 48, no. 4, October 1984, pp. 1434–1438.

4. John Craighead, in preparation; cf. John Craighead, "The Ecocenter," address to Murie chapter of the Audubon Society, Casper, Wyoming, April 28, 1984.

5. Richard R. Knight et al., "Final Report, ad hoc Committee to Investigate the Need and Feasibility of the Supplemental Feeding of Yellowstone Grizzly Bears," December 5, 1983, Table 1, Yellowstone Park files.

6. Knight and Eberhardt, "Population Dynamics"; John Craighead, "The Ecocenter."

7. John Turner et al. "The Great Bear in Our Parks," A Report on Grizzly Bear Management in Our National Parks from the National Park Advisor, Board's Wildlife Committee, April 1985; U.S. Department of the Interior, Fish and Wildlife Service, "Grizzly Bear Recovery Plan" (Denver: Fish and Wildlife Reference Service, 1982); John J. Craighead, "A Proposed Delineation of Critical Grizzly Bear Habitat in the Yellowstone Region," monograph presented at the Fourth International Conference on Bear Research and Management, held at Kalispell, Montana, February 1977; Turner et al., "Report to the Secretary of the Interior." The Fish and Wildlife Service attempted, in the late 1970s, to designate critical habitat for the grizzly, but met so much political opposition that it gave up. Today, government agencies responsible for grizzly management have agreed on an informal designation — "situation one" — for such habitat, but no studies have been done to establish the carrying capacity of this area.

8. John J. Craighead, J. S. Sumner, and G. B. Scaggs, *A Definitive System for Analysis of Grizzly Bear Habitat and Other Wilderness Resources, Utilizing LANDSAT Multispectral Imagery and Computer Technology*, Wildlife-Wildlands Institute Monograph No. 1 (Missoula: University of Montana Foundation, 1982); John Craighead, "Ecosystem Habitat Analysis," Wildlife-Wildlands Institute, 1985.

9. Thomas McNamee, *The Grizzly Bear* (New York: Alfred A. Knopf, 1984), p. 235; Knight and Eberhardt, "Population Dynamics."

10. For the effects of fire-suppression policies on wildlife habitat, see Chapter 9.

11. Charles Kay, "Why Grizzly and Black Bears Are Declining in Yellowstone National Park — A New Explanation," 1984, p. 2. Researchers have found similar depletion of grizzly habitat by grazing ungulates near Glacier Park (Charles Jonkel, "The Blackfeet Reservation: Dilemmas in Range and Grizzly Management," *Western Wildlands*, vol. 11[1], Spring 1985, pp. 25–27).

12. Interview.

13. Interview.

14. *Billings Gazette*, April 10, 1984.

15. John Craighead, "The Ecocenter."

16. Ibid.

17. Tony Povilitis, "On the Citizens' Proposal to Save the Yellowstone Grizzly Bear," Campaign for Yellowstone's Bears, Casper, Wyoming, April 28, 1984.

18. Interview.

19. Interview.

20. Interview.

21. Interview.

22. *Livingston Enterprise*, February 18, 1983.

23. Entry under "Nature," *Oxford English Dictionary;* Rackstraw Downs, "Henri Rousseau and the Idea of the Naive," *New Criterion*, vol. 3, no. 9, May 1985, pp. 15–24; "Nature, Philosophical Ideas of," *Encyclopedia of Philosophy* (New York: Macmillan, 1967); Alfred North Whitehead, *Concept of Nature* (Cambridge: Cambridge University Press, 1920); and Clarence J. Glacken, *Traces on the Rhodian Shore: Nature and Culture in Western Thought from Ancient Times to the End of the 18th Century* (Berkeley: University of California Press, 1967).

24. Plato, *The Republic*, translated with Introduction and Notes by Francis MacDonald Cornford (Oxford, Oxford Unversity Press, 1945), pp. 119–143.

25. For the Greeks, the question was: Is a thing *physis* or *technē* — nature or art? Aristotle viewed the universe using the analogy of a vegetable: All things that could

grow were natural. The Roman and church natural-law traditions viewed anything that was innate as natural. To express their theories of the state, seventeenth- and eighteenth-century philosophers such as Hobbes, Locke, and Rousseau postulated a mythical "state of nature."

26. Martin V. Melosi, *Garbage in the Cities: Refuse, Reform and the Environment, 1880–1980* (College Station: Texas A&M University Press, 1981), pp. 5–6.
27. McNamee, *The Grizzly Bear* p. 235.
28. Thomas Bulfinch, *Bulfinch's Mythology* (New York: Avenel, 1978), pp. 322–323.
29. Letter from Hornocker to editor of the *Wildlifer*, no. 195, November–December, 1982.
30. Interview.
31. Interview.
32. "U.S. Department of Agriculture, U.S. Department of the Interior, and states of Idaho, Montana, Wyoming, and Washington Memorandum of Agreement to Revise and Expand the Interagency Grizzly Bear Committee," executed April 1983.
33. Peterson to Povilitis, April 25, 1984.
34. Eno to Povilitis, March 21, 1984.
35. Plaza to Chapter Presidents, Conservation Chairs. Newsletter Editors, April 20, 1984; Turner et al., "Report."
36. Interview with Regenstein. Paradoxically, it was Reed who had appointed Knight to lead the Interagency Grizzly Bear Study Team in 1973.
37. Interview.
38. Meagher to Reed, August 5, 1982. This was precisely the time when Knight, aware of Eberhardt's projections, had expressed his concern about the grizzly to Wauer.
39. Reed to Russell W. Peterson, Durward L. Allen, Joseph J. Hickey, Rupert Culter, A. Starker Leopold, C. Eugene Knoder, Charles M. Loveless, and Glenn Paulson. The article to which Reed refers is presumably Tom McNamee's, "Breath-holding in Grizzly Country," *Audubon*, vol. 84, no. 6, November 1982.
40. Minutes of the Interagency Grizzly Bear Research Subcommittee, Yellowstone National Park, August 11, 1983, p. 8.
41. Chairman, ad hoc Committee for Population Analysis to Chairman, Interagency Grizzly Bear Steering Committee, January 18, 1983; U.S. Department of the Interior, U.S. Fish and Wildlife Service, Region 6 news release, May 13, 1983; Associated Press release, May 13, 1983.
42. "Errata Sheet," undated, but obtained by this writer in June 1983; *High Country News*, July 8, 1983; *Livingston Enterprise*, December 6, 1983; *Bozeman Daily Chronicle*, December 6, 1983.
43. The ad hoc Committee on Supplemental Feeding was created on February 16, 1983. Its members were: Gary Brown, John Craighead, Mary Meagher, Larry Roop, Chris Servheen, and Richard Knight.
44. Knight et al., "Final Report, ad hoc Committee to Investigate Supplemental Feeding of Grizzly Bears," December 5, 1983, p. 1 and Table 1.
45. Ibid., p. 1.
46. Ibid., p. 2.
47. Ibid., pp. 4–5.
48. Ibid., p. 5.
49. On March 30, shortly after they had begun deliberations, Meagher and Brown sent a memorandum to Knight stating reasons for their opposition to supplemental feeding.
50. Interview.
51. Hank Fischer, "Should We Feed the Grizzlies?" *Defenders*, March–April 1984.
52. Eno to R. Brent Erickson, Office of Senator Alan K. Simpson, August 21, 1984.
53. National Park Service, *Final Environmental Impact Statement, Grizzly Bear Management Program*, July 1983, Yellowstone National Park; National Park Service, "Bear Management Area Use Adjustment," Yellowstone National Park, 1984; Director, Rocky Mountain Region, to the Superintendents of Glacier and Yellowstone, March 29, 1984; *Livingston Enterprise*, July 1, 1983.
54. Although the seriousness of these threats is not to be discounted, they have been

exaggerated. In fact, habitat around the park is better today than it was, for example, in 1910. In that earlier year, the population of Park County, Montana, was greater than it is today; hunting was unregulated; 100,000 sheep were grazing near Gardiner in an area that is now closely protected by the Forest Service; arsenic from the gold mines at Jardine had poisoned all the trout in the Yellowstone River as far as Yankee Jim Canyon (a stretch of river that now provides fine fishing); and towns that no longer exist, such as Cinnebar and Electric, sat cheek to jowl with the park boundary.

55. Most of these steps were taken during spring 1983.
56. *Livingston Enterprise,* October 6, 1982.
57. Wildlife Society, "Resolution on the Yellowstone Grizzly," *Northwest Wildlifer,* January 1985.
58. *New York Times,* October 21, 1984.
59. Concern is widespread among environmentalists in the Yellowstone region that if supplemental feeding were reintroduced, and was obviously successful, the arguments for protecting wilderness in the northern Rockies would be greatly weakened. See Chapter 20.
60. Letter to the Editor of the *Cody Enterprise,* written August 23, 1984.
61. Turner et al., "Report."
62. *Great Falls Tribune,* February 20, 1983; *Livingston Enterprise,* May 23, 1983; *Livingston Enterprise,* June 21, 1983; *Livingston Enterprise,* June 24, 1983; Interagency Grizzly Bear Study Team, "Grizzly Bear Mortalities, 1983," Yellowstone Park files. The last cub to survive was bear number 95, caught in Yellowstone and transported to Missoula to be used in experiments before being destroyed.
63. *Livingston Enterprise,* June 27, 1983; Stephen Herrero, *Bear Attacks* (New York: Nick Lyons, 1985), pp. 70–71; Mark Henckel, "No. 15: Good Bear Gone Bad," *Billings Gazette* July 24, 1983; Mark Henckel, " 'Dum-Dum' was No Brash Bear," *Billings Gazette,* July 24, 1983.
64. *Billings Gazette,* September 3, 1983; *Livingston Enterprise,* September 9, 1983; Interagency Grizzly Bear Study Team, "Individual Bear Histories."
65. *Life,* August 1984.
66. *Billings Gazette,* August 24, 1983; *Billings Gazette,* August 25, 1983; *Livingston Enterprise,* September 8, 1983; *Billings Gazette,* September 17, 1983.
67. *Billings Gazette,* August 25, 1983.
68. National Park Service, "Grizzly Bear Mortalities, 1983."
69. *Livingston Enterprise,* August 2, 6, and 21, 1984.
70. Interagency Grizzly Bear Study Team, "Individual Bear Histories"; Interagency Grizzly Bear Study Team, "1984 Mortalities."
71. Interagency Grizzly Bear Study Team, "1984 Mortalities."
72. Dick Hartman and Clynn Phillips, "Report on ReAcT Meeting," held at Brinkerhof Lodge, Jackson, Wyoming, September 13, 1984. These minutes confirm, nearly verbatim, what he told me (and apparently other reporters) in a recent interview (*Casper Star-Tribune,* May 20, and September 30, 1984).
73. Hartman and Phillips, "Report."
74. Knight et al., "Yellowstone Grizzly Bear Investigations," Table 7, page 32; Richard Knight, David J. Mattson, and Bonnie M. Blanchard, "Movements and Habitat Use of the Yellowstone Grizzly Bear," November 1984, Yellowstone Park files, p. 88.
75. John Craighead, "Yellowstone Grizzly Bear Ecocenter (Role of Food Reserves in the Biology of Yellowstone Grizzlies)," in preparation, 1985.
76. *Casper Star Tribune,* September 28, 1984.
77. Jonkel to LeRoy Ellig, Regional Supervisor, Montana Department of Fish, Wildlife and Parks, dated July 7, 1983.
78. Ibid., Charles Jonkel, "Man Is Not a Natural Prey for Grizzly Bears," *Missoulian,* November 2, 1983; *Missoulian,* October 2, 1983.
79. Herrero, *Bear Attacks,* p. 78.
80. Interview, September, 1982.
81. *Livingston Enterprise,* May 4, 1984.
82. Interview with Knight, 1985; *Billings Gazette,* March 25, 1985; *Casper Star-Tribune,*

March 25, 1985; Greater Yellowstone Coalition, "Yellowstone Grizzly Policy," December 3, 1984.

83. The figure for the fuel cost of trucking garbage was obtained from Steven Ioebst of the Yellowstone maintenance staff in October 1983.

84. Gary Brown and Mary Meagher to members of ad hoc Committee on Supplemental Feeding, March 3, 1983.

13. GRANT VILLAGE AND THE POLITICS OF TOURISM

1. National Park Service, *Environmental Assessment for the Development Concept Plan for Grant Village, Yellowstone National Park* (Denver: Denver Service Center, June 1979), pp. 15–37.

2. Replogle, *The Bannock Indian Trails of Yellowstone;* Aubrey L. Haines, *The Yellowstone Story* (Yellowstone: Yellowstone Library and Museum Association, 1977), vol. 2, p. 228.

3. Yellowstone Superintendent Robert Barbee told me this story.

4. Haines, *Yellowstone Story*, vol. 2, p. 479; Conrad L. Wirth, *Parks, Politics and the People* (Norman: University of Oklahoma Press, 1980), p. 238 ff.

5. Haines, *Yellowstone Story*, vol. 2, pp. 347 ff.; Alfred Runte, *National Parks: The American Experience* (Lincoln: University of Nebraska Press, 1979), pp. 44–45, 83, 91–93, 99–101, 113.

6. John Muir, "The Wild Parks and Forest Reservations of the West," *Atlantic*, vol. 81, January 1898; Runte, *National Parks*, p.172.

7. Runte, *National Parks*, p. 103.

8. Horace Albright, "Research in the National Parks," *Scientific Monthly*, June 1933.

9. For the evolution of the ranger force see Chapter 14.

10. Runte, *National Parks*, p. 111.

11. Haines, *Yellowstone Story*, pp. 478–479, 484–485.

12. Runte, *National Parks*, pp. 157–158.

13. Haines, *Yellowstone Story*, vol. 2, pp. 372–373.

14. Wirth, *Parks, Politics and People*, pp. 238, 240.

15. Wirth, *Parks, Politics and People*, p. 260. Their estimate of eighty million people visiting the national parks by 1966 turned out to be conservative. The actual visitation by 1966 was 133,081,000.

16. Garrison, *Making of a Ranger*, p. 258.

17. Runte, *National Parks*, p. 173.

18. Garrison, *Making of a Ranger*, pp. 276 ff.; Haines, *Yellowstone Story*, p. 375.

19. Haines, *Yellowstone Story*, p. 375.

20. Wirth, *Parks, Politics and People*, pp. 15, 61; Haines, *Yellowstone Story*, pp. 370, 378.

21. Haines, *Yellowstone Story*, pp. 370 ff. The Nichols family mortgaged their family ranch in the Gallatin Valley, Montana to finance construction at Canyon Village.

22. Haines, *Yellowstone Story*, p. 377.

23. Ibid., p. 378.

24. Garrison, *Making of a Ranger*, pp. 279–280. The role of Theodore Wirth, Jr., in designing portions of Grant Village was confirmed to me by his father in an interview in 1984.

25. Jackson Hole Preserve, Inc., *Preliminary Survey of Yellowstone Park Company,* Cresap, McCormick and Paget, Harris, Kerr, Forster & Company, June–September, 1965 (obtained through the Freedom of Information Act).

26. For a portrait of Laurance Rockefeller's role with the conservation movement, see Peter Collier and David Horowitz, *The Rockefellers: An American Dynasty* (New York: Holt, Rinehart and Winston, 1976), pp. 383 ff.

27. Jackson Hole Preserve, Inc., *Preliminary Survey.*

28. Haines, *Yellowstone Story*, vol. 2, p. 374; National Park Service, *Draft General Management Plan, John D. Rockefeller, Jr., Memorial Parkway, Wyoming* (Denver: National Park Service, 1980), p. 1. The parkway was authorized by Public Law 92-

404 on August 25, 1972. A bypass road for Fishing Bridge was also planned in 1968 and a "Fishing Bridge Bypass Committee" established (National Park Service, "Fishing Bridge By-Pass Committee Report," September 21, 1972, Yellowstone Park files).

29. The effect of the West Thumb bypass was described to me by Jonas, who was stationed there during the time of construction as a seasonal ranger. The effect of the marina on the lakeshore and the original design of sewage treatment were confirmed in several interviews with park engineers and maintenance staff. (See also *Environmental Assessment, Grant Village*, pp. 33 ff. and *Master Plan for Yellowstone National Park* (Denver: Midwest Regional Office, National Park Service, 1973, p. 5).

30. The controversy over the design of Glacier Park sewage treatment was described to me by Riley McClellen, research scientist in Glacier.

31. Garrison, *Making of a Ranger*, p. 280.

32. Runte, *National Parks*, p. 177.

33. Edward Abbey, *Desert Solitaire: A Season in the Wilderness*, p. 52.

34. The *Master Plan* was completed in 1973, and approved in 1974 (National Park Service, *Master Plan*).

35. *Master Plan.*

36. Ibid.

37. Ibid.

38. Ibid.

39. The National Historic Preservation Act was passed in 1966. See Lee, *Family Tree of the National Park System*, p. 74.

40. Williams, "*Giardia* and the Waterborne Transmission of Giardiasis"; David G. Stuart et al., "Effects of Multiple Use on Water Quality of High-Mountain Watersheds: Bacteriological Investigations of Mountain Streams," *Applied Microbiology*, vol. 22, no. 6, 1971, pp. 1048–1054; William G. Walter et al., "Microbiological and Chemical Studies of an Open and Closed Watershed," *Journal of Environmental Health*, vol. 301, no. 157, 1967; Thomas K. Haack and Gordon A. McFeters, "Microbial Dynamics of an Epilithic Mat Community in a High Alpine Stream," *Applied and Environmental Microbiology*, vol. 43, no. 3, March 1982; Sidney A. Stuart et al., "Aquatic Indicator Bacteria in the High Alpine Zone," *Applied and Environmental Microbiology*, vol. 31, no. 2, February 1976; Kenneth L. Temple et al., "Survival of Two Enterobacteria in Feces Buried in Soil Under Field Conditions," *Applied and Environmental Microbiology*, vol. 40, no. 4, October 1980.

41. This account was given to me by an authoritative source, who wished to remain anonymous.

42. Jonas to Barbee, September 28, 1983.

43. Interviews with Jonas and John Craighead, 1984–1985. Until the late 1960s a boat dock at Fishing Bridge attracted fishermen, fish, and therefore bears. See Department of the Interior, Bureau of Sport Fisheries and Wildlife, *Summary Report, 1967, Fishery Management Program, Yellowstone National Park, Wyoming*, Albuquerque, New Mexico, February 9, 1968, p. 10.

44. Department of the Interior, Fish and Wildlife Service, *Annual Project Report, Yellowstone Fishery Investigations, Yellowstone National Park, Calendar Year 1974*, July 11, 1975, p. 30. The Pelican Creek trap had fallen into disrepair by 1967. See Department of the Interior, Bureau of Sport Fisheries and Wildlife, *Annual Project Report, Calendar Year 1972*, May 11, 1973, p. 13.

45. Cole to Anderson, August 23, 1971, Table 2; National Park Service, *Fishing Bridge and the Yellowstone Ecosystem: A Report to the Director*, Yellowstone National Park, 1984, Table 12, pp. 64–65; Cole to Anderson, September 21, 1972.

46. Garrison, *Making of a Ranger*, pp. 258–259. Yellowstone passed the two-million mark in attendance in 1965, yet in 1979 the total was only 1,875,169. Indeed, despite the popular belief that visitation is constantly increasing, it has remained roughly constant for the past two decades.

47. National Park Service, *Final Environmental Statement, Yellowstone Master Plan* (Denver, Midwest Regional Office, 1974), pp. 6, 30.

48. Ibid., pp. 28 ff.

49. Interview with Anderson. In 1966 the Nichols family sold the Yellowstone Park

Company to Goldfield Enterprises, which later became the General Host Corporation. In October 1966, the company signed a thirty-year contract with the National Park Service to operate concessions in Yellowstone. See Haines, *Yellowstone Story,* p. 423; National Park Service, *Yellowstone National Park Concessions Management Review of the Yellowstone Park Company* (Denver: National Park Service, February–October, 1976), pp. 7–8; Laventhol & Horwath, Certified Public Accountants, "Concessions Operations Plan, Yellowstone National Park, June 1979 (obtained through the Freedom of Information Act), pp. I-4 to I-10.

50. Interview with Townsley.
51. Everyone, it seems — both those in favor of the project and those opposed — credit (or blame) Townsley with the building of Grant Village.
52. National Park Service, *Concessions Management Review,* pp. 133 ff.
53. Department of the Interior and Department of Agriculture, *Greater Yellowstone Cooperative Regional Transportation Study.* See, in particular, "Final Report, Assessment of Alternatives" (February 1979), "Visitor Use Portion" (April 1978), "Regional Assessment, Socioeconomic Report" (June 1978), and "Regional Assessment, Summary Report" (December 1978).
54. Hersler to W. A. and John Warnock, October 19, 1982.
55. Interviews with Whalen, Turner, and Jarvis.
56. Interviews with Hocker.
57. National Park Service, "Fiscal 1980 Construction Program, Justification Material," pp. 1058 and 1063, *Budget for the Fiscal Year 1980* (U.S. Government Printing Office, 1979). See 1978 testimony by Director Whalen before Congressman Yates's Subcommittee on Interior and Related Agencies, for fiscal 1980 budget, pp. 781–788.
58. Galen Buterbach, Regional Director of the Fish and Wildlife Service, interview.
59. *Environmental Assessment for Grant Village,* June 1979.
60. Ibid., pp. 13, 43.
61. Ibid., pp. 11, 13. Notice that it also claims that "a small number of endangered gray wolves are believed to inhabit the park" (p. 4).
62. Ibid., p. 41.
63. Ibid., p. 12.
64. The status of this appropriations request at this time was explained to me by C. Bruce Sheaffer, Assistant Chief of the Budget Division, National Park Service, and Dale Snape, former officer in charge of National Park Service Oversight, Office of Management and Budget (interviews, 1963–1965).
65. John A. Townsley and L. E. Surles, "Finding of No Significant Impact," executed February 8, 1982.
66. From records of Budget Division, National Park Service, Washington, D.C.
67. Snape visited Yellowstone in August 1979 and discovered by accident that the second stage of planning for Grant Village was already under way. According to Snape he sought out Townsley and had long conversations with him on the project. Snape thought it an inappropriate use of federal dollars, and told Townsley so.
68. Interviews with Bowser, 1983–1984.
69. National Park Service, "Preinvitation Notice," October 3, 1979 (obtained through the Freedom of Information Act); National Park Service, "Development / Study Package Proposal, Grant Village," January 19, 1979; James C. Gritman, Acting Regional Director, Region 6, U.S. Fish and Wildlife Service, Denver, "Section 7 Consultation on Grant Village, Yellowstone Park," October 31, 1979.
70. Gritman, "Section 7 Consultation."
71. Ibid.
72. Ibid.
73. Robert M. Ballon, Acting Area Manager, Fish and Wildlife Service, Billings, Montana, to Regional Director, National Park Service, Rocky Mountain Region, December 12, 1980 (obtained through the Freedom of Information Act). Throughout 1980 and 1981 the Fish and Wildlife Service was uneasy about Park Service intentions at Grant and Fishing Bridge (Wally Steucke, Area Manager, Fish and Wildlife Service, Billings, to Townsley, February 18, 1981; Roy C. Slatkavitz, Chief, Division of Park Planning, to Regional Director, Rocky Mountain Region, November 25,

1980; Regional Director, Region 6, Fish and Wildlife Service, Denver, to Regional Director, National Park Service, Denver, October 2, 1980 and May 3, 1980).

74. National Park Service, "Yellowstone National Park — Wyoming, Reprogramming for Concessioner Buyout," May 2, 1980; Larry E. Meierotto, Assistant Secretary of the Interior, to Congressman Sidney R. Yates, Chairman, Subcommittee on Interior and Related Agencies, Committee on Appropriations, House of Representatives, May 7, 1980; J. Robinson West, Assistant Secretary of the Interior, to Mark O. Hatfield, Chairman, Committee on Appropriations, U.S. Senate, February 15, 1983; Senator Robert C. Byrd, Chairman, Subcommittee on the Department of Interior and Related Agencies, U.S. Senate, to Meierotto, March 28, 1980; Yates and Joseph M. McDade, Rankin Minority Member, Subcommittee on Interior and Related Agencies, House of Representatives, to Meierotto, March 27, 1980; Associate Director, Park Service Administration, to Regional Director, Rocky Mountain Region, May 13, 1980 (all documents obtained through the Freedom of Information Act).

75. Executive Offices of the President, Office of Management and Budget, "Presidential Allowances, Department of Interior, National Park Service, 1981 Budget," November 15, 1979. (A request for this document through the Freedom of Information Act was denied; but I managed to obtain a copy from other sources.)

76. Interview with Herbst. That he reversed his decision is documented by the chronology presented to Congressman Yates (Reprogramming for Concessioner Buyout, May 2, 1980).

77. Interview with Surles; Larry E. Meierotto to Sidney R. Yates, May 7, 1980.

78. Mary E. Reiter, Contracting Officer, Denver Service Center, National Park Service, to Associate Director, Administrative Services, National Park Service, Washington, D.C., December 12, 1980; contracting officer, National Park Service, to Kober Construction, Billings, Montana, February 4, 1981 (obtained through the Freedom of Information Act). This letter apparently confirmed an earlier notification. It states, in part, "you are notified to proceed with the work on Wednesday, January 21, 1981."

79. Interviews with Joe Cutter.

80. Interviews with Snape.

81. Interview with Davidge; National Park Service, "Issue Paper: Grant Village," April 14, 1981 (obtained through the Freedom of Information Act).

82. Ibid. On April 8, 1981, the *Billings Gazette* quoted Montana's Senator Baucus as favoring construction. "Today," he said, "we are paying the price for that deferred maintenance."

83. Interviews with MacDonald and Bevinetto.

84. Russell E. Dickenson, Director, National Park Service, "Recommendation — Development of 100 Motel Units Grant Village — Yellowstone National Park," decision, April 14, 1981.

85. The buyout of General Hosts took place April 1, 1980; TW Services, Inc., then took over operation of these concessions on an interim basis. A five-year contract between TW Services, Inc., and the Park Service was signed November 1, 1981, and was finally approved February 18, 1982.

86. Public Law 89-249, 89th Congress, H.R. 2091, October 9, 1965, "Relating to the Establishment of Concession Policies in the Areas Administered by National Park Service and for Other Purposes"; Don Hummell, Chairman, Conference of National Park Concessioners, *The Concession System in National Parks* (undated).

87. National Park Service, contract with TW Services, Inc., Yellowstone National Park, Contract no. CC 1570-2-0001, executed February 18, 1982. I talked with several people with long experience in concessions management who expressed uneasiness about this proposal. Within the Park Service itself, among ordinary rangers, worry is considerable over this potential conflict of interest.

88. Interview with Hummell, 1984.

89. The most active of these groups is the Committee to Protect Our Yellowstone Heritage, founded by W. A. Warnock, John A. Warnock, and Bill Warnock.

90. Dennis Madden, Compliance Historian, Wyoming State Historic Preservation Office, to Mark Junge, Chief of the State Historic Preservation Trust, April 11, 1984, and

forwarded by Governor Hershler to Superintendent Barbee April 19, 1984; Hershler to Lorraine Mintzmyer, Regional Director, Park Service, Rocky Mountain Region, February 16, 1983; Randall A. Wagner, Director of the Wyoming Travel Commission to Dick Hartman, State Planning Coordinator for the Wyoming State Clearinghouse, January 21, 1983.

91. Dan Wenk, Landscape Architect, Yellowstone, to Superintendent of Yellowstone, May 3, 1984.
92. The shift of visitor services from Fishing Bridge to Grant Village would result in a transfer in tax revenues from the Cody district of Wyoming to the Jackson district. Spokesmen for the Cody district were reluctant to lose this tax base. Hamilton Stores, its President, Terry Povah told me, had just recently spent $500,000 renovating its store at Fishing Bridge, and was reluctant to see it torn down.
93. The first steps were taken to comply with the 1979 Fish and Wildlife Service Section 7 consultation "no-jeopardy" decision by dismantling Fishing Bridge in late 1982, and were pursued vigorously by Superintendent Barbee shortly after he arrived in 1983 (Associate Director, Cultural Resources Management, National Park Service, Washington, D.C., to Regional Director, Rocky Mountain Region, December 1, 1982; Barbee to G. Ray Arnett, Assistant Secretary of the Interior for Fish, Wildlife and Parks, October 23, 1983).
94. All Barbee found were the 1968–1972 memoranda from Cole and plant ecologist Don Despain to Anderson about the environmental impact of the proposed Fishing Bridge bypass road; this evidence did not make a strong case for closing Fishing Bridge. "These types of vegetational units," Cole had written in 1972, referring to the habitat surrounding Fishing Bridge, "i.e., meadows and open mature lodgepole pine, are not in short supply in Yellowstone" (to Anderson, September 21, 1972).
95. R. Knight, B. Blanchard, and D. Mattson, "Influences of the Fishing Bridge Area on the Yellowstone Grizzly Bear Population," January 26, 1984, Yellowstone Park files.
96. *Missoulian,* February 6, 1984.
97. *Fishing Bridge and the Yellowstone Ecosystem, 1984;* Senators Malcolm Wallop and Alan K. Simpson, and Congressman Dick Cheney, to Russell Dickenson, Director, National Park Service, requesting an Environmental Impact Statement on the closure of Fishing Bridge, September 10, 1984; Dickenson's reply, November 1, 1984.
98. The National Parks and Conservation Association woke up to the dangers of Grant Village very late. In a letter to Barbee dated February 7, 1983, Jarvis urged that "before the Grant Village area is chosen, an evaluation of the impact of such construction on the fragile remaining grizzly population must be completed." This letter might have been useful if it had been written ten years earlier.
99. Interview.
100. National Park Service, "Completion Report for construction of lodging units at Grant Village," April 26, 1984, Yellowstone Park files; National Park Service, "Newsletter," April 4, 1985. As mentioned in chapter 12, a boy was mauled by a grizzly in the Grant Village campground during summer 1984.
101. National Park Service, *Draft General Management Plan, John D. Rockefeller, Jr., Memorial Parkway, Wyoming,* January 1980. The plan was approved by the Acting Regional Director, James B. Thompson, on May 5, 1982. That no Environmental Impact Statement was done on the project was confirmed to me by William W. Shenk, Assistant Superintendent of Grand Teton National Park, in a letter dated August 7, 1984.

14. GUMSHOES AND POSY PICKERS

1. *Washington Star,* February 26, 1936; *New York Herald Tribune,* March 1, 1936; *Livingston Enterprise,* February 26, 1936; *Missoula Sentinel,* February 26, 1936. I pieced together the story of this accident from interviews with Conrad Wirth and Wright's daughters, Mrs. Pamela Wright Lloyd and Mrs. R. A. Brichetto.
2. Interviews with Thompson.

3. Horace M. Albright, "Research in the National Parks," *Scientific Monthly,* June 1933.

4. Alfred Runte, *National Parks: The American Experience* (Lincoln: University of Nebraska Press, 1979), pp. 102–104.

5. Albright, "Research in the National Parks." On September 30, 1916, just a month after Congress had created the Park Service, twenty-two enlisted men working for the cavalry in Yellowstone were discharged from the army and took immediate employment as park rangers. See Haines, *Yellowstone Story,* vol. 2, p. 289.

6. Interviews with Raymond E. Moran, Division of Standards Development, Office of Personnel Management, 1984.

7. C. Frank Brockman, "Park Naturalists and the Evolution of National Park Service Interpretation through World War II," *Journal of Forest History,* vol. 22, no. 1, January 1978; C. Frank Brockman, "Chronology of Interpretation in the National Park Service," draft of private research project, Yellowstone Park Library, October 1976, pp. 38 ff.

8. Ibid.; Harold C. Bryant and Wallace W. Atwood, Jr., *Research and Education in the National Parks* (Washington, D.C.: U.S. Government Printing Office, 1932).

9. Skinner, "Predatory and Fur-Bearing Animals of Yellowstone," 1927, p. 165.

10. Bryant and Atwood, "Research and Education in the National Parks," and Brockman, "Chronology of Interpretation"; Lowell Sumner, "Biological Research and Management in the National Park Service: A History," *George Wright Forum,* vol. 3, no. 4, Autumn 1983; Joseph Grinnell, "What Is a Natural Balance for Wild Life in the National Parks, and How Can It Be Maintained?" address delivered at the National Parks Superintendents' Conference, February 17, 1928. See also Louis C. Cramton, *Early History of Yellowstone National Park and Its Relation to National Park Policies* (Washington, D.C.: U.S. Government Printing Office, 1932).

11. Wright's father, George Tennant Wright, was a Yankee sea captain who married the heiress of a large agricultural empire in El Salvador. George Wright was orphaned at an early age and raised by an aunt in Oakland, California. He met Grinnell while studying forestry at Berkeley.

12. See Chapter 3.

13. Wright, Thompson, and Dixon, *Fauna Number 1;* Wright and Thompson, *Fauna Number 2.*

14. Sumner, "Biological Research."

15. Wright, Thompson, and Dixon, *Fauna Number 1,* pp. 3–4.

16. Albright, "Research in the National Parks."

17. Interviews with Grater, Cahalane, and Sumner, 1983–1984. See also Sumner, "Biological Research and Management in the National Parks."

18. Sumner, "Biological Research."

19. Interview with Cahalane.

20. Interviews with Cahalane, Sumner, and Thompson. See also Sumner, "Biological Research."

21. Interview with Cahalane; Sumner, "Biological Research." By World War II, the only biologists in the national parks were Cahalane, Dixon, and Adolph Murie. They worked for the Fish and Wildlife Service, but were paid from Park Service funds.

22. Interview with Cahalane.

23. Sumner, "Biological Research"; Brockman, "Chronology of Interpretation."

24. Sumner, "Biological Research"; Wirth, *Parks, Politics and People,* pp. 261–262. The goal of Mission 66 was $786,545,600; it raised $1,035,225,000.

25. A. Starker Leopold et al., "Wildlife Management" (see Chapter 4).

26. William J. Robbins et al., "A Report by the Advisory Committee to the National Park Service on Research of the National Academy of Sciences"; for full bibliographic data, see Chapter 1, note 13.

27. Ibid.

28. Ibid.

29. Ibid.

30. Ibid.

31. Ibid.

32. Ibid.
33. Department of the Interior, Office of the Secretary, press release, "National Academy of Sciences Recommends Positive Research Program for National Park Area; Secretary Udall Orders Corrective Steps," October 18, 1963; Director, National Park Service, to Secretary of the Interior, 1963.
34. Interview with Moran, 1984; U.S. Civil Service Commission, Bureau of Policies and Standards, Standards Division, *Park Ranger Series*, GS-453, Section 5105, Title 5, United States Code, December 1957.
35. U.S. Civil Service Commission, *Guide for the Evaluation of Professional Positions Engaged in Interpretive Work*, December 1962.
36. Brockman, "Chronology of Interpretation."
37. J. Carlisle Crouch et al., "Report of Field Operations Study Team," 1966; Associate Director, National Park Service, to Regional Directors, November 15, 1966.
38. U.S. Civil Service Commission, Bureau of Policies and Standards, Standards Division, *Park Management Series*, GS-025, June 1969; National Park Service, Single Agency Qualification Standard, *Park Management Series*, GS-025, August 1969.
39. Crouch et al., "Field Operations Study"; Office of Personnel Management, *Qualifications Standards — Park Technician Series*, GS-026, August 1969; National Park Service, *Park Management (Superintendent Guidelines — GS-025)*, June 1975.
40. Grant W. Sharpe, *Interpreting the Environment* (New York: John Wiley, 1976), p. 615.
41. Ibid., pp. 616 ff.
42. Interview with Utley.
43. National Park Service, "Interpretation and Visitor Service Guideline," NPS release no. 2, March 1980.
44. Sharpe, *Interpreting the Environment*, p. 615.
45. Freeman Tildon, *Interpreting Our Heritage* (Chapel Hill: University of North Carolina Press, 1957), p. 30.
46. From interviews with Sprugel, Sumner, and Linn. See also Sumner, "Biological Research — A History."
47. Sumner, "Biological Research — A History." The story of how the resource-manager position and the research centers in Everglades and St. Louis Bay were created was explained to me by Linn and by Dr. James Kushlan, a former biologist in Everglades.
48. Part of the reason for not creating a line item for research was that doing so would make it more visible and thus more vulnerable to reductions in funding. By supporting research indirectly, the money could be "hidden."
49. U.S. Civil Service Commission, *Research Grade-Evaluation Guide*, June 1964; U.S. Civil Service Commission, *Research Grants Grade-Evaluation Guide*, April 1968; National Park Service, *Research Personnel Evaluation Plan*, September 25, 1979.
50. Interviews with Bearss.
51. Confirmed by Linn and Kushlan.
52. Anderson to Stephanie Wood, Jackson Hole Environmental Action Society, July 18, 1972; Metcalf to Secretary Reed, January 22, 1975; Director Dickenson to Regional Director, October 18, 1973; Reed to Metcalf, February 14, 1975; Merrill A. Beal, Regional Director, to Anderson, October 20, 1973; National Park Service, "Status of the Yellowstone Research Center," May 9, 1975 (all obtained through the Freedom of Information Act). Instead of building a research center in Yellowstone, the Park Service created, in Jackson Hole, a Cooperative Park Study Unit under the direction of the University of Wyoming. The contract with the University, however, restricted their role to "interpretation"; that is, they could not advise management.
53. Resource managers belong to the 025 series.
54. Bernard Shanks, *This Land Is Your Land: The Struggle to Save America's Public Lands* (San Francisco: Sierra Club Books, 1984), p. 223; Bernard Shanks, "National Park Service Chases Squirrels of Political Popularity," *High Country News*, vol. 10, no. 20, October 1978.
55. As we have seen in our story, the Craigheads and now Knight are examples of researchers around whom the Park Service built a wall.

56. National Park Service, *Final Environmental Assessment, Grant Village,* 1974, p. 42.
57. Interview.
58. James A. Kushlan, "External Threats and Internal Management: The Hydrologic Regulation of the Everglades," 1983; According to Kushlan, this report was suppressed by the Park Service (interview, 1984).
59. In Everglades, Kushlan told me, the Park Service spent $1.5 million a year for seven years with "no results." That is, the results of research were rejected. Continuous efforts were made to keep researchers from publishing, and eventually they were forbidden outright to do so. Recently, according to Kushlan and others, the qualified scientists left, and their positions have been filled by individuals without professional degrees who have been promoted far beyond the level to which their qualifications should lift them.
60. James A. Kushlan and Terri Jacobsen, "Management of Alligator Nest Flooding," Report SFRC/84/0, January 1984. (According to Kushlan, this piece too was suppressed by the Park Service.)
61. Shanks, *This Land Is Your Land,* p. 221; Daniel H. Henning, "National Park Wildlife Management Policy: A Case Study in Decentralized Decision-Making," *Northwest Science,* vol. 46, no. 2, 1972.
62. The public is told, repeatedly, how much "independent" research goes on in our national parks. But in practice, a member of Yellowstone's research office told me recently, the decision on who should work is "delegated to the field level."
63. Throughout my research on this subject, I found many people — both in the Park Service and outside — who expressed the worries I have relayed here. But in nearly all cases, they would speak only off the record.
64. Shanks, *This Land Is Your Land,* p. 222.
65. Of the twenty-nine works by Yellowstone biologists cited by Houston in the *Northern Yellowstone Elk,* for example, by my count five appeared in refereed journals and only sixteen were published at all.
66. National Park Service, "Research Conducted in Yellowstone National Park, 1981," Yellowstone Park, 1981; National Park Service, "Ongoing Biological Research," Yellowstone Park, March 1982; Glen Cole, "Role of Research in National Park Natural Areas," March 11, 1968, Yellowstone Park files; National Park Service, "Information for Cooperating Researchers," Information Paper no. 9, November 1982, Yellowstone National Park; National Park Service, "Reminders to Researchers," Information Paper no. 11, Yellowstone Park, May 1982; Carl P. Russell, "A Concise History of Scientists and Scientific Investigations in Yellowstone National Park," parts 1 and 2, Department of the Interior, National Park Service, 1935; Robert M. Linn, ed., *First Conference on Scientific Research in the National Parks,* National Park Service Transactions and Proceedings Series, no. 5, 1979.
67. Interview.
68. Interview.
69. Instead of a research division there is an Associate Director for Natural Resources in Washington. But this office has no supervisory responsibility. Park Service science is organized along feudal lines. Each regional office has nearly absolute control of the budget for science in its area. In some fields the work is entirely delegated to one region (for history, it is the Denver office). There is simply no overall authority for science in the national parks.
70. The proposed budget for training resource managers for fiscal year 1985 was $851,000 (from budget material supplied by Carole Bickley of the Office of the Associate Director for Natural Resources, 1984). See also National Science Foundation, *Annual Survey of Federal Funds for Research and Development, National Park Service, Department of the Interior, Fiscal years 1983, 1984, 1985,* NSF form 818, 1984; Memorandum to Directorate, Field Directorate, and WASO Division Chiefs from Assistant Director, Financial and Data Systems, "Subject: Information: Progress of the Fiscal 1985 Budget — Appropriations Stage," October 17, 1984 (obtained through the Freedom of Information Act).
71. *Arizona Daily Star,* September 2, 1984. In fiscal 1985, the budget for acid rain and air quality was $4,324,000; in addition, much of the $17,677,000 budgeted for "ser-

vicewide science programs" went to acid rain and air-quality monitoring. The budget for other research was $3,279,000.

72. Interview with John Reed, Chief, Biological Resources Division, Office of the Assistant Director for Natural Resources, and with William Supernaugh, formerly with the Division of Natural Resource Management, 1984.

73. Dickenson to Regional Directors and Superintendents, March 29, 1982.

74. This information was obtained in extensive interviews with division chiefs in Washington and elsewhere in the Park Service. See also National Park Service, "Occupational Listing, Servicewide," as of FY-84, pay period 07, December 30, 1983, and National Park Service, *1980 National Park Service Science and Technology Directory* (Washington, D.C., Research Evaluation Division, National Park Service, January 1980).

75. National Science Foundation, *Annual Survey for Research and Development;* National Science Foundation, Division of Science Resources Studies, *Federal R&D Funding, by Budget Function,* Fiscal Years 1983–1985 (Washington, D.C.: National Science Foundation, March 1984); National Science Foundation, *Federal Funds for Research and Development,* vol. 33, detailed statistical tables (Washington, D.C.: National Science Foundation, September 1984).

76. J. Robert Stottlemyer, "Evolution of Management Policy in the National Parks," *Journal of Forestry,* vol. 79, January 1981; Douglas L. Gilbert, *Natural Resources and Public Relations* (Washington, D.C.: Wildlife Society, 1971), pp. 230 ff.

77. Shanks, *This Land Is Your Land,* pp. 223–224.

78. Dickenson to Davidge, August 23, 1982.

79. Donald G. Brauer, Chief Engineering and Science Occupations Branch, Standards Development Center, to Director, National Park Service, transmitting *Tentative Standards for Park Ranger Series,* GS-025 (Classification only), November 30, 1981. See also, Office of Personnel Management, *Tentative Standards for Park Ranger Series,* GS-025 (Qualification only), February 26, 1982.

80. Interview with Supernaugh.

81. Chairman, 025 Task Force to Director, National Park Service, February 26, 1982; National Park Service, "Qualification Standard," 1982, National Park Service files, Washington, D.C.

82. Ibid.

83. Interview.

84. Present status confirmed by qualified sources with the National Park Service in Washington, D.C.

85. "The Changing Rangers," *Newsweek,* July 25, 1983.

86. Ibid.

87. Shanks, *This Land Is Your Land,* p. 225.

15. THE DEEP HOLE GAP

1. Nathaniel P. Langford, *The Discovery of Yellowstone Park* (Lincoln: University of Nebraska Press, 1971), p. 43.

2. A. M. Pitt and R. A. Hutchinson, "Hydrothermal Change Related to Earthquake Activity at Mud Volcano, Yellowstone National Park, Wyoming," *Journal of Geophysical Research,* vol. 87, no. B-4, pp. 2762–2766. The seismic swarm continued from May through November 1978. On October 23 and November 7, more than 100 "earthquakes" arrived per minute.

3. William R. Keefer, *The Geologic Story of Yellowstone National Park,* Geological Survey Bulletin 1347 (Yellowstone: Yellowstone Library and Museum Association, 1971), pp.71, 87; T. Scott Bryan, *The Geysers of Yellowstone* (Boulder: Colorado Associated University Press, 1979), p. 13; Robert B. Smith and Robert L. Christiansen, "Yellowstone Park as a Window on the Earth's Interior." *Scientific American,* vol. 242, no. 2, February 1980.

4. J. W. Barlow, *Letter from the Secretary of War Accompanying an Engineer Report of a Reconnaissance of the Yellowstone River in 1871,* Senate Ex. Doc. No. 66, 42nd

Congress, 2nd Session, April 18, 1872, 30; Langford, *The Discovery of Yellowstone Park,* p. 44.

5. Bryan, *Geysers of Yellowstone,* pp. 9–10. During summer 1984 the intervals between eruptions of Old Faithful were the greatest ever recorded: an average of 76.33 minutes. At this writing the interval has decreased to an average of 73 minutes (interview, R. A. Hutchinson).

6. Osborne Russell, *Journal of a Trapper* (Lincoln: University of Nebraska Press, 1955), p. 100. Cf. Doane's comment in 1870: "As a field for scientific research, it promises great results" (*Letter from the Secretary of War Communicating the Report of Lt. Gustavus C. Doane upon the So-called Yellowstone Expedition of 1870,* 41st Congress, 3rd Session, Ex. Doc. No. 51).

7. Doane, *Letter from the Secretary of War;* Runte, *National Parks: The American Experience* (Lincoln: University of Nebraska Press, 1979), pp. 36–39.

8. Hiram M. Chittenden, *The Yellowstone National Park* (Norman: University of Oklahoma Press, 1961), p. 27; Aubry L. Haines, *The Yellowstone Story* (Yellowstone: Yellowstone Library and Museum Association, 1977), vol. 1, pp. 43, 47.

9. Walter W. deLacy, "A Trip up the Snake River in 1863," Contributions to the Historical Society of Montana, vol. 1 (Helena: Rocky Mountain, 1876), p. 106; Haines, *Yellowstone Story,* vol. 1, p. 100.

10. Haines, *Yellowstone Story,* vol. 1, p. 53; Chittenden, *The Yellowstone National Park,* p. 48.

11. Haines, *Yellowstone Story,* vol. 1, pp. 53, 85 ff.

12. Ibid., p. 53.

13. Ibid., pp. 64, 140–141; Alfred Runte, *Trains of Discovery: Western Railroads and the National Parks* (Flagstaff: Northland Press, 1984), pp. 16–35.

14. Ibid., pp. 142, 349; "The Yellowstone Scientific Expedition," *Helena Daily Herald,* July 11, 1871.

15. Haines, *Yellowstone Story,* vol. 1, p. 155.

16. Ibid., pp. 166–167.

17. Ibid., p. 169; "Dodgson, Charles Lutwidge," entry in *Oxford Companion to English Literature,* ed. by Sir Paul Harvey (Oxford, Clarendon Press, 1937).

18. Barlow, *Letter from the Secretary of War,* p. 25.

19. Haines, *Yellowstone Story,* p. 180–181; Robert L. Christiansen and H. Richard Blank, Jr., "Quaternary and Pliocene Volcanism of the Yellowstone Rhyolite Plateau Region of Wyoming, Idaho and Montana," Geological Survey Professional Paper 729, Yellowstone National Park Library, 1975, vol. 1, pp. 12–15. A revised version of this paper will soon be published under Christiansen's name alone.

20. C. Frank Brockman, "Evolution of Interpretation in the National Park Service," Yellowstone Park Library, 1976; Harold C. Bryant and Wallace W. Atwood, *Research and Education in the National Parks* (Washington, D.C.: U.S. Government Printing Office, 1932), pp. 41–42, 62; Christiansen and Blank, Quaternary and Pliocene Volcanism," pp. 14–15; George D. Marler, "Some of Yellowstone's Geological Story," *A Manual of General Information on Yellowstone National Park* (Yellowstone: National Park Service, 1963); J. P. Iddings, "Obsidian Cliff, Yellowstone National Park," *Seventh Annual Report* (U.S. Geological Survey: U.S. Government Printing Office, 1888), pp. 249–295.

21. F. V. Hayden, *Preliminary Report of the United States Geological Survey of Montana and Portions of Adjacent Territories* (Washington, D.C.: U.S. Government Printing Office, 1872); W. H. Holmes, "Report on the Geology of the Yellowstone National Park," *A Report of the Exploration in Wyoming and Idaho for the Year 1878,* U. S. Geological and Geographic Survey of the Territories, Twelfth Annual Report, part 2, Yellowstone, 1879; J. P. Iddings, "The Rhyolites," *Geology of the Yellowstone National Park,* U.S. Geological Survey monograph 32, part 2, W. H. Holmes, ed. (Washington, D.C.: U.S. Government Printing Office, 1899); Keefer, *The Geologic Story of Yellowstone National Park.*

22. Marler, "Some of Yellowstone's Geological Story," p. 70.

23. Christiansen and Blank, "Quaternary and Pliocene Volcanism," pp. 14 ff.

24. Ibid., pp. 15–16; Francis R. Boyd, "Welded Tuffs and Flows in the Rhyolite Plateau

of Yellowstone Park, Wyoming," *Geological Society of America Bulletin*, vol. 72, March 1961, pp. 388–389. I also received help in piecing together this part of the story from Christiansen and Boyd.

25. Christiansen and Blank, "Quaternary and Pliocene Volcanism," pp. 15–16; Boyd, "Welded Tuffs," pp. 388–389; Richard Foster Flint and Brian J. Skinner, *Physical Geology* (New York: John Wiley, 1974), pp. 313–314.

26. Christiansen and Blank, "Quaternary and Pliocene Volcanism," pp. 15–16; Boyd, "Welded Tuffs," pp. 388–389; Flint and Skinner, *Physical Geology*, pp. 314–315. The importance of Pelée and Katmai in advancing thinking on welded tuffs was explained to me by both Boyd and Christiansen. In addition, Boyd cited the work of P. Marshall in New Zealand (P. Marshall, "Acid Rocks of the Taupo-Rotorua Volcanic District," *New Zealand Transactions of the Royal Society*, vol. 64, 1935).

27. Francis R. Boyd, Jr., "Geology of the Yellowstone Rhyolite Plateau," Ph.D. thesis presented to Harvard University in May 1957, Yellowstone National Park Library, pp. 92 ff.

28. Ibid., Boyd, "Welded Tuffs," pp. 393 ff.

29. Peter Francis and Stephen Self, "The Eruption of Krakatau," *Scientific American*, vol. 249, no. 5, November 1983; Stephen H. Schneider and Randi Londer, *The Coevolution of Climate and Life* (San Francisco: Sierra Club, 1984), pp. 277–282.

30. Interview.

31. Boyd, "Geology of the Yellowstone Rhyolite Plateau," p. 85.

32. Ibid., p. 85; Boyd, "Welded Tuffs," p. 410. Boyd, however, was aware that future research might show the date of the eruptions to be even later.

33. Hayden, *Preliminary Report for 1972*, p. 81.

34. William A. Jones, *A Report upon the Reconnaissance of Northwestern Wyoming Made in the Summer of 1873 by Captain William A. Jones, Corps of Engineers* (Washington, D.C.: U.S. Government Printing Office, 1873), p. 34.

35. Edmund Christopherson, *The Night the Mountain Fell: The Story of the Montana-Yellowstone Earthquake* (West Yellowstone, Montana: Yellowstone Publications, 1960), p. 10.

36. Ibid., p. 5; William A. Fischer, *Earthquake! Highlights of Yellowstone Geology with an Interpretation of the 1959 Earthquakes and Their Effects in Yellowstone National Park* (Yellowstone: Yellowstone Library and Museum Association, 1976), pp. 1–2.

37. Christopherson, *Night the Mountain Fell*, p. 6.

38. Ibid., p. 19.

39. Marler, "Geological Story," p. 79; Fischer, *Earthquake!* pp. 25–27, 47 ff.

40. Fischer, *Earthquake!* pp. 5–7. At first the quake was measured as reaching 7.1 on the Richter scale; but recently, after recalculation, it has been found to be 7.5.

41. Marler, "Geological Story."

42. Hayden, *Preliminary Report for 1872*, p. 82; Fischer, *Earthquake!* Table 2, opposite page 24.

43. Warren Hamilton, "Late Cenozoic Tectonics and Volcanism of the Yellowstone Region, Wyoming, Montana and Idaho," Billings Geological Society, 11th Annual Field Conference, September 7–10, 1960, pp. 92–105.

44. Interview with Hamilton, 1985; G. M. Richmond, "Glacial Geology of the West Yellowstone Basin and Adjacent Parts of Yellowstone National Park," U.S. Geological Survey professional paper 435-T, 1964. See also Kenneth L. Pierce, *History and Dynamics of Glaciation in the Northern Yellowstone Park Area*, Geological Survey professional paper 729-F (Washington, D.C.: U.S. Government Printing Office, 1979).

45. Hamilton, "Late Cenozoic Tectonics."

46. Ibid.

47. Ibid.

48. Campbell died before work on the book began, but his close friend, John Good, supplied me with some of his concerns of this period. See Arthur B. Campbell, "Current U.S.Geological Survey Research in Yellowstone National Park," *Abstracts of the Northwestern Section of the U.S. Geological Society*, Bozeman, Montana, 1964.

49. Interviews with Warren Hamilton, John Good, and Robert L. Christiansen.

50. Campbell, "Current Geological Survey Research." See Robert L. Christiansen, "Technical Letter NASA-1-4, A distinction between bedrock and unconsolidated deposits on 3–5µ infrared imagery of the Yellowstone rhyolite plateau" (Houston, Texas: Manned Spacecraft Center, National Aeronautics and Space Administration, May 1968), and Robert L. Christiansen and H. Richard Blank, Jr., *Volcanic Stratigraphy of the Quaternary Rhyolite Plateau in Yellowstone National Park*, Geological Survey professional paper 729-B (Washington, D.C.: U.S. Government Printing Office, 1972). See also the quarterly reports of these teams, Yellowstone Park Library, 1966–1967.

51. Interview with Fournier; Keefer, *Geologic Story of Yellowstone*, p. 73.

52. Christiansen and Blank, "Quaternary and Pliocene Volcanism," p. 277.

53. Smith and Christiansen, "Yellowstone Park as a Window"; Christiansen and Blank, "Quaternary and Pliocene Volcanism; Christiansen and Blank, "Stratigraphy of Quaternary Rhyolite."

54. Christiansen and Blank, "Quaternary and Pliocene Volcanism," pp. 271 ff.

55. Ibid., pp. 271 ff.

56. Ibid., pp. 10, 311, 146. For a similar prediction, see Robert L. Christiansen, "Yellowstone Magmatic Evolution: Its Bearing on Understanding Large-Volume Explosive Volcanism," *Explosive Volcanism: Inception, Evolution, and Hazards*, ed. by Committee of Geophysics Study, National Research Council (Washington, D.C.: National Academy Press, 1984).

57. These competing theories were explained to me by Christiansen and Warren Hamilton. See Christiansen and Blank, "Quaternary and Pliocene Volcanism," pp. 295 ff.; Smith and Christiansen, "Yellowstone National Park as a Window."

58. R. B. Smith et al., "Yellowstone Hot Spot: New Magnetic and Seismic Evidence," *Geology*, September 1974; R. B. Smith et al., "Yellowstone Hot Spot: Contemporary Tectonics and Crustal Properties from Earthquake and Aeromagnetic Data," *Journal of Geophysical Research*, vol. 82, no. 26, September 10, 1977; J. R. Pelton and R. B. Smith, "Recent Crustal Uplift in Yellowstone National Park," *Science*, vol. 206, December 7, 1979; Smith and Christiansen, "Yellowstone National Park as a Window"; R. B. Smith and Lawrence W. Braile, "Crustal Structure and Evolution of an Explosive Silicic Volcanic System at Yellowstone National Park," in *Explosive Volcanism*.

59. Smith et al., "Yellowstone Hot Spot: Contemporary Tectonics," p. 3670. See also Smith et al., "Yellowstone Hot Spot: New Magnetic and Seismic Evidence."

60. Smith and Braile, "Crustal Structure and Evolution of an Explosive Silicic Volcanic System," p. 98.

61. Smith and Christiansen, "Yellowstone Park as a Window."

62. Ibid.; Smith et al., "Yellowstone Hot Spot: Contemporary Tectonics," pp. 3673 ff.

63. That is to say, within five kilometers it is too warm to break. See Smith and Christiansen, "Yellowstone Park as a Window"; Smith et al., "Yellowstone Hot Spot: Contemporary Tectonics," p. 3668. The continental crust, in relation to the oceanic crust, is quite thick on the average, reaching in places a depth of forty-four miles or more. See B. Clark Burchfiel, "The Continental Crust," *Scientific American*, vol. 249, no. 3, September 1983.

64. Smith and Christiansen, "Yellowstone Park as a Window"; Smith et al., "Yellowstone Hot Spot: Contemporary Tectonics," pp. 3673–3675.

65. Smith and Braile, "Crustal Structure and Evolution of an Explosive Silicic System," p. 103. See also Smith and Pelton, "Recent Crustal Uplift."

66. Wayne L. Hamilton, "Applying Lake Level Gauging Records to the Investigation of Uplift within the Yellowstone Caldera, Yellowstone National Park," *Abstracts of the American Geophysical Union*, Spring Meeting, Cincinnati, Ohio, 1984; A. L. Bailey and Wayne L. Hamilton, "Submerged Shorelines of Yellowstone Lake as Indicators of Uplift," address presented at symposium on Geophysics and Geology of Yellowstone, American Geophysical Union, Trout Range Branch, 1984; D. Dzurisin et al., "Recent Crustal Uplift in Yellowstone National Park," *Red Book of the U.S. Geological Society* (Washington, D.C.: U.S. Government Printing Office, 1984).

67. Christiansen, "Yellowstone Magmatic Evolution," p. 94.

68. Interview with Toon; Smith and Blank, "Quaternary and Pliocene Volcanism," p. 311; O. B. Toon, "Sudden Changes in Atmospheric Composition and Climate," *Patterns of Change in Earth Evolution,* ed. by H. D. Holland and A. F. Trendall (Berlin: Springer Verlag, 1984); Committee on the Atmospheric Effects of Nuclear Explosions, National Research Council, *The Effects on the Atmosphere of a Major Nuclear Exchange* (Washington, D.C.: National Academy Press, 1985).

69. Interview with Christiansen. For a calculation of the combined megatonnage of the two superpowers, see Jonathan Schell, *The Fate of the Earth* (New York: Avon, 1982), p. 54.

70. *Wyoming State Tribune,* July 20, 1981; *Bozeman Chronicle,* July 21, 1981; *Idaho Statesman,* July 22, 1981.

71. Yellowstone Park geologist Wayne L. Hamilton told me about these events.

72. deLacy, "A Trip up the Snake River," p. 124.

73. E. M. Shoemaker, ed., *Continental Drilling* (Washington, D.C.: Carnegie Institution, 1974); Geophysics Research Board and Division of Earth Sciences, National Academy of Sciences, National Research Council, *Solid-Earth Geophysics: Survey and Outlook,* publication 1231 (Washington, D.C.: National Academy of Sciences, 1964); Committee on Ocean Drilling, National Research Council (Washington, D.C.: National Academy Press, 1982); Willard Bascom, *A Hole in the Bottom of the Sea: The Story of the Mohole Project* (New York: Doubleday, 1961).

74. Interviews with Robert S. Andrews, Senior Staff Officer, Continental Scientific Drilling Committee, Board of Earth Sciences, National Research Council; U.S. Geodynamics Committee, National Research Council, *Continental Scientific Drilling Program* (Washington, D.C.: National Academy Press, 1979).

75. Geodynamics Committee, *Continental Scientific Drilling,* pp. 7–8.

76. Ibid., especially Appendix E. See also Continental Scientific Drilling Committee, *Mineral Resources: Research Objectives for Continental Scientific Drilling* (Washington, D.C.: National Academy Press, 1984).

77. Geodynamics Committee, *Continental Scientific Drilling Program,* pp. iii, 7 ff.

78. Continental Scientific Drilling Committee, *A National Drilling Program to Study the Roots of Active Hydrothermal Systems Related to Young Magmatic Intrusions* (Washington, D.C.: National Academy Press, 1984).

79. Continental Scientific Drilling Committee, *Mineral Resources,* p. 1; Geodynamics Committee, *Continental Scientific Drilling Program,* p. 97.

80. Continental Scientific Drilling Committee, *A National Drilling Program,* p. 4.

81. Ibid., p. 98.

82. U.S. Department of Agriculture, Forest Service, *Final Environmental Impact Statement of the Island Park Geothermal Area* (Missoula, Montana: Forest Service, 1980), pp. 111–112. Paradoxically, this EIS was written with the help of the Park Service, whose geologists are well aware of the threat that geothermal exploration poses for Yellowstone.

83. Ibid., p. 111.

84. Ibid., pp. 90 ff.

85. Ibid., pp. 90 ff.

86. Ibid., pp. 90 ff.

87. Ibid., pp. 90 ff.

88. The cost was estimated in 1979 (Geodynamics Committee, Continental Scientific Drilling Committee, p. 122, Table C-1).

89. Ibid., p. 98.

90. Ibid., pp. 149–150.

91. Ibid., p. 121.

92. Ibid., pp. 121–122.

93. Interviews with Varley and Fournier. In 1984 the CSDC wrote that "it is urged that this area continue to be considered for possible research drilling in the future" (*A National Drilling Program,* p. 4).

94. Interview with Hamilton.

95. Interviews with Warren Hamilton, Robert B. Smith, and others, 1984–1985.

96. Ye. A. Kozlovsky, "The World's Deepest Well," *Scientific American,* vol. 251, no. 6,

December 1984; USSR Ministry of Geology, "The Kola Super-Deep Borehole," 1984.
97. U.S. Senate, "Senate Resolution 439 — Relating to the Continental Scientific Drill-ing Program," *Congressional Record,* vol. 130, no. 111, September 10, 1984.
98. Ibid.
99. Interview with Abell.
100. Calvin M. Kaya, "Reproductive Biology of Trout in a Thermally Enriched Environ-ment: The Firehole River of Yellowstone National Park, Final Report," prepared for the U.S. Department of Energy under Contract No. EY-76-S-06-2228, Task Agreement 2, Bozeman, Montana, June 30, 1978.
101. W. A. Berggren and John A. van Couvering, eds., *Catastrophes and Earth History: The New Uniformitarianism* (Princeton: Princeton University Press, 1984).
102. Brockman, "Chronology of Interpretation."
103. The story of the Scaup Lake salamander is an interesting one. A similar species was mentioned by deLacy when traveling through the park in 1863. Townsley gave Smith permission to dynamite the lake against the advice of his staff. "Are we going to let a lousy little salamander stop science?" he was reported to have asked them. The dynamite was detonated July 16, 1980. On that day, rangers were encouraged to keep the visiting journalists far enough away from the lake that they could not hear the explosion. Subsequent searches for the salamander by the Fish and Wildlife Service proved fruitless. They concluded, "Absence of salamanders . . . is curious, but does not provide irrefutable evidence that they no longer occur in Scaup Lake" (Fish and Wildlife Service, *Fishery and Aquatic Management Program in Yellowstone National Park,* Report for Calendar Year 1980, May 1981).

16. THE NEW PANTHEISTS

1. Northern Rockies Action Group, "Groups NRAG Worked with in 1982–1983"; NRAG Activity Reports, October 1983–February 1984 and March 1984–July 1984.
2. Willis W. Harmon, "Biographical Summary," Futures Seminar Material — 1, July 18–19, 1983; Willis W. Harmon, "Five 'Forcing Functions' Shaping the Future," Futures Seminar Material — 2, July 18–19, 1983.
3. Institute of Noetic Sciences, *Newsletter,* vol. 10, no. 2, Fall 1982, Futures Seminar Material — 3, July 18–19, 1983.
4. *Noetic Sciences Newsletter,* pp. 3 ff.; Tarrytown Group, *Tarrytown Letter,* February 1983.
5. Lynn White, Jr., "The Historical Roots of Our Ecologic Crisis," *Science,* vol. 155, no. 3767, March 10, 1967, pp. 1203–1207.
6. Ibid.
7. Ibid.
8. Ibid.
9. René Dubos, *A God Within* (New York: Scribner's, 1972), pp. 157–161.
10. White devotes a paragraph to explaining why the Eastern Orthodox countries were not as destructive as the Western Christian ones. The Orthodox religion, he sug-gested, "was essentially artistic rather than scientific," and thus created an atmosphere in which science could not flourish. But science flourished for centuries in the Arab world. Were the Arabs any less enlightened in their treatment of nature than the Orthodox countries?
11. Dubos, *A God Within,* pp. 159 ff. For an example of the threats to nature in Nepal, see *The Barun Valley Report* (Missoula: Wildlife-Woodlands Institute, 1985). The wildlands in this country are being destroyed, as are many in the third world, by poor farming practices.
12. John Muir, *Our National Parks* (Madison: University of Wisconsin Press, 1981), p. 74.
13. Stephen Fox, *John Muir and His Legacy: The American Conservation Movement* (Boston: Little, Brown, 1981), pp. 29 ff.
14. Ibid., p. 12.
15. Ibid., p. 5.

16. Carl Bode, ed., *The Portable Emerson* (New York: Penguin, 1981), pp. xiii, xx.
17. Curiously, trancendentalism was based on a misunderstanding of the philosophy of Kant. In his *Critique of Pure Reason* the German philosopher had argued that what was transcendental — the "thing in itself" — could never be known. Yet Emerson and other transcendentalists tried to do just what Kant said was impossible: to know the "thing in itself." See Emerson, "Nature," in Bode, *The Portable Emerson*, pp. 8–50; Perry Miller, ed., *The American Transcendentalists: Their Prose and Poetry* (Baltimore, Johns Hopkins University Press, 1957), pp. 49–69.
18. Miller, *The American Transcendentalists*, pp. 69–73, 308–329; Henry David Thoreau, *The Natural History Essays* (Salt Lake City: Peregrine Smith, 1980); Edward Abbey, *Down the River* (New York: E. P. Dutton, 1982), pp. 13–48.
19. H. D. Thoreau, "Walking," in H. D. Thoreau, *The Natural History Essays*, p. 112; H. D. Thoreau, *Walden* (New York: Carlton House, 1854); Fox, *John Muir*, p. 363; Laurence Stapleton, ed., *H. D. Thoreau: A Writer's Journal* (New York: Dover, 1960); H. D. Thoreau, *A Week on the Concord and Merrimack Rivers* (Boston: Houghton Mifflin, 1961).
20. Fox, *John Muir*, pp. 218–249, and pp. 358 ff.
21. Ibid., ch. 7, 11; Sigurd F. Olson, "The Emergent God," *Reflections from the North Country* (New York: Alfred A. Knopf, 1982), p. 164.
22. Fox, *John Muir*, ch. 11; John McPhee, *Encounters with the Archdruid* (New York: Farrar, Straus & Giroux, 1971), p. 84; Robinson Jeffers, "Intellectuals," *The Selected Poetry of Robinson Jeffers* (New York: Random House, 1927), p. 458.
23. Fox, *John Muir*, ch. 11.
24. Ibid., p. 80; William Everson, *Archetype West: The Pacific Coast as a Literary Region* (Berkeley: Oyez Press, 1976).
25. Susan L. Flader, *Thinking Like a Mountain* (Lincoln: University of Nebraska Press, 1974), p. 18; Fox, *John Muir*, p. 19; McPhee, *Archdruid*, p. 86.
26. Fox, *John Muir*, pp. 368–371.
27. Roderick Nash, *Wilderness and the American Mind* (New Haven: Yale University Press, 1967), p. 133.
28. Ibid., pp. 134–138; Fox, *John Muir*, pp. 110–111.
29. Carolyn Merchant, *The Death of Nature: Women, Ecology and the Scientific Revolution* (San Francisco: Harper and Row, 1983), pp. 236 ff.
30. Nash, *Wilderness and the American Mind*, pp. 134–135.
31. Ibid., pp. 135 ff.
32. Ibid., pp. 161 ff.; Fox, *John Muir*, pp. 139–147.
33. Fox, *John Muir*, p. 142.
34. Nash, *Wilderness and the American Mind*, ch. 12, 13; Fox, *John Muir*, 289–290.
35. Pantheism is really less of a religion than a philosophy, for it is an attempt to provide a completely systematic account of the universe, making it a *deterministic* philosophy: everything happens as it was meant to be. For a good exposition of pantheism, see Stuart Hampshire, *Spinoza* (London: Penguin, 1951).
36. Fox, *John Muir*, pp. 362–363.

17. THE SUBVERTED SCIENCE

1. Stephen Fox, *John Muir and His Legacy: The American Conservation Movement* (Boston: Little, Brown, 1981), p. 292.
2. Susan L. Flader, *Thinking Like a Mountain* (Lincoln: University of Nebraska Press, 1974), p. 5.
3. A. G. Tansley, "The Use and Abuse of Vegetational Concepts and Terms," *Ecology*, vol. 16, no. 1, July 1935.
4. Joseph M. Petulla, "Toward an Environmental Philosophy: In Search of a Methodology," *Environmental Review*, vol. 2, 1977.
5. Tansley, "The Use and Abuse of Vegetational Concepts."
6. Howard T. Odum, *Systems Ecology: An Introduction* (New York: John Wiley, 1983), p. 17.

7. Aldo Leopold, *A Sand County Almanac* (Oxford: Oxford University Press, 1949), p. 215.
8. Ibid., p. 203.
9. Ibid., p. 204.
10. Tansley, "The Use and Abuse of Vegetational Concepts."
11. Odum, *Systems Ecology*, p. 17.
12. R. L. Lindeman, "The Trophic-Dynamic Aspect of Ecology," *Ecology*, vol. 23, no. 4, October 1942.
13. Interview.
14. Vincent Schultz and Alfred W. Kleinert, Jr., eds., *Radioecology: Proceedings of the First National Symposium on Radioecology, Held at Colorado State University, Fort Collins, Colorado, September 10–15, 1961* (New York: Reinhold, 1963); F. Ward Whicker and Vincent Schultz, *Radioecology: Nuclear Energy and the Environment* (Boca Raton, Fla.: CRE Press, 1982), 2 vols; J. E. Lovelock, *Gaia: A New Look at Life on Earth* (Oxford: Oxford University Press, 1979), ch. 3, "The Recognition of Gaia," pp. 33 ff.
15. Ibid.
16. Stephen H. Schneider and Rundi Londer, *The Coevolution of Climate and Life* (San Francisco: Sierra Club, 1984), pp. 249–252; W. A. Berggren and John A. van Couvering, eds., *Catastrophes and Earth History*.
17. John A. Wiens, "On Understanding a Non-Equilibrium World: Myth and Reality in Community Patterns and Processes," Donald R. Strong, Jr., et al., *Ecological Communities: Conceptual Issues and the Evidence* (Princeton: Princeton University Press, 1984), p. 439.
18. Ibid., pp. 440, 451.
19. Peter M. May, "An Overview: Real and Apparent Patterns in Community Structure," in Strong et al., *Ecological Communities*, pp. 3–4.
20. Wyatt W. Anderson, "Achieving Synthesis in Population Biology," in Charles E. King and Peter S. Dawson, eds., *Population Biology: Retrospect and Prospect* (New York: Columbia University Press, 1983), pp. 191–192.
21. May, "An Overview."
22. Strong et al., *Ecological Communities*, p. viii.
23. Ibid., p. ix.
24. Petulla, "In Search of a Methodology."
25. Between 1979 and 1981, on a grant from the Exxon Education Foundation, I studied the transformation in college undergraduate curricula between 1945 and 1980. (Alston Chase, *Has Success Spoiled America? The Decline of General Education*, 1982 report to the Exxon Education Foundation; Alston Chase, *Group Memory: A Guide to College and Student Survival in the 1980s* [Boston: Atlantic–Little, Brown, 1980], ch. 5, "Curriculum: The Patina of Reform," pp. 110–131.)
26. Strong et al., *Ecological Communities;* Robert H. Whittaker, *Communities and Ecosystems* (New York: Macmillan, 1975); Eugene P. Odum, *Fundamentals of Ecology* (Philadelphia: Saunders, 1971).
27. C. P. Snow, *The Two Cultures: And a Second Look* (New York: New American Library, 1964).
28. Petulla, "In Search of a Methodology"; E. C. Pielou, *Mathematical Ecology* (New York: John Wiley, 1977), p. v.
29. Strong et al., *Ecological Communities*, p. viii.
30. Theodore Roszak, *Where the Wasteland Ends: Politics and Transcendence in Postindustrial Society* (New York: Doubleday, 1972), p. 367.
31. Olson, *Reflections from the North Country*, p. 42.

18. THE HUBRIS COMMANDOS

1. Alfred Runte, *National Parks: the American Experience* (Lincoln: University of Nebraska Press, 1979), p. 90.
2. Ibid., pp. 90–91.

3. Joseph Sax, *Mountains Without Handrails: Reflections on the National Parks* (Ann Arbor: University of Michigan Press, 1980), p. 14.
4. M. R. Montgomery, *In Search of L. L. Bean* (Boston: Little, Brown, 1984); U.S. Department of the Interior, Fish and Wildlife Service, Bureau of Sport Fisheries and Wildlife, *National Survey of Fishing and Hunting, 1970,* Resource Publication 95 (Washington, D.C.: U.S. Government Printing Office, 1970), p. 1.
5. J. Baldwin, "Equipment for the New Age," *Outside,* vol. 10, no. 1, January 1985.
6. Lee, *Family Tree,* p. 64; Shanks, *This Land Is Your Land,* pp. 239 ff.; Madison-Gallatin Alliance, *The Lee Metcalf Wilderness Proposal* (Bozeman, Montana: Madison-Gallatin Alliance, June 1980).
7. Stephen Fox, *John Muir and His Legacy: The American Conservation Movement* (Boston: Little, Brown, 1981), p. 315; Russell W. Peterson, President, National Audubon Society, "1983 Report to Members," *Audubon Action,* vol. 1, no. 4, October 1983; *Newsletter,* Yellowstone Basin Sierra Club Group, 1982.
8. *Audubon,* vol. 83, no. 1, January 1981, p. 110; *Sierra,* vol. 68, no. 1, January/February, 1983, pp. 49 ff.; *Defenders,* vol. 56, no. 1, April 1981, insert.
9. Ronald F. Lee, *Family Tree of the National Park System* (Philadelphia: Eastern National Park and Monument Association, 1972), p. 88; Conrad L. Wirth, *Parks, Politics and People* (Norman: University of Oklahoma Press, 1980), p. 261.
10. David Johst, "Does Wilderness Designation Increase Recreation Use?" Department of Agriculture, National Forest Service, Bozeman, Montana, February 21, 1984; Gallatin National Forest, "Wilderness Area Use Statistics," National Forest Service, Bozeman, Montana, October 26, 1984; John T. Stanley, Jr., "Sierra Club Wilderness Impact Study: Conclusions and Specific Findings," *Recreational Impact on Wildlands, Conference Proceedings,* Ruth Ittner et al., eds. (Seattle: National Forest Service, Northwest Region, 1979), pp. 189–194; Stephen F. McCool, "Does Wilderness Designation Lead to Increased Recreational Use?" University of Montana School of Forestry, 1983.
11. Shaller, interview; National Park Service, "Backcountry Statistics," 1983, Yellowstone National Park files.
12. National Wildlife Service, Bureau of Sport Fisheries and Wildlife, *National Survey of Fishing and Hunting, 1970,* Resource Publication 95 (Washington, D.C.: U.S. Government Printing Office, 1971); National Wildlife Service, Bureau of Sport Fisheries and Wildlife, *National Survey of Fishing and Hunting, 1960,* Circular 120 (Washington, D.C.: U.S. Government Printing Office, 1961); and *National Survey of Fishing, Hunting and Wildlife-Associated Recreation, 1975* (Washington, D.C.: U.S. Government Printing Office, 1977); Fish and Wildlife Service, *Fishery and Aquatic Management Program in Yellowstone National Park,* Technical Report for Calendar Year 1982, Yellowstone National Park, May 1983, p. 6, Table 3; Laura Loomis, "Park Crowds Are Pushing the Limits," *National Parks,* vol. 59, nos. 1–2, January/February, 1985, pp.11–17.
13. Recently in Montana Trout Unlimited has been lobbying to increase stream access for the public. Whatever the social justice of these efforts, there can be little doubt that more fishermen on the streams does not help the trout.
14. Ittner et al., *Recreational Impact on Wildlands.*
15. Elizabeth L. Horn, "Without a Trace — A Wilderness Challenge," Ittner et al., *Recreational Impact on Wildlands,* pp. 248–250.
16. Daniel O. Holmes, "Experiments on the Effects of Human Urine and Trampling on Subalpine Plants," Ittner et al., *Recreational Impact on Wildlands,* pp. 79–88.
17. Roderick Nash, *Wilderness and the American Mind* (New Haven: Yale University Press, 1967), p. 141.
18. Bill Devall, "John Muir and His Legacy," book review in *Humboldt Journal of Social Relations,* vol. 9, no. 1, Fall/Winter, 1981–1982.
19. The "*R* value" or insulating property of wood is, builders say, "an *R* an inch." By comparison, six inches of fiberglass offers an *R* value of 19.
20. Joan Haines, "Black Powder Buffs," *Bozeman Daily Chronicle Extra,* September 15, 1982, and, for more information on the buckskinners, see *Buckskin Report,* published in Big Timber, Montana.

21. Interview with David Baird.
22. William L. Bryan, "Preventing Burnout in the Public Interest Community," *NRAG Papers,* vol. 3, no. 3, Fall 1980, p. 23.
23. Bryan, "Preventing Burnout," p. 8; Northern Rockies Action Group, "Training for Citizen Action in the Eighties," Helena, Montana, undated; Bruce P. Ballenger, *Membership Recruiting Manual,* a training publication of the Northern Rockies Action Group, Helena, Montana, February 1981.
24. H. Ross Toole, *The Rape of the Great Plains* (Boston: Atlantic Monthly Press, 1976).
25. Rosak, *Where the Wasteland Ends,* p. 370.
26. Murray Bookchin, "An Open Letter to the Ecology Movement," quoted from John Baden and Richard Stroup, "Saving the Wilderness," *Reason,* July 1981, pp. 26–36.
27. Thomas S. Kuhn, *The Structure of Scientific Revolutions,* 2nd ed. (Chicago: University of Chicago Press, 1970).
28. Alan Drengson, "Shifting Paradigms, from the Technocratic to the Person-Planetary," *Environmental Ethics,* vol. 2, no. 3, Fall 1980), pp. 221–240.
29. Bill Devall, "New Age and Deep Ecology: Contrasting Paradigms," Mimeographed, 1981; George Sessions, "Shallow and Deep Ecology: A Review of the Philosophical Literature," Colloquium on the Humanities and Ecological Consciousness, *Earthday X,* Denver, April 21–22, 1980; Willis W. Harmon, *An Incomplete Guide to the Future* (Stanford: Stanford Alumni Association, 1976), p. 28; Fritjof Capra, *The Turning Point: Science, Society and the Rising Culture* (New York: Simon and Schuster, 1982), p. 30; Marilyn Ferguson, *The Aquarian Conspiracy: Personal and Social Transformation in the 1980s* (Los Angeles: J. P. Tarcher, 1980), p. 29; George Sessions and Bill Devall, *Deep Ecology: Living as if Nature Mattered* (Salt Lake City: Gibbs M. Smith, 1985).

19. THE CALIFORNIA COSMOLOGISTS

1. George Sessions, *Ecophilosophy Number 2,* May 1979, p. 16.
2. Theodore Rozak, *Where the Wasteland Ends: Politics and Transcendence in Postindustrial Society* (New York: Doubleday, 1972), pp. xv–xvi.
3. Fritjof Capra, *The Tao of Physics: An Exploration of the Parallels Between Modern Physics and Eastern Mysticism,* 2nd ed. (New York: Bantam Books, 1984), p. xvii.
4. M. C. Richards, *The Crossing Point,* quoted in Marilyn Ferguson, *The Aquarian Conspiracy: Personal and Social Transformation in the 1980s* (Los Angeles: J. P. Tarcher, 1980), p. 60.
5. Bill Devall, "John Muir and His Legacy," *Humboldt Journal of Social Relations,* vol. 9, no. 1, Fall/Winter 1981–1982; Rosak, *Where the Wasteland Ends,* pp. xxiv, 417.
6. Peter Gunther, "Man-Infinite and Nature-Infinite: A Mirror-Image Dialectic," 1980, unpublished.
7. Jeremy Rifkin, *Entropy* (New York: Viking Press, 1980), p. 239.
8. Ehrlich, *The Population Bomb,* p. 156.
9. Jacob Needleman, *The New Religions,* quoted in Ferguson, *Aquarian Conspiracy,* pp. 60–61.
10. Ferguson, *Aquarian Conspiracy,* pp. 132, 137.
11. Devall, "John Muir."
12. Rosak, *Where the Wasteland Ends,* p. 160.
13. Gary Snyder, "The Four Changes," *The Environmental Handbook,* Garrett de Bell, ed. (New York: Ballantine, 1970); Bill Devall and George Sessions, "Deep Ecology and Paradigm Change," 1983, unpublished.
14. Devall, "John Muir"; Bill Devall, "New Age and Deep Ecology: Contrasting Paradigms," Mimeographed, 1981; Devall, "Muir Redux: From Conservation to Ecology," 1982, unpublished; Conrad Bonifazi, *The Soul of the World: An Account of the Inwardness of Things* (Washington, D.C.: University Press of America, 1978).
15. Devall, "New Age and Deep Ecology"; Tom Regen, "On the Nature and Possibility of an Environmental Ethic," 1979, unpublished; Devall, "John Muir"; Peter Singer,

Animal Liberation (New York: Random House, 1975); Tom Regen and Peter Singer, eds., *Animal Rights and Human Obligations* (New York: Prentice-Hall, 1976); John Rodman, "The Liberation of Nature?" *Inquiry*, vol. 20, Spring 1977; *Inquiry*, vol. 22, nos. 1 and 2, 1979.

16. Edward Johnson, "Animal Liberation versus the Land Ethic," *Environmental Ethics*, vol. 3, no. 200, 1981; Christopher Stone, *Do Trees Have Standing? Toward Legal Rights for Natural Objects* (New York: Avon Press, 1975); Roderick Nash, "Do Rocks Have Rights?" *The Center Magazine*, November/December, 1977; Richard Watson, "Ultra-Anti-Anthropocentrism," 1982, unpublished; John Passmore, *Man's Responsibility for Nature: Ecological Problems and Western Traditions* (New York: Scribner's, 1974); and William Leiss, *The Domination of Nature* (New York: George Braziller, 1972).

17. Sessions, "Shallow and Deep Ecology."

18. Hans Jonas, "Gnosticism," *Encyclopedia of Philosophy*, ed. by Paul Edwards (New York: Macmillan, 1967).

19. Joseph L. Henderson and Maud Oakes, *The Wisdom of the Serpent: The Myth of Death, Rebirth, and Resurrection* (New York: Collier, 1963); William I. Thompson, *The Time Falling Bodies Take to Light* (New York: Saint Martin's, 1981); Allan W. Watts, *The Supreme Identity: An Essay on the Oriental Metaphysics and Christian Religion* (New York: Noonday, 1957).

20. Rosak, *Where the Wasteland Ends*, p. 346; Bahro quotation from Fritjof Capra and Charlene Spretnak, *Green Politics* (New York: E. P. Dutton, 1984), p. 56.

21. Institute for Noetic Sciences, *Newsletter*, Fall 1982; Capra, *The Tao of Physics*, pp. 116–117, 129; John Rodman, "Theory and Practice in the Environmental Movement," *The Search for Absolute Values in a Changing World*, International Cultural Foundation, 1978.

22. Devall, "John Muir"; *Ecophilosophy Newsletter, IV*, May 1982.

23. Carolyn Merchant, *The Death of Nature: Women, Ecology and the Scientific Revolution* (San Francisco: Harper and Row, 1983), p. xvii.

24. Thompson, *Falling Bodies*, p. 252; Ferguson, *Aquarian Conspiracy*, p. 389.

25. Ginsberg speech quoted from Devall, "John Muir"; remarks by Rubins, Kurpnik, and Thomas quoted from Ferguson, *Aquarian Conspiracy*, pp. 58, 206–207.

26. Devall, "New Age and Deep Ecology"; Rodman, "The Liberation of Nature?"

27. Ferguson, *Aquarian Conspiracy*, p. 100.

28. Willis W. Harmon, "Hope for the Earth: Connecting Our Social, Spiritual and Ecological Visions," *Newsletter*, Fall 1982; Tarrytown Group, *Tarrytown Letter*, February 1983; Willis W. Harmon, *An Incomplete Guide to the Future* (Stanford: Stanford Alumni Association, 1976), pp. 89 ff.; Ferguson, *Aquarian Conspiracy*, p. 61.

29. Ferguson, *Aquarian Conspiracy*, p. 412.

30. Harmon, *An Incomplete Guide*, p. 32.

31. Ferguson, *Aquarian Conspiracy*, pp. 89–90.

32. Ibid., p. 412; Harmon, "Hope for the Earth."

33. Arne Naess, "The Shallow and the Deep: Long-Range Ecology Movements," *Inquiry*, vol. 16, 1973; Sessions, "Shallow and Deep Ecology"; Devall, "New Age and Deep Ecology"; Jill Engledow, "A Long-Term Look at Rights," *Ecophilosophy Newsletter, VI*, May 30, 1983; Linda Centell, "Deep Ecology," *Union*, Arcata, California, March 31, 1983; Stephan Bodian, "Simple in Means, Rich in Ends: A Conversation with Arne Naess," *Ten Directions*, Zen Center, Los Angeles, Summer/Fall 1982; Bill Devall and George Sessions, *Deep Ecology: Living as if Nature Mattered* (Salt Lake City: Gibbs M. Smith, 1985).

34. Raymond F. Dasmann and Peter Berg, "Reinhabiting California," *Reinhabiting a Separate Country*, Peter Berg, ed. (San Francisco: Planet Drum Foundation, 1978); Raymond F. Dasmann, "National Parks, Nature Conservation and 'Future Primitive,'" *Ecologist*, vol. 6, no. 5, 1976; Sessions, "Shallow and Deep Ecology."

35. Engledow, "A Long-Term Look at Rights"; *Ecophilosophy Newsletter, II*, May 1979; Sessions, "Shallow and Deep Ecology"; Peter Berg, *Reinhabiting a Separate Country*.

36. Devall, "New Age and Deep Ecology"; Capra and Spretnak, *Green Politics*, pp. 193 ff.; James Ogilvy, "From Command to Co-Evolution: Toward a New Paradigm for

Human Ecology," *Ecological Consciousness,* ed. by Robert Schultz (Washington, D.C.: University Press of America, 1981).

37. Ferguson, *Aquarian Conspiracy,* p. 412; Bill Devall, "New Age and Deep Ecology"; Lovelock, *Gaia,* p. 127.
38. J. Peter Vajik, *Doomsday Has Been Cancelled* (Culver City, California: Peace Press, 1978), p. 61; Harmon, *An Incomplete Guide,* p. 144; Harmon, "Hope for the Earth."
39. Ferguson, *Aquarian Conspiracy,* p. 29.
40. Alexis de Tocqueville, *Democracy in America* (New York: Anchor, 1969), p. 9.
41. Greenpeace, "The Greenpeace Philosophy," brochure; Nash, "Do Rocks Have Rights?"
42. Capra and Spretnak, *Green Politics,* p. 56.
43. Ibid., p. 200.
44. NRAG, *Training Citizen Action in the Eighties:* "NRAG'S Anti-Nuclear Program, First Year Summary," 1983; "NRAG's Economic Justice Program," 1983; "Groups NRAG Worked with in 1982–1983"; "NRAG Activity Reports," October 1983– February 1984 and March 1984–July 1984.
45. Sessions, "Shallow and Deep Ecology."
46. Russell W. Peterson, "Think Globally, Act Locally," *Audubon Action,* vol. 1, no. 2, February 1983; "The Sierra Club Activist Program," brochure; "The World after Nuclear War," Conference on the Long-Term Worldwide Biological Consequences of Nuclear War, October 31–November 1, 1983; "On the Fate of the Earth," Rocky Mountain Regional Conference, October 20–23, 1983.
47. Rick Reese, *Greater Yellowstone: The National Park and Adjacent Wildlands* (Helena, Montana: Montana Magazine, 1984), p. 95.

20. THE LAND ETHIC

1. Taped at the second annual convention of the Greater Yellowstone Coalition, June 15, 1984. See also Franz J. Camenzind, "The Greater Yellowstone Ecosystem," January 26, 1984, Report to the Greater Yellowstone Coalition.
2. Bryan to author, November 9, 1984.
3. Greater Yellowstone Coalition Brochure, 1984.
4. Greater Yellowstone Coalition, "Progress Report," October 1, 1984.
5. Minutes of "React" meeting, sponsored by the Wyoming Governor's Office, held at Brinkerhof Retreat, Jackson, Wyoming, September 13, 1984.
6. See "Progress Report," October 1, 1984, and Greater Yellowstone Coalition, *Threats to Greater Yellowstone,* 1984 (Bozeman, Montana: Greater Yellowstone Coalition, December 1, 1984).
7. *Threats to Greater Yellowstone.*
8. Taped by author at meeting, June 16, 1984. See also Greater Yellowstone Coalition, "Yellowstone Grizzly Policy," December 3, 1984.
9. Greater Yellowstone Coalition, "Draft Ungulate Policy Statement," 1985.
10. Greater Yellowstone Coalition, "Draft Geothermal Policy," 1985; Greater Yellowstone Coalition, *Threats to Greater Yellowstone.*
11. Rick Reese, *Greater Yellowstone: The National Park and Adjacent Wildlands* (Helena, Montana: Montana Magazine, 1984), pp. 22–28.
12. Sessions, *Ecophilosophy Number 2;* Sigurd F. Olson, *Reflections from the North Country* (New York: Alfred H. Knopf, 1982), p. 41; Hank Fischer, "Is Wilderness Bad for Wildlife?" *Defenders,* vol. 59, no. 5, September/October, 1984.

21. YELLOWSTONE ELEGY

1. Paul Schullery, *Mountain Time* (New York: Nick Lyons/Schocken, 1984), p. 69.
2. Ibid., p. 94.
3. *Livingston Enterprise,* February 1, 1982.

EPILOGUE

1. James A. Kushlan, John C. Ogden, and Aaron L. Higer, "Relation of Water Level and Fish Availability to Wood Stork Reproduction in the Southern Everglades, Florida," (Tallahassee, Florida: U.S. National Park Service, 1975); interview with William B. Robertson, Jr., biologist, National Park Service, Everglades; Science Support Staff, National Park Service, "Fiscal 1987 Briefing Statement," February 21, 1986, unpublished; William Porter Bradley, "History, Ecology, and Management of an Introduced Wapiti Population in Mount Rainier National Park, Washington," Ph. D. thesis, University of Washington, 1982; Carl E. Gustafson, "Wapiti Populations in and Adjacent to Mount Rainier National Park: Archaeological and Ethnographic Evidence," Washington State University, 1983, unpublished; Ed Starkey, "Elk of Mount Rainier National Park: A Review of Existing Information," National Park Service, 1984, unpublished; Paul Schullery, "A History of Native Elk in Mount Rainier National Park," 1983, unpublished; National Park Service, "The Problem of Feral and Exotic Animals in the National Park System," 1986, unpublished; Director, National Park Service, to Regional Directors, Alaska Area Director, Manager, Denver Service Center, "State of the Parks: A Report to the Congress on a Servicewide Strategy for Prevention and Mitigation of Natural and Cultural Resource Management Problems," January 14, 1981.
2. William D. Newmark, "A Land-Bridge Island Perspective on Mammalian Extinctions in Western North American Parks," *Nature*, volume 325, January 29, 1987.
3. National Park Service, State of the Parks, A Report to Congress, May, 1980; *Arizona Daily Star*, September 2, 1984.
4. Conservation Foundation, "Executive Summary, National Parks for New Generation: Visions Realities, Prospects," April 24, 1985; National Parks Conservation Association, "No Park Is an Island," reprinted from *National Parks & Conservation Magazine*, 1979.
5. The Wilderness Society, National Parks and Conservation Association, Sierra Club and National Audubon Society, "Toward a Premier National Park System," 1983, unpublished.
6. Ibid.
7. Linn to author, January 25, 1987.
8. Burch to author, January 22, 1987.
9. Reed to author, February 22, 1987.
10. Ibid.
11. See Alfred Runte, *National Parks: The American Experience* (Lincoln: University of Nebraska Press, 1979).
12. Ronald A. Foresta, *America's National Parks and Their Keepers* (Washington, D.C.: Resources for the Future, 1984), p. 121.
13. Results of 1985 poll given to me by Park Service Chief Scientist Dr. Richard Briceland on January 24, 1986.
14. Interview with Dr. Richard Briceland; National Park Service, Fiscal Year 1987 Budget Summary and Justification Material.
15. Fiscal Year 1987 Budget Summary and Justification Material.
16. Acting Director, National Park Service, to Regional Directors, "Research and Resource Management Program Changes," January 21, 1986; Associate Director, Natural Resources, to Regional Chief Scientists, "Recent Actions Affecting the Research/Natural Resource Program," February 4, 1986.
17. William R. Jordan III, Michael E. Gilpin, and John D. Aber, *Restoration Ecology: Ecological Restoration as a Technique for Basic Research* (Cambridge: Cambridge University Press: to be published).
18. William R. Jordan III, Editor, *The Arboretum* (Madison: University of Wisconsin-Madison, 1981), p. 1.
19. Daniel H. Janzen, *Guanacaste National Park: Tropical Ecological and Cultural Restoration* (San Jose, Costa Rica: Editorial Universidad Estatal A Distancia, 1986).

20. William R. Jordan, III, "Restoration and Reentry of Nature," *Restoration & Management Notes*, vol. 4, no. 1, Summer, 1986.
21. Ibid.
22. George M. Wright, Joseph S. Dixon, and Ben H. Thompson, *Fauna of the National Parks of the United States* (Washington, D.C.: U.S. Government Printing Office, 1933), pp. 3 ff.
23. A Starker Leopold, et. al., "Wildlife Management in the National Parks," Report of the Advisory Board on Wildlife Management to Secretary of Interior Udall. March 4, 1963.
24. Ronald F. Lee, *Family Tree of the National Park System* (Philadelphia: Eastern National Park & Monument Association, 1972), pp. 61-62. For an example of the tendency of research to ignore social studies under the new organization, see, Office of Natural Science Studies, National Park Service, *Proceedings of the Meeting of Research Scientists & Management Biologists of the National Park Service* (Washington, D.C.: National Park Service, 1968).
25. Acting Director, National Park Service, to Regional Directors, January 21, 1986.
26. "The Grand Illusion," *Newsweek*, July 28, 1986.
27. Dangermond to author, February 10, 1987.
28. National Park Service, *Twelve-Point Plan: The Challenge, the Actions* (Washington, D.C.: U.S. Government Printing Office, 1986), pp. 1 and 11.
29. Ibid., p. 11.
30. Charlotte Curtis, "Trouble in Yellowstone," *New York Times*, June 3, 1986.
31. Daniel Rosenheim, "A Push to Privatize Parkland," *San Francisco Chronicle*, December 15, 1986.
32. Ibid.
33. Ibid.
34. William J. Robbins, *A Report by the Advisory Committee of the National Park Service on Research of the National Academy of Sciences-National Research Council* (Washington, D.C.: National Academy of Sciences—National Research Council, August 1, 1963), p. 53; National Park Service, Fiscal Year 1987 Budget Summary and Justification Material.
35. See Theodore Sudia, "National Parks and Domestic Affairs," *The George Wright Forum*, vol. 4, no. 4, 1986, pp. 15-27.
36. Frederick Jackson Turner, *The Frontier in American History* (New York: Henry Holt and Company, 1920).

Index

Abbey, Edward: *Desert Solitaire*, 207–208
Abell, Bruce, 290
Absaroka mountains, 268, 271
academic freedom, 248, 251–253
accessibility to wilderness, 327–329
acid rain, 256, 361
Adams, Ansel, 302–303, 341, 346; as pantheist, 304, 349
Advisory Board on Wildlife Management, 33–35, 340
Advisory Committee on Natural History Studies, 247–248
African buffalo, 55–56, 67
Agriculture, U.S. Department of, 193, 238, 239; 1949 Yearbook of, 111
Alaska, 64, 269
Albright, Horace, 20, 21, 125, 126, 127, 235; on National Park Service, 234, 237
Albright, Horace, Training Center, 143
Allen, Durward, 154, 166, 175, 178, 341
Allen, E. T., 267, 268
American Association for the Advancement of Science (AAAS), 298
American Automobile Association, 204
American Indian Movement, 45
American Ornithological Union, 120
American Society of Mammalogists, 125
Anderson, Bob, 365
Anderson, Jack, 38, 51, 74; and bear management, 156, 157, 162; and Grant Village, 214; rejects research lab, 249; and wolf transplant, 129, 130, 134, 141
Anderson, Terry, 388
Anderson, Wyatt W., 320
andesite, 268, 290
Andrews, Robert S., 289
Andrus, Cecil, 178, 220, 222, 341
angel dust, used on bears, 190, 192
Animal Friends of Yellowstone (book), 137
animism, 304, 347, 349
antelope: decline in, 80, 103, 371; Indians hunt, 97, 99; killed, 193; *see also* pronghorn antelope

anthropologists, 257, 374
Appalachian Mountain Club, 327
archaeologists, 79, 199, 257
Arnett, G. Ray (Assistant Secretary of the Interior), 140
Arnold, F. T., 66
aspen, 24, 27, 79; decline of, 63, 68–69, 79, 371; overgrazing of, by elk, 86–87
Association of Park Rangers, 381
attendance at Yellowstone, 213–214, 331, 332, 374
Audubon (magazine), 109, 181
Audubon Society, 378
automobile: banned in Yosemite, 207; efforts to discourage, 209–211, 214; increases park use, 202–203; lobbies on, 213
"aversion" experiments for bears, 191–192

backpacking, 328–329, 331–332
Baden, John, 388
Bahro, Rudolf, 351
Bailey, Elizabeth, 188
Bailey, John, 78–79
Bailey, Vernon, 66, 67, 122–123
Baines, James: *The Whole Earth Papers*, 357
Baird, John, 336
balance of nature, 26, 29–30, 34–35; based on "biotic pyramid,"; 314–315; doubt in, 318–319
Bang's disease. *See* brucellosis
Bannock Indians, 108, 109, 198; decline of, 103, 106, 113
Barbee, Robert, 65–66, 74, 77, 227–229, 257
Barlow, Arthur, 110
Barlow, J. W., 15, 263, 267
Barmore, William, 52, 64–65, 143
Barnes, Victor, and black bear study, 148, 153, 163
Baronett, Collins Jack, 85, 198
Barrett, Stephen, 96–97
Battelle Memorial Institute, 142
Baucus, Max, 225, 297

"Bayer 1470" (Rampon), 157
Bear Attacks (Herrero), 191
bear-human encounters, 146–148, 155, 166, 187–190, 218, 231
bear management, 149–154, 161, 165–166, 168, 175–178, 182–186; "control actions" for, 155–158, 186–189, 191–193; *see also* black bears; grizzly bears
Bearss, Edwin, 248–249, 251
Beartooth-Absaroka Wilderness, 331–333, 364
beavers, 11–13; decline in, 28, 63, 80, 371; destroyed, 128, 193; and wolves, 135–136; at Yancey's Hole, 85–86
Beetle, Alan, 64
behavior modification for bears, 191–192
benign neglect. *See* noninterference
Beowawe Geysers, Nevada, 285
Berg, Peter, 356
Berkholter, Bob, 130
Bevinetto, Anthony, 225
Bianchi, Don, 191
Bickley, Carol, 257
Big Game Ridge, erosion on, 78
bighorn sheep. *See* sheep, bighorn
Billings Gazette, 50
Biography of a Grizzly (Seton), 167–168
Biological Survey, U.S., 21, 239; and predator control, 66–67; and extermination of wolves, 120–123; and Wright, 238
biologists: and bear policy, 193; doubt community ecology, 320; vs. geologists, 291–292; oppose Grant Village, 199, 211; role of, in Park Service, 235–239, 251
"bioregional politics," 356
BioScience, 179
"biospherical egalitarianism," 356
"biotic pyramid," 25, 314
biotic systems, 33, 40
Birds and Mammals of Mount McKinley National Park (Dixon), 238
birdwatchers, 332
bison. *See* buffalo
Bitterroot Valley, 14, 135
black bears: decline of, 142–143, 163, 371; killed, 156, 160, 184, 193; as problem, 144–149
Blackfoot Indians, 96, 103, 105, 106
Blank, H. Richard, 276, 277, 279, 281
Blank, Tim, 186
Blonson, Gary, 89
blowout danger, 287, 288

bluegrass, 28, 78, 87
bobcats, 24, 124, 128, 193
Bob Marshall Wilderness, 162
Bode, Carl, 301
Bohlen, Curtis, 153, 178, 341
Bonine, Bob, 296
Bookchin, Murray, 342, 359
Boone and Crockett Club, 18–19, 125, 178, 341
Boutelle, F. A., 17
Bowen, N. L., 267
Bowser, Richard, 219–220
Boyd, Francis, 269–272
Bozeman, Montana, 211
Brand, Stewart, 357
Bray, Olin, and black bear study, 148, 153, 163
Brett, Lloyd, 66
Brewster, William, 305
Briceland, Richard, 257, 382
Bridger, Jim, 15, 61, 265–266
British Ecological Society symposium on competition for food, 319
Brockman, Frank, 235
Brophy, Robert, 346
Brouwer, H. A., 267
Brower, David, 303–305, 324, 346, 361
Brown, Gary, 82, 176, 183, 186
Brown, James H., 316–317
browse, 12–13; disappears, 22, 26–28, 63, 68, 87
brucellosis in buffalo, 75, 83
Bryan, William, 296, 304, 310, 338–340
buckskinners, 336
Buckskin Report, 336
Buddhism, 303, 304, 349
buffalo, 14; African, 55–56, 67; brucellosis in, 75, 83; decline of, 16, 103; and herd reduction, 31–32, 35, 130, 193; Indians hunt, 97, 99–102, 145; proliferation of, 83–84, 88–89, 115; restoration of, 18, 22, 30
Bunnell, Fred, 185
Burch, William, 378
Burlington Railroad, 200, 203
burning: by Indians, 94–97, 111; and natural-burn policy, 70; encourages new growth, 63; Wright supports, 238; *see also* fires
Burrows, C. J., 54–55
Butte, Montana, 339

Cache Creek, 271
Cahalane, Victor H., 126, 237, 239–240
Cain, Stanley, 247, 248
California as center of radical thought, 345–347, 373

California Cosmology, 347–355; and environmental activists, 373; *see also* new politics; new religion; new science
California Tides foundation, 295
Camenzind, Franz, 363
Cammerer, Arno, 125–126, 127
Campaign for Yellowstone Bears, 175
Campbell, Arthur B., 275
campsites as "sacrifice areas," 333
Canada, wolves in, 133
Canadian Wildlife Service, 58, 64, 134
Canyon, 205, 208
Canyon Creek fish restoration project, 76
Canyon Hotel, 205
Canyon Village, 204, 205
Capra, Fritjof, 343, 344, 346, 352; and *Green Politics*, 359; as New Age/Aquarian, 357
caribou, 58, 74
Carnegie Geophysical Laboratory, 267, 268
Carnegie Institute, 283
carnivores, 57–58, 64; *see also* predation; predators
"carrying capacity" ("K") of range, 62, 73–74, 90, 171–174
Carson, Rachel: *Silent Spring*, 42, 43–44, 303, 311–312
Carter, James E., 199, 224; opposes Grant Village, 222, 223, 227
Casper *Star-Tribune*, 190
Catlin, George, 14, 96
Caughley, Graeme: on Kaibab, 69; self-limiting theory of, 153–156
Cavalry, U.S.: and Indians, 113; runs Yellowstone, 16, 17, 18, 66, 121
Cedar Lake Bog, Minnesota, study, 316
Celestine Pool, 281–282
census: aspen, 79; bear, 146, 148, 170; deer, 88; efficiency of, 67, 68; elk, 17, 66–67, 73, 83
Chamberlain, Allen, 327–328
Channel Islands National Monument, 377
Chapman, Scotty, 80
Chatham, Russell, 49
Chicken Ridge, 78
Chippewa Indians, 104
Chittenden, Hiram, 14, 61, 264
chlamydia, 81–82
Christianity: as cause of ecologic crisis, 298–300; and need for reformation, 345; rejected, 301–304, 310, 351
Christiansen, Robert L., 276, 277, 279, 280–281
Citizens' Alliance for Progressive Action, 360

Civilian Conservation Corps (CCC), 237
Clark, William, 145, 264
Cleveland, Grover, 5
Cody, Wyoming, 228
Cody *Enterprise*, 186
Coeur d'Alene Indians, 95, 96
coevolution, 357–358
Co-Evolution Quarterly, 357, 359
Coffman, John, 238
Cohen, Mark N., 102
Cole, Glen: and bear problem, 143, 148, 151, 158, 160, 165; and caribou, 74; and elk problem, 73; and Fishing Bridge, 229; on moose, 255; and "natural control" policy, 38–42, 46–48, 52, 130; and theory of natural regulation, 59–60, 62–70; and reintroduction of wolves, 129–132, 134, 139
Colter, John, 264, 329
Committee on Rare and Endangered Wildlife Species, 150
Committee on the Yellowstone Grizzlies. *See* National Academy of Sciences: Committee on the Yellowstone Grizzlies of
Commoner, Barry, 43, 367; and "Third Law of Ecology," 44
communicators, interpreters as, 245–247
community ecology, 313–314, 316; doubts on validity of, 318–320; fails as true science, 321
"compensatory mechanisms," 160, 163, 171; and self-regulation, 318
competition among ungulates, 29, 62–63, 68
Comstock, Theodore, 15, 61
concessionaires: typical franchise contract of, 226; oppose Grant Village, 199, 205–206, 223; funded by railroads, 201
Condon, David de L., 11
Congress, U.S.: and deep-drilling project, 289–290; environmental legislation of, 43, 306; and Grant Village, 216–217, 222, 225–227; and Park Service, 200, 213, 258, 261; and predator control, 66, 121–122; passes Yellowstone Park Act, 3, 113, 266–267
Conservation Foundation, 296, 378
continental plate: drilling proposed in, 283; movements of, 274, 278; thickness of, 279–280
Continental Scientific Drilling Committee (CSDC), 284, 285, 287–290
Continental Scientific Drilling program, 289–290
"control actions" for bears, 155–158, 160, 162–165, 186–189, 191–193

Cook, Charles W., 265
Cooke, Jay, 266
Cooke City, dumps at, 164, 185, 186
Cooperative Wildlife Study Units, 247,
253, 256
cougars. *See* mountain lions
Council on Environmental Quality, 43,
162, 341
coyote extermination, 23, 24, 121, 123,
126, 127, 193, 292
Craighead, Frank, 64, 131, 252; grizzly
study by, 148, 149–153, 157; his *Track
of the Grizzly,* 179; *See also* Craighead
brothers
Craighead, John, 30, 61, 64, 128, 131,
212; on bear decline, 158, 163, 164,
168, 171, 189–190; proposes ecocenters, 174–176; and ecosystem habitat
analysis, 172; on feeding committee,
182–184; *see also* Craighead brothers
Craighead brothers: propose feeding centers, 174–177, 182–184; grizzly bear
study by, 148, 149–153, 157; vs. Park
Service, 151–154, 157–165, 179, 291,
296; support for, 158, 160, 169
crash of ungulates, 24–26, 67
craters, volcanic, 270, 271, 274, 277
Cross, Walker, 188
Crow Indians, 99, 103, 105, 106
Crystal Creek elk trap, 31–32
Cundall, Alan W.: *Hamilton's Guide to
Yellowstone,* 5
Cunningham, Bill, 296
Curie depth, 279–280
Cutter, Joe, 223–224
cyanide for predators, 128

Dangermond, Peter, 387
Dasmann, Raymond F., 134, 346, 356
Davidge, Rick, 224–225, 258
Day, A. L., 267, 268
DDT, 44, 312
Death Gulch, 167–169
Death Valley National Monument, 377
Deckard Flat, 49–50
deep drilling, 283–291, 366
"deep ecology," 356–358, 364
Deep Sea Drilling Project, 283
deer, 15, 68; *see also* mule deer; red deer;
white-tail deer
Defenders of Wildlife, 158, 159, 184; ex-
peditions of, 331
Defense, U.S. Department of: deep-drill-
ing projects of, 284; tests of, 278, 291
deLacy, Walter W., 11, 96, 106, 265, 282

density-dependence mechanisms, 83–84,
319
Desert Solitaire (Abbey), 207–208
deSmet, Pierre, 96, 265
Despain, Don, 60, 63, 229
deterioration of Yellowstone, 114–115,
256–257, 372–375
Devall, Bill, 334, 343–348, 352, 354, 356
Diablo Canyon nuclear reactor, 346, 347
Dickenson, Russell, 140, 225, 228, 229,
256; and Park Service policies, 257,
258–260
diet of animals, 12–13
Dilley, Willard E.: *Wildlife in Yellowstone and Grand Teton National Parks,*
5
Dingell, John, 75
direct reduction of elk, 27, 31–33, 36–37,
44–47
Dirks, Richard A., 68
diseases: of animals, 75, 81–82, 83, 90;
backcountry, 211, 333; that kill Indians,
104–105
Dixon, Joseph, 236; *Birds and Mammals
of Mount McKinley National Park,* 238
Doane, Gustavus C., 92, 115, 264, 267
Dome Mountain, 61, 65, 72
doming at Yellowstone, 280–281
Doomsday Has Been Cancelled (Vayk),
357
Dorr, George, 305
Douglas, William O., 112, 303, 349
Douglas fir, 14, 24
"downhill equilibrium" in "irruptive sequence," 26
Drengsen, Alan: *Environmental Ethics,*
343
Drugs: and self-transcendence, 354–355;
used on animals, 157, 188–189, 190,
192
Dum-Dum (grizzly bear), 187
dumps. *See* garbage
Dunmire, Robert, 137
Dunraven, Earl of, 11, 15, 61
Dzurisin, Daniel, 280

earth: as ecosystem, 317–318; as living
organism, 357
earthquakes, 272–275, 279
Eastern cultures, 352
Eastern religions, 302, 303, 345–346
Eastman, Gordon, 129–130, 133
Eberhardt, Lee, 142, 170–171, 173
"ecocenters," 174–176, 183
"ecological sensibility," 352, 354
Ecological Society of America, 57, 125

ecology, 42–48, 113, 115, 312, 320–321;
crisis in direction of, 342–343; fails as
true science, 321–325, 353; and geology, 290–291; of Yellowstone, 12–13,
29; *see also* community ecology
Ecology of the Coyote in the Yellowstone
(Murie), 238
economic development, 339, 364–367
ecophilosophers, 343
Ecophilosophy (journal), 356
ecosystem: concept of, 33–35, 312–313;
difficulty of defining, 316–318, 363,
368; doubts about, 318–320; earth as,
317–318; environmentalists' idea of,
324–325; and fire, 94; human activity as
part of, 314, 315; Indians' effect on,
105–107; and study of nature, 313–314;
universe as, 318; Yellowstone as, 40–
42, 55–56, 62–73, 90, 364–367
ecosystem habitat analysis, 172
ecosystems management of Park Service,
33–35, 38–41, 46, 129; and bear problem, 149–154, 166; and true "ecosystems management," 367; and
noninterference, 250, 367; policy of,
fails, 368; and self-regulation, 318–319;
see also natural regulation
education: replaced by interpretation,
245–247; and work of naturalists, 235,
239, 260–261
egalitarianism, 346, 358–359; "biospherical," 356
Ehrlich, Paul, 43, 345
electric shock, 192
elk, 3, 5, 14–17, 71–74, 240, 366; effect
of, on other animals, 80–84, 173; effect
of, on vegetation, 28, 77–80, 86–87; as
food for bears, 189–190; herd reduction
of, 27–30, 31–37, 240; Indians hunt,
97, 99, 102, 103; killed, 49–51, 193; migration of, 38, 61–62, 65, 72–73; "natural control" theory and, 38–39, 46–48,
52–53; natural regulation theory and,
59–70, 129, 130; trampling by, 28, 77
Emerson, Ralph Waldo, 301–303, 305
endangered species, 138–141, 150, 162,
217, 220, 230
Energy, U.S. Department of, 284, 289
Engledow, Jill, 356
Eno, Amos, 180, 184, 230
Entropy (Rivkin), 345
environmental activists, 359, 360, 373; *see
also* environmentalists; environmental
movement
Environmental Center conference "On
the Fate of the Earth," 361
Environmental Ethics (Drengson), 343

Environmental Impact Statements: on
deep drilling, 285–286; on dump clos-
ings, 152, 157; on Fishing Bridge, 229;
on Grant Village, 217, 218; on Park
Service Master Plan, 137, 213–214
Environmental Information Center, 339
environmentalists: and bear policy, 177–
178, 193; dilemma of, 308–310; and
Grant Village, 207, 216–217; and motorists, 204; and predator control, 125;
and religion, 299–300, 302–304, 343;
and science, 304–305, 322–325; and
Yellowstone, 361–362; *see also* environmental activists; environmental movement
environmental movement: awakening of,
43–47, 302–303, 312; growth of, 330; in
northern Rockies, 295–298, 326–327;
political activism and, 337–342, 359–
360; and recreation industry, 329–330,
332; spiritualism of, 300–305, 308–310,
322–325; split in, 340–343, 358–361;
promotes wilderness use, 331; *see also*
environmental activists; environmentalists
environmental organizations, mainline:
become bureaucracies, 340–341; and
grass-roots politics, 360–361; revolving
door of, with government, 341; "Shallow Ecology" of, 360
Environmental Policy Act, 157, 217, 218
Environmental Protection Agency, 43
Erickson, Glenn, 81, 88
erosion, 12, 22, 27, 28, 77–78, 286
Estey, Harold J., 140
Evelyn, John: *Silva: A Discourse of Forest Trees and the Propagation of Timber
in His Majesty's Dominions*, 306
Everglades National Park, 247, 249, 251–
252, 257, 377
Everson, William, 304, 346
Evoy, Jeffrey A., 279
Excelsior geyser, 263
exclosures, 27, 47, 78, 86–87, 291
Executive Order 11643 (Nixon), 157
explorations of Yellowstone, 264–272
extermination: of bears, 155, 184, 193; as
control policy, 27, 31–37, 49–51, 193;
in 1800s, 16; of predators, 23–25, 119–
128; vs. preservation, 239

Fabich, Hank, 191
farm lobbies, 121, 122, 127, 139, 140
*Fauna of the National Parks of the United
States* (Wright and others), 236
fauna series, 236–238, 247; fauna one and
two, 236, 237; fauna three, 238; fauna
four and five, 238

Federal Register, 76
feeding centers for bears, 174–177, 183–184
Fenner, C. N., 267
Ferguson, Marilyn, 343, 346, 353–358
Field Natural History school, 234
Field Operations Study Team (FOST), 243–244, 245, 249, 260
Final Environmental Statement for the Yellowstone Master Plan (1974), 137, 152, 213–214
Finley, Robert B., Jr., 161–162, 179
Firehole, 96
Firehole geyser basin, 263, 267, 272, 276
Firehole River, 290
"fire hunting," 92, 94
fires, 68–69, 87, 92–97; forest, 111, 113; *see also* burning
firing line, 49–51, 77
First World Conference on National Parks, 42
Fischer, Hank, 184, 367
fish, 17–18, 44, 75–76, 212
Fish and Game, Montana Department of, 30, 39, 50, 71–72, 77, 191
Fish and Wildlife Service, U.S.: Ad Hoc Fisheries Task Force of, 76; and bears, 159, 162, 172, 184–185; and fish restoration project, 75–77; fish-trapping studies of, 212; and Grant Village, 199, 217, 220–222, 227; in Interior Department, 239; Division of Predator and Rodent Control of, 128, 129, 139; and wolves, 138, 139
fisher, 124, 193
fishermen, 332
Fishing Bridge: and Grant Village, 198, 206, 208–210, 221, 227–229; as habitat for grizzly bear, 154, 189, 211–213
flow rock, 269, 270
Flynn, James W., 77
Folsom, David E., 265
food supply and natural control, 54–56, 59
Ford Foundation, 340
Foresta, Ronald A., 381
forest management and "sustained yield" vs. protection, 306–307
"forest reserves," 306–307
Forest Service, U.S., 72, 111, 285, 306, 333, 365
forests of Yellowstone, 14, 114
"fortress mentality" of Park Service, 254–256
Foshag, W. V., 267
Foss, Arnold, 64, 77

Fournier, Robert, 276, 288, 289
Fowler, Charles W., 64–65
Fox, Stephen, 302, 304–305, 310, 312
foxes, 124, 193
franchise contract, 226
Fredenhagen, Brigitta, 188
Freedom of Information Act, 65
Friends of the Earth, 303, 351, 359
Frison, George, 97, 99
fuel crisis, 213, 223, 283–284, 339
fumaroles, 262–263
Fund for Animals, 159, 162
fur trappers, 4, 11, 45, 264
"future primitive," 348, 356

Gaab, Joe, 80, 131
Gallatin mountain range, 268
Game and Fish, Wyoming Department of, 192
garbage: as bear food, 145–147; and phase-out of dumps, 149–154, 164, 166, 171, 174, 182, 185, 186, 192, 208, 334; naturalness of, 176–178
Garbage in the Cities (Melosi), 177
Gardiner, Montana, 3, 59, 83; bears at, 164, 186; firing line at, 49–51, 88
Garrison, Lemuel (Lon), 28–30, 42–43, 60, 64; on elk, 33; and Mission 66, 204, 205, 207; on scientists, 251; and wolves, 128
Gates, Marshall, 119, 130
General Hosts Corporation, 206, 214
genetic pools, 75–77
Geological Survey, U.S., 266, 267, 292; and deep drilling, 289, 291; studies earthquakes, 273–277; and NASA study, 275–277
Geologists, 262–263, 267–275, 279–281; vs. biologists, 291–292
geology of Yellowstone, 290–291
geothermal development, 285–287, 364
geothermal system of Yellowstone: activity of, 273–277; features of, 262–265, 268, 282; *see also* thermal areas
German Greens, 351, 359, 360
geysers, 109, 262–265, 268; activity of, 273, 274, 277, 281–282; deep drilling destroys, 285–286; and "geyser gazers," 227
Ghost Ranch workshop, 283
giardiasis, 211, 333
Gibbon River, 18, 371
Ginsberg, Allen, 353
Glacier Bay National Park, 377
Glacier–Bob Marshall Wilderness area, 139

Glacier National Park, 172, 173, 207, 247

Gnosticism, 350–351

Golley, Frank, 57

Good, John 275

Goose Lake, 18

Gore, Sir George, 14

Grand Canyon National Game Preserve, 24–25

Grand Canyon National Park biologists' meeting, 143–144, 149

Grant, U. S., 111, 204

Grant, Ulysses S., 5, 204, 267

Grant Village, 173, 188, 197–199, 209–210; in grizzly bear habitat, 211–213, 217–218, 221, 228–231; objections to, 219–220, 222; stages in development of, 204–208, 223–230; Townsley pushes, 214–217

grasses, 12; and burning, 63, 93–94; decline of, 22, 53–55, 57, 114; native vs. exotic, 27–28, 29, 78, 87, 371; efforts to restore, 35

Grater, Russ, 237, 261

Graves, Henry S., 19, 61, 73

gravitational acceleration, 279–280

grayling, 75–76

Great Bannock Trail, 98, 103

Greater Yellowstone Coalition (GYC), 169, 185, 361, 363–367

Greater Yellowstone Cooperative Regional Transportation Study, 215–216

Green Book, 39–40, 47, 129

Greening of America, The (Reich), 42

Greenpeace, 359

Green Politics (Capra and Spretnak), 359

Green River, Wyoming, 99

Griffin, Robert P., 153

Grinnell, George B., 112

Grinnell, Joseph, 22, 125, 128, 236, 237, 252

Grizzly Bear Recovery Program, 168

grizzly bears, 15, 41, 144; control-action deaths of, 155–158, 160, 162–165, 186–193, 231; decline of, 142–144, 163, 170–171, 182–183; and Fishing Bridge, 221, 227, 229–230; and garbage dumps, 149–154, 182, 334; and Grant Village, 154, 189, 211–213, 217–218, 221, 228–230; as problem, 145–149, 154, 184–186; as threatened species, 75, 162, 164, 168, 217; *see also* bear management; Craighead brothers

Grizzly Bear Steering Committee, 159

Gros Ventre Indians, 104, 105

ground squirrels destroyed, 128, 193

Gruell, George E., 78

Guadalupe Mountains National Park, 377

Guanacaste National Park, Costa Rica, 383

Gunter, Peter, 345

Habeck, J. R., 68–69, 90

Haeckel, Ernst, 312

Hague, Arnold, 267, 268, 269

Haines, Aubrey L., 4–5, 67, 137, 265

Haleakala National Park, 377

Haley, James A., 158

Hamilton, Warren, 273–275, 289

Hamilton, Wayne, 168–169, 280, 288–289

Hamilton's Guide to Yellowstone (Cundall and Lystrup), 5

Hamilton Stores, 164, 197, 228

Handler, Philip, 159–160

Hansen, Clifford, 249

Harmon, Willis W., 297, 343, 346, 351–352; on global future, 355, 357–358; on transcendence, 354

Harper's Ferry center for interpretation, 246

Harris, Mary Hazell (Executive Director of Defenders of Wildlife), 159

Harris, Moses, 121

Harry, Bryan: *Wildlife in Yellowstone and Grand Teton National Parks,* 5

Hart, Philip, A., 158

Hartzog, George, 36, 37, 39

Hassrick, Buzzy, 190

Hawaii Volcanoes National Park, 377

Hayden, Ferdinand V., 266, 267, 271–273, 291; expedition of, 15, 61, 96, 114

Hayden Valley, 15, 132, 188, 371

Haynam, Karen Craighead, 136–137

Heap, D. F., 15, 263

Heartline Fund, 296

heat flow, 276–277

Hebgen Lake, 272–273, 279, 364

helicopters: bears dropped from, 155–156, 187–188, 192; drive elk, 31

helium, 3, 262

Henning, Daniel H., 252

Herbst, Bob, 165, 178, 222, 341

Herrero, Stephen: *Bear Attacks,* 191

Hershler, Ed, 140, 216, 228

Hetch Hetchy dam, 307–308, 327, 346

Hickerson, Harold, 104

"high-lining," 86, 371

hikers, 210–211, 333; *see also* backpacking

Hinduism, 347, 349

historians, 245, 248–249, 258

historical research, 34–35, 90–91, 244

Hobbs, Thomas, 140

Hocker, Phil, 217

Holmes, William H., 267, 268, 269
Homestake Mining Company, 364
Hornocker, Maurice, 64, 128, 138, 169, 179, 291–292
hot-spot theory on volcanism, 278
hot springs, 262, 263, 273, 277; benefits of, 290
House of Representatives, U.S., 267; *see also* Congress, U.S.
Houston, Douglas, 52, 60, 74, 79–80, 108, 138; and buffalo problem, 83–84; on carrying capacity of range, 73–74; on deer, 87–88; and natural regulation, 61–70; his *Northern Yellowstone Elk,* 89
Howe, Robert, 28, 77–78, 82, 251
Hultkrantz, Ake, 109
human activity as part of ecosystem, 314
humanism, 345
humanists vs. scientists, 324–325
human-wildlife encounters, 286; with bears, 146–148, 155, 166, 187–190, 218, 231
Hummell, Don, 226
Hunter, David, 297
hunters: and bears, 162, 184, 185; Indian, 92, 94, 97–103, 105; increase in numbers of, 332; and drive for public hunting, 33, 36–37, 47–48
hydrothermal system, 273, 276, 277, 280, 285, 287; *see also* thermal areas

Ice Ages, 92, 100; in Yellowstone, 274
Idaho, 339
Idaho, University of, 64
Idaho Conservation League, 295, 339
Iddings, J. P., 267, 268, 269
Imperial geyser, 263
"inanition," 84
Indians, 5, 29, 382, 384; decline of, 103–105; use of fire by, 92–97, 111; history of, in Yellowstone, 105–109, 114–115; hunting by, 97–103; intertribal buffer zones of, 103, 104; view of, 110–115
"industrial tourism," 207
inhumanism, 304, 347
Institute of Biological Sciences, 179
"intensive management," 237
Interagency Grizzly Bear Committee (IGBC), 180, 182, 191–192
Interagency Grizzly Bear Study Team (IGBST), 142, 159, 161, 169, 171; Steering Committee of, 142, 180; subcommittee on population analysis of, 182, 185
Interior, U.S. Department of, 5, 23, 43, 239, 390; and bear policy, 159, 179, 180,

193; deep-drilling project of, 284; creates feeding committee, 182–184
interpretation: as apology for park policy, 251, 257; replaces education, 245–247
Interpretation and Visitor Service Guideline, 246
Interpreting Our Environment (Park Service textbook), 245
Interpreting Our Heritage (Tildon), 246
intuition, 302, 304–305, 346
"irruptive sequence," 26
Island Park, Idaho, 274, 285–286, 364

Jackson, W. H., 114, 266
Jackson Hole Preserve, Inc., 206
James, Edwin, 94–95
Janzen, Daniel H., 383
Jarvis, Destry, 217, 230, 378
Jeffers, Robinson, 303, 346, 351; and inhumanism, 304
Jefferson, Thomas, 94
Jefferson River Valley, 14
John D. Rockefeller, Jr., Memorial Parkway (JODR), 206–207, 230–231, 366
Johnson, Edward, 349
Johnson, Huey, 295
Johnson, "Liver-eating," 337
Jonas, Robert J., 64, 131, 179, 261; beaver study of, 11–13, 28, 86; on Grant Village, 212; on wolves, 134; on Yellowstone, 255
Jones, William A., 15, 61, 272
Jonkel, Charles, 188, 190
Jordon, William R., III, 383
Judeo-Christianity, 298–300, 304, 309, 310, 350
jumps, 99–100
June, James, 155

"K." *See* "carrying capacity" of range
Kaibab North Plateau, 24–26, 69
Kansas City Star, 265
Katmai, Mount, 269, 270
Kay, Charles, 79, 80, 86, 87, 173, 179
Keating, Kimberly Allen (Kim), bighorn study, 81
Kennedy, George C., 269, 270
Kerwin, David A., 281
Kittams, Walt, 27–28, 78, 132, 251, 291
Klingelhofer, George, 296
Knight, Richard, 142–143; and bear problem, 159, 161, 175; bear research of, 162–163, 167, 170–171, 173, 189, 192; and analysis of Fishing Bridge, 229; as scapegoat, 181–182
Knowles, Joseph, 334

Kober Construction Company, 223, 225
Kotchek, Seymour, 260
Krupnik, Lou, 353
Krutch, Joseph Wood, 303, 304
Kuhn, Thomas S., 342, 343
Kushlan, James, 251–252

Lacey Act, 5, 16, 121, 264
Lahren, Larry, 65
Lake (village), 198, 214
Lake Bear Logs, 154
Lamar River, 28, 44, 78, 87
Lamar Valley, 59, 119, 134
Land and Water Conservation Fund Act
    of 1965, 43
"land ethic" of Leopold, 314–315, 325,
    368
LANDSAT satellite and habitat analysis,
    172
Lane, Franklin K., 124
Langford, Nathaniel Pitt, 105, 106, 115,
    262, 291; on Bridger, 265–266; expedi-
    tion of, 15, 61, 92; and creation of Yel-
    lowstone Park, 266–267
Last Chance Peacemakers Coalition, 360
Lava Creek Campground, 106
lava flows, 268, 269
law enforcement, 202; emphasis on, 249–
    250, 253; ranger experts in, 234–236,
    243, 260
Lemhi Indians, 96, 103
Leopold, Aldo, 383; on intuition, 304–305;
    his theory of "irruptive sequence," 26;
    "land ethic" of, 314–315, 325, 368; his
    Sand County Alamac, 25, 42, 314, 315;
    founds Wilderness Society, 303
Leopold A. Starker, 6, 113, 247, 248, 346,
    387; and bear problem, 143–154, 165,
    166; "Leopold Report" of, 33–35, 39–40,
    46, 60, 129, 178, 240; and Park Service,
    43, 341
Leopold Report, 33–35, 39–40, 46, 60, 129,
    178, 240, 385, 387, 390, 392
Lewis, Henry T., 94, 95, 111-112
Lewis, Meriwether, on buffalo, 145
Lewis and Clark, 94–96, 103–105
Lewis Lake, 14, 76
Lewis River, 14
lightning and "natural-burn" policy, 79–80
Lindeman, R. L., 316
Lindesfarne Association, 352–353
Lindsley, Chester, 21
Linduska, Joseph P., 151, 341
Linn, Robert M., 378; on bears, 149, 151,
    153; and ecosystems, 143, 144; on power
    in Park Service, 248, 261; on wolves,
    134–135

live-trapping, 31–32, 37, 47
Livingston *Enterprise*, 20
lobbying: by automobile and RV indus-
    try, 213, 228; by environmental organi-
    zations, 340–341; by farm groups, 121,
    122, 127, 139, 140
Locke, John: *Second Treatise on Civil
    Government,* 110
Lost Creek, beavers at, 85–86
Louis XI, King of France, 386
Lovelock, James, 317, 357, 358
Ludlow, Robert, 15, 61
lynx, 124, 128, 193
Lystrup, Herbert T.: *Hamilton's Guide to
    Yellowstone,* 5

MacDonald, Patty, 225
McGee, Gale, 37
McLaughlin, John S., 37, 38, 64, 128
McNamee, Thomas, 173, 178, 181
McPhee, John, on Brower, 304, 305
Madison River, 271, 272–273
mail-order houses, 329–330
management. *See* ecosystems manage-
    ment; park management; people man-
    agement; wildlife management
Management and Budget, U.S. Office of,
    199, 217, 219, 222, 223, 227
Manns, Timothy (Tim), 156
*Manual of General Information on Yel-
    lowstone National Park* (1964), 129
Marcoux, Ron, 83
marina at Grant Village, 205, 207, 230
marten, 124, 193
Martin, Paul S., 101
Marting, Leeda, 296
Master Plan. *See* National Park Service
    Master Plan
Mather, Stephen T.: as Director of Park
    Service, 201, 234–235; on extermina-
    tion of animals, 21, 124, 125, 217
Mathiessen, Peter, on Indians, 110
May, William, 187
Meagher, Mary, 43, 60, 143; on bears,
    165, 175, 179; on beavers, 255; on buf-
    falo, 59; succeeds Cole, 74; and elk
    problem, 65–66, 82–83; on Indians,
    107; and Knight, 181; on trapping, 75
Mealey, Steven, 189
Mebane, Alan, 137
Mech, David: *The Status of the Wolf in
    the United States,* 120
Meek, Joe, 264
megafauna of North America, 100–101
Melosi, Martin V.: *Garbage in the Cities,*
    177
Merchant, Carolyn, 346, 352

Merriam, C. Hart, 120
migration, elk, 38, 61–62, 65, 72–73
Miller, Kenneth, 260
Milwaukee Railroad, 201, 203
mineral deposits, 283–285, 288, 339
minimum-impact camping, 333–334
Mirror Plateau, 131
Mission 66, 203–208; Steering Committee of, 204
Missoula, Montana, 335, 339
Missouri River, 14, 145
M-99 Etorphine, 190
Molly Islands, 124, 237
Montana, 37, 47–48, 135, 339; and bears, 156, 162, 191; Department of Fish and Game of, 30, 39, 50, 71–72, 77, 191; and Grant Village, 223, 225
Montana, University of, 252
Montana Citizens to End the Arms Race, 360
Montana Cooperative Wildlife Research Unit, 30, 252
Montana Environmental Information Center, 295, 296
Montana Wilderness Society, 296
moose, 17; Cole on, 255; decline of, 68, 80–81, 83; Indians hunt, 97, 102, 103
Moran, Raymond E., 244
Moran, Thomas, 267
Morris, Ben, 155–156
Morton, Rogers C. B., 159–160
Morton, Thomas, 94
Mother Earth News, 335
Mott, Maryanne, 296
Mott, William Penn, Jr., 260, 387
mountain lions, 15, 47; control program for, 21, 22, 23, 24; destroyed, 121, 123, 126, 128, 193; and natural control, 52–53, 58; reintroduced, 35, 138
mountains of Yellowstone, 268, 280–281
Mount Cook National Park, 53
Mount Rainier National Park, 377
Mudd, Tom, 296
mud pots, 264
Mud Volcano, 262–263; doming at, 280
Muir, John, 5, 12, 112, 201, 346; on forest use, 306, 307; as pantheist, 300–301, 304, 305, 349; vs. Pinchot, 307–308; and tourism, 328
mule deer, 3, 87–88, 102; decline of, 21, 81, 103, 371; hunted by Indians, 98; mismanagement of, 24–25
Mullan (army engineer), 14, 95
Mummy Cave, 99, 102
Murie, Adolph, 23, 61; coyote study by, 126–128, 292; his *Ecology of the Coy-*

ote in the Yellowstone, 238; his *Wolves of Mount McKinley,* 238
Murphy, Patricia, 186
Myers-Hindman site, 65
mysticism, 304–305, 351

Naess, Arne, 356
NASA studies, 317; of Yellowstone, 275–277, 291
Nash, Roderick, 346, 349, 350
National Academy of Sciences, 155, 159, 257; Committee on the Yellowstone Grizzlies of, 160–161; and forest reserves, 306; National Research Council of, 240, 283, 284
National Association of Audubon Societies, 127
National Association of Primitive Riflemen: *Buckskin Report,* 336
National Audobon Society, 169, 171, 174, 178, 180, 185, 217; Citizen Mobilization Campaign of, 361; growth of, 330, 340–341; tours of, 331
National Conference on "The World After Nuclear War," 361
National Environmental Policy Act of 1970, 43
National Historic Preservation Act of 1966, 43, 210
National Outdoor Leadership School, 331
National Park Concessioners, 226
National Parks and Conservation Association (NPCA), 136, 217, 227, 378, 388
National Park Service: animals killed by, 193–194, 233; and bear problem, 143–144, 147–154; chain of command in, 153, 222, 249; and Congress, 200, 213, 258, 261; in conflict with Craigheads, 151–154, 157–165, 175–176, 177–179; created, 6, 16, 113, 234; and deterioration of Yellowstone, 256–257, 372; and elk problem, 19–23, 27–35; and environmentalists, 43–48, 341; Forestry Division of, 238; and Grant Village, 205–208, 217–230; image of, 233, 261; and Indians, 45–46; Division of Interpretation of, 239; Natural Sciences Advisory Committee of, 153–154; Office of Natural Science Studies of, 43; organization of, 243–247, 258–260; vs. preservationists, 126–128; promotes park use, 200, 203–205; and research, 60, 161–162, 239–240, 247–253, 258; Division of Research and Education of, 267; and scientists, 235–238, 250–258; twofold

National Park Service (cont.)
agenda of, 201–203; Wild Life Division of, 126, 237–239, 261; and wolves, 129–141; *see also* ecosystems management; rangers
National Park Service Act, 6, 201, 379
National Park Service Master Plan, 208–214; Environmental Impact Statement on, 213–214, 229
National Park Service, "State of the Parks Report," 378
National Parks Preservation Act, 185
national parks reorganization, 39
national park system: attendance in, 200–203, 331; growth of, 330; becomes playground, 332; reorganization of 1964, 386
National Park System Advisory Board, 154, 166, 178, 240; Wildlife Committee of, 172, 186
National Parkways Comprehensive Guide to Yellowstone National Park, 107, 137
National Register of Historic Places, 228
National Research Council. *See* National Academy of Sciences: National Research Council of
National Science Foundation, 240, 247, 289
National Wild and Scenic River System, 43
National Wilderness Preservation System, 70
National Wildlife Federation, 330, 361
natural control, policy of, 38–42, 46–48, 52–53, 56, 130; *see also* self-limiting theory of natural control
natural-fire regime, 79–80, 173
"naturalists," 235, 243; biologists as, 239; become communicators, 245–247; as educators, 260–261; and friction with rangers, 243–244
natural regulation, 59–70, 77, 113–114, 366–367; and elk problem, 73–77, 82–84; and park development, 198–199, 208–209; becomes people management, 250; and self-regulation, 318–319; testability of, 89–91; *see also* noninterference with wildlife
Natural Resources Defense Council, 361
Natural Resources Management Plan of 1983, 137, 138
Nature, 377
nature as religion, 301–305, 348–351
Nature Conservancy, 178, 341
Needleman, Jacob, 345–346
Nelson, Allen, 71–72

Nelson, E. W., 19, 61, 73
Nelson, Gaylord, 341
"neolithic conservatives," 334, 340
Nettleton, A. B., 266
"New Age/Aquarian Conspiracy," 356–358
Newmark, William, 377
new philosophy of nature, 341–360
new politics, 347–348, 353–355
new religion, 347–349; parallels Gnosticism, 350–351
"New Resource Economists," 361, 388
new science, 347, 351–353
Newsweek, 266, 387
New York Times, 142, 185
New York Zoological Society, 125
New Zealand: thar in, 53–56, 67; geysers ruined in, 285
Nez Perce Indians, 108–109, 113
Nichols family, 205–206
Nixon, Richard M., 157, 359
no-action alternatives for planners, 218–219, 228–229
noetic sciences, 297
noninterference with wildlife, policy of, 75, 250, 348–349, 358, 367, 373; *see also* natural regulation
Norris, Philetus W., 16, 106, 109, 113
Northern Pacific Railroad, 200, 203, 265, 266
Northern Plains Resource Council, 339
Northern Rockies Action Groups (NRAG), 296–298, 326; work of, 338–340, 360
Northern Rockies Foundation, 295, 296
Northern Rocky Mountain wolf (*canis lupus irrematis*), 138–139; *see also* wolves
Northern Yellowstone Elk, The (Houston), 89
nostalgia craze, 336–337
Novakowski, Nicholas, 134
nutritional stress and disease, 82

Obsidian Cliff, 98, 268
occult, 305
Odum, Howard T., 314, 316
Office of Ecological Studies, proposed, 391
Office of Management and Budget, U.S., 199, 217, 219, 222, 223, 227
Office of Personnel Management, U.S., 243, 244; and Park Service guidelines, 258–260
Office of Science and Technology, U.S., 290
O'Gara, Bart, 36, 64, 136

Old Faithful geyser, 206, 208, 210, 211, 216, 263; and Grant Village, 227–228
Olsen, Arnold, 32–33
Olson, Sigurd, 46, 112, 178, 208, 247, 248, 303; and clairvoyance, 305; advises Park Service, 341; on wilderness, 324, 367
Olympic National Park, 377
Omaha Indians, 105
organicism, 303, 304
Osgood, John, 186
otters, 124
"025 Task Force," 259–260
Ouspensky, Peter, 303
Outward Bound, 331
overgrazing: of deer, 24–25; of elk, 18, 22, 23, 28, 62, 173
overkill by Indians, 102–103
overnight accommodations in Yellowstone, 204–205, 208–209, 228

panpsychism, 349
pantheism, 304–305, 308–309, 318, 334, 340, 349
"paradigm," 342–343, 356
"paradigm change," 342
Park Improvement Company, 16
park management, 19–20, 33–35, 234; and Robbins Report, 240–250; *see also* wildlife management
Park Management Superintendent Guidelines, 258–260
Park Service. *See* National Park Service
"Park Specialists" (new ranger category), 243–244
pedestaling, 77, 86
Peek, James M., 57, 58, 64, 68, 80, 179, 255–256
Pelican Creek, 212
pelicans, 124, 193, 237
Pelton, J. R., 280
Pengelly, W. Leslie, 30, 60, 64, 65, 86, 87, 179
people management, 44–45, 250, 260, 367, 374
personal transcendence, 353, 354
Personnel Management, U.S. Office of, 243, 244; and Park Service guidelines, 258–260
pesticides, 44, 238, 311–312
Peter (career Park Service officer), 141
Peterson, Russell W., 162, 180, 341
Petulla, Jospeh M., 313, 321
Phillips, Elizabeth, 186
Phillips, Frank, 186
Phillips, Jerry, 165

Pike, Drummond, 295
Pimlott, Douglas, 58
Pinchot, Gifford, 306–308
pine: lodgepole, 14, 93, 114; whitebark, 173, 190
pinkeye, 81–82, 90
Pitchstone Plateau, 219
Planet Drum Foundation, 356
playgrounds, parks as, 332
Plaza, Polly, 180
Pleasant Valley Hotel, 85
Pleistocene epoch, 100–101
"Pleistocene variations," 55
Pliocene epoch, 271–272
poaching, 16, 184, 185, 254
pocket gophers, 128
Point Foundation, 338
poisoning: of bears, 157; of fish, 76; of predators, 128
political activism, 337–342, 353–355, 359
Political Economic Research Center, 388
politics: and environmental groups, 340–341, 361; Grant Village and, 216–217, 221–227, 229, 230; and Greater Yellowstone Coalition, 365–366; influence policy, 47–48, 51; radicalism in, 338–341, 358–359; *see also* new politics
pollution, 207, 217–218, 230, 335, 339
Population Status of Large Mammal Species in Yellowstone National Park, 1982, 137
Povah, Terry, 228
Povah, Trevor S., 164
Powder River Basin Resource Council, 339
Powell, John Wesley, 96
predation, natural, 25–26, 59, 62, 64–65, 113–114; *see also* predators
predators, 384; in Africa, 56; control of, 21–23, 24–26, 29, 66–67; destroyed, 121–128; and elk, 40, 61, 66, 69, 236–237; and natural control, 52–54, 56–58, 69; reintroduced, 34, 129–141
preindustrial cultures, 303, 334–335, 348, 351
preservation, 5–6; vs. extermination, 239; through noninterference, 308, 367; and agenda of Park Service, 201–202; vs. promotion, 202, 208; and protection, 44–45, 308; *see also* protection
preservationists: as humanists, 324–325; vs. progressives, 307–309; and railroads, 201–202; *see also* environmentalists
Preservation Principle of Regan, 348
Pressler, Larry, 289–290
primitive chic, 334–337, 340

Pritchard, Paul, 388
private enterprise, 224, 225, 226–227
pronghorn antelope, 21, 35–36
propagating-rift theory on continental
  plate movements, 278
propane cannons, 187, 192
prospectors, 45, 265, 282
protection: of animals, 5–6, 234, 239; and
  forest reserves, 306–309; and preservation, 308; and ethic of protectionism,
  309; vs. "sustained yield" management,
  306–307; of visitors, 234, 249–250, 260,
  374; *see also* preservation
publication of research, 251–252, 255
public hunting, 36–37, 39, 47–51, 74, 88;
  see also hunters
public-interest groups, 338–342, 359
"punctuated equilibrium," 291
Pyne, Stephen J., 93, 95, 96, 112
pyroclastic flows, 269, 271

Quiet Crisis, The (Udall), 42

Rabbit Creek dump, 146, 154, 164
radiation, 219
radio collars, 30; on bears, 148, 152, 157–
  158; on elk, 72; use of, forbidden, 47,
  75, 81, 291
railroads, 200–203, 205, 380
Rainbow Point Campground, 187
Rampon, 157
Randall, Dick, 155
range. *See* range conditions; winter
  range
range conditions: and disease, 82; and
  fires, 68–69, 93–97; and Park Service
  policies, 202; "rest-rotation" for, 58;
  see also grasses
rangers: classification of, 244, 259–260;
  "fortress mentality" of, 254–256; and
  Grant Village, 199; image of, 233; and
  friction with naturalists, 243–244; run
  Park Service, 248, 249, 253, 256; as police force, 202, 234–236, 250, 260; vs.
  scientists, 248, 249–253; standards for,
  243–244, 258–260, 386, 390–391; *see also*
  National Park Service
Rasmussen, D. Irwin, 25
Ratliff, Ronald, 281
Raynolds, William F., 265
Reagan, Ronald, 199, 223, 289–290, 359,
  361
recreational activities, 332–333
recreational-vehicle (RV) lobbies, 213,
  228
recreation industry, 329–330, 332
red deer, 104

Red Lodge, Montana, 211, 333
Red Mountain, 78
Reed, John, 379
Reed, Nathaniel, 75, 136, 138, 341; and
  bear policy, 157, 158, 159, 178, 180–
  185
Reed, Scott W., 296
Reese, Rick, 185, 361–362, 365–367
Regan, Tom: "Preservation Principle,"
  348
Regenstein, Lewis, 158–159, 179,
  180
Reich, Charles A.: *The Greening of
  America,* 42
reinhabitation, 356
religion: ecology as, 323–325; exotic, 347;
  nature as, 301–305; *see also* new religion
reprogramming of funds, 219
research: funding for, 248, 258; vs. geology, 291–292; historical, 34–35, 90–91,
  244; independent, 161, 236, 252–256;
  "mission-oriented," 60, 253; and Park
  Service, 161–162, 239–240, 247–253,
  258; publication of, 251–252, 255; and
  Robbins Report, 240–242; science centers for, 247–249; *see also* fauna series
Research-Grade Evaluation, 248–249, 257
resource management: funding for, 256;
  and Reagan administration, 361–362;
  standards for, declining, 258–260
resource managers, 247, 249, 257, 260–
  261
"rest-rotation" method of range control,
  58
"resurgent caldron cycle," 277
Reynolds, Harry, Jr., 134, 155
Reynolds, Harry, III, 64, 179
Reynolds, William F., 15
rhyolite, 268–270, 274, 290
"rhyolite plateau," 271
Richards, M. C., 344–345
Richmond, Gerald, 274
Rivkin, Jeremy: *Entropy,* 345
road-building program, 204, 206, 209–
  210
Robbins, William J. 230, 248
Robbins Committee, 161
Robbins Report, 240–242, 244, 247–248,
  250, 251, 253, 258
Rock Creek: tidal wave in, 272; West
  Fork of, contaminated, 333
Rockefeller, John D., Jr., Memorial
  Parkway (JODR), 206–207, 230–231,
  366
Rockefeller, Laurance, 206
Rockefeller Brothers Fund, 296, 338, 340

Rocky Mountain National Park, 199–200
Rocky Mountains, 268, 338–340
Rodman, John, 346, 349, 352, 354
Rogers, Edmund, 127, 128
Roop, Larry, 163–164, 190
Roosevelt, Theodore, 3–4, 18, 61, 145, 252, 306
Roosevelt Lodge, 11, 107
Rosak, Theodore, 323, 342, 344–346, 348, 351
Ross, Alexander, 14
Rot and Grün, 351
Roush, Jon, 296
rubber bullets, 192
Rubin, Jerry, 353
Rush, W. M., 22, 61, 252
Russell, Osborne, 4, 96, 106, 263

safety, visitor, 234, 249–250, 260, 374
sagebrush, 22, 94, 97, 114
sagebrushers, 202–203
"Sagebrush Rebels," 361
Saint Louis Bay, 247, 257
Sand County Alamanac, A (Leopold), 25, 42, 314, 315
Sargent, Leonard, 296
"saturation trapping" of bears, 181
Sax, Joseph, 328
Scaup Lake, 291
Schaller, George, 57
Schleyer, Bart, 190
Schullery, Paul, 88, 107–108, 165, 341
Schwinden, Ted, 297
science, 245–247, 253–258, 304–305, 332, 351; *see also* new science; scientists
Science, 89
Science and Technology, U.S. Office of, 290
science research centers, 247–249, 255
scientists: concept of ecosystem of, 313–319; and environmentalists, 322–325; independent, 236, 252–256; vs. rangers, 248–254; *see also* biologists
Scribner's Monthly, 267
Seaquist, Bobbi, 373
"seasonals," 261
Seater, Stephen, 158, 159, 179
Second Treatise on Civil Government (Locke), 110
seismic studies of Yellowstone, 279–281
self-limiting theory of natural control, 53–56, 59–60, 83–84
self-realization, 353–356
self-regulation, 313, 316–319
self-sufficiency ethic, 329, 333, 335
self-transformation, 354–356
Selway, Idaho, elk in, 68

Selway-Bitterroot wilderness, 139
Senate, U.S., 132, 225, 267; *see also* Congress, U.S.
seral succession of vegetation, 93–97; climax, 90
Sernylan, 190
Servheen, Christopher, 168, 175, 179, 181–182, 190
Sessions, George, 343, 344, 346, 360, 367
Seton, Ernest Thompson: beaver study by, 85; his *Biography of a Grizzly,* 167–168
sewage plant, 205, 207, 217–218, 230
Shaller, George, 331–332
"Shallow Ecology," 360
Shanks, Bernard, 250, 252, 258, 261
sheep, bighorn, 14, 21, 103, 371; and elk, 81–82; Indians hunt, 98
Sheepeater Indians, 99, 105, 106, 108
Shepard, Paul, 346
Shinn, Dean A., 96
Shiras, George III, 17, 24
Shoesmith, Merlin, 64, 131
Shoshone Indians, 103, 104, 106, 108
Shoshone Lake, 14, 76
Shoshone National Forest, 189
Shoshone River, 99
Should Trees Have Standing? (Stone), 349
Sideler, Richard, 130
Sierra Club, 18, 21, 213, 217, 296, 300, 378; Activist Program of, 361; as bureaucracy, 340–341; growth of, 330; Legal Defense Fund of, 349; outings of, 331
Sierra of California, 331
Silent Spring (Carson), 42–44, 303, 311–12
Silk, Tom, 296
Silva: A Discourse of Forest Trees and the Propagation of Timber in His Majesty's Dominions (Evelyn), 306
Silvaculture, 306
simple life, 335–337
Sinclair, A. R. E., 55–56, 89
Sioux Indians, 103, 104
Six Mile Creek, elk at, 71–72
Skinner, Milton, 12, 21–22, 61, 235–236, 291
smallpox, 104–105
Smith, Robert B., 280, 289
Snake River, 14
Snape, Dale, 224
Snyder, Gary, 346, 348, 356
social engineering, 202, 209–210
Soda Butte Creek, 67
soil of Yellowstone, 290
South Dakota School of Mines and Technology, 290
South Florida Research Center, 252

Specimen Ridge petrified forest, 265
Sports Afield, 31
Spretnak, Charlene: *Green Politics,* 359, 360
Sprinkle, Ron, 373
Sprugel, George, 247–248
Squirrel Island, 311–312
Stanford University, 387
starvation: of bears, 170–173, 191; of elk, 19–20, 74, 84; in Kaibab, 24–25
Status of the Wolf in the United States, The (Mech), 120
Stearns, Allen E., 104–105
Stearns, E. Wagner, 104–105
Steinhart, Peter, 109
Stockman, David, 224
Stone, Christopher: *Should Trees Have Standing?,* 349
Stone, Richard, 165
Stottlemyer, J. Robert, 258
Strickland, Dale, 192
Sumner, Jay, 131
Sumner, Lowell, 237–240
Superior, Lake, trout of, 76
Supernaugh, William, 259
supplemental feeding, 175–178, 182–184; *see also* "ecocenters"; feeding centers
Surles, L. E. (Buddy), 223
survival programs, 331, 333
Swan Lake Flat, 102
Swenson, Jon, 71–72, 73, 82

tags, 30, 47, 48, 75
Tansley, A. G., 312–316, 319
Taoism, 303, 304
Taylor, Dale, 108
"technicians," 244, 259, 260
technology: distrust of, 304–305, 309–310, 323–324; and exploration, 275–276, 279, 283–284, 287–289; and New Age/Aquarians, 357, 358; and recreation industry, 329–334, 337
Tertiary period, 55, 268, 271
thar, Caughley's study of, 53–56
thermal areas, 264, 265, 268, 278, 285–286, 290; *see also* geothermal system; hydrothermal system
Thomas, Irving, 353–354
Thompson, Ben, 233–234, 236–239, 385
Thompson, William Irwin, 352–353
Thoreau, Henry David, 301, 302, 350; his *Walden,* 302
Thumb Bay, 197, 204
Tilden, Freeman: *Interpreting Our Heritage,* 246
Toll, Roger, 124, 232
Toon, Owen, 281

tourism, 201–204; as factor in decisions, 52, 214, 216, 228; and fuel crisis, 213, 223; and Indians, 109; as motive for expansion, 226–227; necessity for, 328; vs. protection, 208–210
Tower Junction, 11, 67, 198
Townsley, John, 74, 140, 175; and Grant Village, 214–217, 222–227, 230
Track of the Grizzly (Craighead), 179
Train, Russell, 341
tranquilizers, 71, 155, 156, 187, 190, 191
transcendence, personal, 353, 354
transcendentalism, 302–304
transplant, wolf, 129–141
trappings of animals, 31–32, 37, 47, 75, 190, 191; by Indians, 99–100; "saturation," 181
trees, 24, 27; decline in, 114
Trees (Department of Agriculture Yearbook), 111
Tri-State MX Coalition, 360
"trophic levels," 56–57
trout, 17–18, 75–76; poisoning of, 76; popularity of, 332; spawning streams of, closed, 218, 221, 230
Trout Creek dump, 154, 164, 212–213
Trout Unlimited, 178, 341
Trust for Public Lands, 295, 361
tuff, welded, 269–271
Turner, Frederick Jackson, 336
Turner, William, 216–217
TW Services, 86, 226–227
"twelve-point plan," 387

Udall, Stewart, 33–35, 39, 112, 240–242; his *Quiet Crisis,* 42
undulant fever. *See* brucellosis
UNESCO, 74–75
ungulates: and beavers, 12–13; Caughley's theory on, 53–56; competition among, 62–63, 68; crash of, 24–26, 67; destroy range, 29, 57, 62; and natural predators, 25–26, 58, 66; *see also* deer; elk
Union Pacific Railroad, 201, 203
United Indians of All Tribes, 45
Utah, University of, 279–280
Utley, Robert, 245

Varley, John, 74, 288
Vayk, J. Peter: *Doomsday Has Been Cancelled,* 346, 357
vegetation: deterioration of, 13, 27–28, 77–80, 86–87, 371; and seral succession, 90, 93–97; *see also* grasses
Virgin Islands National Park, 377
visitation: as basis for Grant Village, 220;

to national parks, 200–203; at Yellowstone, 213–214, 331, 332, 374
visitor safety, 234, 249–250, 260, 374
visitor-use corridors and enclaves, 208–209
volcanic activity in Yellowstone, 268–272, 274–275; cycles in, 276–277; study of, 276–281
volcanic soil, 267–269
Vore buffalo jump site, 99

Wagner, Randall, 228
Wahb Springs, 167–168
Wakulla Springs symposium on ecology, 320
Walden (Thoreau), 302
Walker, Harry, 155
Wallace, Henry, 24
Wallop, Malcolm, 225
Wall Street Journal, 159
Ward, Jay, 134
Warren, Edward, 85–86
Washakie, Chief of Shoshones, 103
Washburn mountain range, 268, 271
Wasson, Isabel Bassett, 267
water table, 13, 63
Watt, James, 224–225, 227, 359
Watts, Alan, 346
Watts, David, A., 161
Wauer, Roland, 142, 175; memorandum by, 168, 169, 180
weather, 68, 84
Weaver, John, 131–133, 135, 137
Wells, Albert, 296
Wenk, Dan, 228–229
Western Association of Game and Fish Commissioners, 74
Western Solidarity, 360
Western tradition, 345–346, 348
West Thumb, 198, 204, 208, 211; bypass of, 206–207
West Yellowstone, 164, 187–188, 216, 223
Whalen, William, 216, 222
White, John R., 127
White, Lynn, Jr., 298–300, 309–310, 346, 373
white-tail deer: decline of, 21, 29–30, 371; and elk, 62–63; Indians hunt, 98
Whole Earth Catalog, The, 335, 338, 357
Whole Earth Papers, The (Baines), 357
Wiens, John A., 318, 319
Wilcox, Bruce A., 387
Wilcox, R. E., 267
wilderness, 44–47; accessibility of, 327–329; consumptive use of, 332; damage to, 210–211, 332–333; need to expand, 368; and search for nostalgia, 336–337;

use of, promoted, 331
Wilderness Act of 1964, 43–46
Wilderness Society, 46, 178, 185, 213, 295, 303, 330, 341, 361, 378
wilderness system, growth of, 330
"wilderness threshold community," 198, 209, 213–214
Wildlife in Yellowstone and Grand Teton National Parks (Harry and Dilley), 5
wildlife management: early policy on, 16–23; and Green Book, 39–40; and Leopold Report, 33–35; and noninterference, 75, 250, 358, 366–367, 373; and Robbins Report, 240–250; Wright's theory of, 237; *see also* ecosystems management; natural control; natural regulation; park management
wildlife of Yellowstone, 5; and deep drilling, 286; and Indians, 97–100, 102–104; destroyed by Park Service, 193, 233
Wildlifer (magazine), 169
Wildlife Society, 134, 185, 295
Willow Park, 102
willows, 12, 13, 28, 29; decline of, 63, 68–69, 79, 371
Wilson, Woodrow, 307
Wind River Range, 138
winter range: deterioration of, 27–29; Houston's theory on, 61, 65–66; inadequacy of, 19–20, 22, 23, 71–74, 77
Wirth, Conrad L. (Connie), 203, 261
Wirth, Theodore (Ted), 205
Wisconsin glaciation (ice ages), 274
Wisconsin, University of, Arboretum, 383
Wolf Recovery Team, 136, 139, 140
wolverines, 128, 193
wolves: destroyed, 23–25, 29, 119–123, 193, 371; and elk, 58, 64–66; Northern Rocky Mountain, 138–139; reintroduced, 35, 129–141; role of, in natural control, 52–53, 56
Wolves of Isle Royale, The (fauna series), 247
Wolves of Mount McKinley, The (Murie), 238
woodcraft, 329, 333–334
World Heritage Site, Yellowstone as, 74–75
World Wildlife Fund, 178, 341
Wright, Gary, 79, 88, 97, 100, 102–103
Wright, George, 22–23, 61, 125, 126, 232–233, 250, 261, 308, 346, 385; early life of, 234–236; fauna series of, 236–237; research of, 236–237, 252, 291; and Wild Life Division, 237–238

Wyoming: bear policy in, 184, 185; Department of Game and Fish of, 192; and Grant Village, 216, 223, 225–227; wildlife in, 37, 38, 184

Wyoming, University of, 64, 68

Wyoming State Historic Preservation Office, 228

Wyoming Travel Commission, 228–229; *Travel Log* of, 130

Yancey, "Uncle John," 85

Yancey's Hole, 11, 13, 85–87, 90–91

Yannone, Vince, 64

Yellowstone ecosystem, 55–56, 73, 90; need to enlarge, 368; drive to protect, 40–42, 44–47, 363–367

Yellowstone Lake, 197, 198, 204, 271, 280, 287; pollution in, 207, 217–218, 230

Yellowstone Library and Museum Association, 137

"Yellowstone melting anomaly," 279

Yellowstone National Park Concessions Management Review, 215

Yellowstone Park Act, 3, 5, 204, 264, 266–267

Yellowstone Park Company, 205–206, 214; bought out by Park Service, 215–217, 222, 223

Yellowstone plateau, formation of, 268–272

Yellowstone River, 4, 44, 87, 290

Yellowstone Valley, lower, 65

Yosemite National Park, 22, 207, 250, 327–328; Field Natural History school in, 234

Zen Buddhism, 345, 346, 347

"zootic climax" sites, 62; vegetation in, 87, 90